Current Developments in Spatial Planning

Current Developments in
Spatial Planning

Edited by **Marina De Lima**

R CALLISTO
REFERENCE

New York

Published by Callisto Reference,
106 Park Avenue, Suite 200,
New York, NY 10016, USA
www.callistoreference.com

Current Developments in Spatial Planning
Edited by Marina De Lima

International Standard Book Number: 978-1-63239-137-7 (Hardback)

Printed in the United States of America.

Contents

Preface

I am honored to present to you this unique book which encompasses the most up-to-date data in the field. I was extremely pleased to get this opportunity of editing the work of experts from across the globe. I have also written papers in this field and researched the various aspects revolving around the progress of the discipline. I have tried to unify my knowledge along with that of stalwarts from every corner of the world, to produce a text which not only benefits the readers but also facilitates the growth of the field.

Geoscience is developing with a rapid pace and spatial planning is an important unit of it. Number of advanced methods and modeling techniques such as Global Positioning System (GPS), Geographical Information System (GIS) or remote sensing technologies have been introduced and practiced to enhance various facets of spatial planning. The sections of this book discuss the present profile of the theories, data, analytical methods and modeling techniques adopted in several case studies. The book not only deals with the latest developments in urban and spatial planning, but also highlights the different elements of spatial planning and provides access to different case studies in several countries.

Finally, I would like to thank all the contributing authors for their valuable time and contributions. This book would not have been possible without their efforts. I would also like to thank my friends and family for their constant support.

Editor

Part 1

Theoretical Aspects of Spatial Planning

Strategy Planning of Sustainable Urban Development

Katalin Tánczos and Árpád Török
Budapest University of Technology and Economics
Hungary

1. Introduction

Motto: „And what, I said, will be the best limit for our rulers to fix when they are considering the size of the State and the amount of territory which they are to include, and beyond which they will not go?
What limit would you propose?
I would allow the State to increase so far as is consistent with unity; that, I think, is the proper limit." – Plátōn

As the rapid growth of world population and its concentration in cities around the globe takes place, sustainable urban development has constituted a crucial element affecting the long-term outlook of humanity (Auclair, 1997). Besides, in the urbanized areas a high level of GDP (gross domestic product) concentration can be observed. That should make us care for the operational efficiency of the urban areas more and more studiously. Dynamic and continuous horizontal and vertical growth of urban areas makes the question of efficiency more and more important. Handling productivity concentration has been a key issue for economists and solving transportation problems – due to population concentration – has been just as important for engineers for a long time. Representatives of abysmally separated scientific fields tried to solve the efficiency problems of certain areas which can be characterised by lack of capacities, extreme population and productivity density.

To effectuate an uniform, consistent methodology for urban planning – taking into consideration the viewpoints of the land use and the transportation - according to Platon's words, we need to approach the subject by considering complex social and economic aspects. With the desire to achieve urban development that meets the needs of the present without compromising the ability of future generations to meet their needs (WCED, 1987), urban development is required to minimize threats from wasteful use of non-renewable regenerations, to avoid the uncompensated geographical or spatial displacement of environmental costs onto other places, and not to draw on the regeneration base and waste generation capacities to the levels which disrupt dynamic equilibrium of the ecosystem (Burgess, 2000).

The structure of urban areas can be understood as a result of the relationship between transport and land-use. The interaction is known as a two-way process. This process, however, is more complicated than other reciprocal processes that are frequently encountered in everyday life. This is mainly because the various interactions take place over different time scales and involve factors with varying degrees of certainty. Hence, an analysis of the interaction between transport and land-use requires disentangling diverse relationships among the factors and this makes it a difficult task.

For sustainable urban development it is inevitable to constitute policies – based on a useful planning method - that can support sustainable urban development without sacrificing either economic growth or the freedom of movement. Sustainable urban policies have to be based on comprehensive planning with the involvement of different sectors and fields of competence.

To effectuate a uniform, consistent planning methodology a flexible, consistent, and well manageable model needs to be developed, which involves the aspects both of land use and transportation. Beside this, it is important to see that planning methods generally focus on estimating and comparing the effects of one or two selected measures (e.g. infrastructure investments, land-use or fee-collection possibilities). In contrast to the above mentioned process the assignment of the best solution from a given set of measures seems to be more effective.

Based on this assumption we introduce a newly developed optimizing method. With this it is possible to describe individual system components' (e.g. consumers, firms) behaviour, to estimate and compare social effects of the changed urban environment (based on the investigated set of measures) and furthermore to define optimal solution from a social point of view. Hereby we will have the possibility to control the urban environment based on the available set of measures (as a part of the controlling method), the object of the individual system components (energy consumption, costs, benefits - as a part of the modelling method) and the selected social objectives (e.g. operational efficiency, summed up social costs, summed up social benefits, pollution - as a part of the controlling method).

2. Objectives

As it was mentioned above the current method of selecting the development package comprising the most beneficial measures is to select the best measures by comparing their expected impact, after which the development strategy can be formulated based upon the joint implementation of selected measures. The effects of community development plans on the local level are often evaluated solely from a transport perspective. The state of the art, activity chain-based transport models analyse community interests with regards to mobility demand. However, this can lead to the omission of such important aspects as the use of community land in accordance with community interests.

The limiting factor that makes activity-based modelling unsuitable for the analysis of some of the aspects of community development is evident from its very definition.

The definition by Torsten Hägerstrand (Hägerstrand, 1970) states that transport decisions are induced by individual activities responsible for mobility demand. Individuals make these decisions based on changes in the utility function related to transport processes. For example, the farther a given activity is from the starting point of the trip, the less the increase in the utility function, while the comfort level associated with the mode of transport decreases the opportunity cost.

When comparing the above with the general definition of the utility function, a contradiction arises, since the definition states that utility is the property which makes things fit for the purpose of meeting individual needs. Needs inspire and energize action, while directing it at the same time (Hunyady, 2003).

Considering how transport demand is primarily generated by the fact that the production, use and consumption of products and services are separated, the dependence of transport on the demand for the original product or service cannot be ignored (Gubbins, 2003). Therefore, keeping the secondary role of transport in mind, we can come to the conclusion that demand which can be met in its entirety by transport alone rarely arises.

The enforcement of community interests, on the other hand, requires an analysis of the changes in the utility function as comprehensive as possible, since – starting from the dictionary of definitions – interest is "necessary and important, for the benefit of someone or something" (Juhász, 2003).

Taking into account that transport demand is usually not the only constituent in complex human needs, the goal of the chapter is to expand the evaluation procedure based on transport models in use today for supporting community development decisions. Analysing the changes in the utility of the individual (e. g. estimating the effects of consumed goods), the modelling of individual decisions that affect how a community works becomes possible. A further goal of the chapter is to introduce a community development process using the utility and decision theory of economics that analyses the utilization of urban areas, the structure of the transport system and the properties of the economic and social strata in a complex manner, also taking interactions between these aspects into account. Additionally the paper introduces a process to maximize social utility as a function of community development initiatives. This could be an important contribution to the development of a methodology for the selection of the urban development packages that best enforce community interests.

The model development orientations are based on the evaluation of the related literature. This approach makes it possible to analyse the model development processes and the reasons of methodological evolution. Beside this, the detailed literature analysis explains the chosen development orientations of our research.

3. Evolution of urban models

Before the investigation of the development process of urban models, the reasons for the evolution of the methodologies. Ambition to forecast the needs for transport can be originated from modelling of individual needs, which include physiological needs, safety, social needs, the needs for appreciation and self-realization according Maslow's theory (Maslow, 1943). In Maslow's hierarchy model of needs, the lowest level of necessities is the subsistence. Man always provided the goods for subsistence by production. Hence the efficiency of production directly influences the level of complacency. So it is understandable that urban models aim to represent the operation efficiency of urban system and to forecast traffic structure, which directly effects productivity.

3.1 Evolution of transport models

In modern history, the industrial revolution of the nineteenth century led to a very dynamic development on both micro and macro level. The economic growth generated by the industrial revolution coupled with the rapid expansion of transport needs generated bottlenecks in the transport system. Thus it is not surprising that the history of the modelling started in the nineteenth century. The well-known gravity model applied in transport modelling based on one of the most elemental rule of the classical physics was probably first formulated by Carey (Carey, 1873).

After the First World War the economic growth further escalated, the effect of which gave a higher priority to infrastructure development. The formulation of metropolitan areas led to the restructurization of transport needs. The main problem of the spatially concentrated urban areas became the rapid intensification of transport demand. After the Second World War the economic growth could continue, which resulted further increases in transport needs. It is not surprising that the first complex transport models, which contain the bases of the urban transport planning methodologies used nowadays, developed in 1950's for the first time. The well-known six-step-model (data collection, forecast, goals, network proposals, testing network proposals, evaluation) was first introduced by Dr. J. Douglas Caroll in the Detroit Regional Traffic Study. The final method was worked out and applied also by Dr. J. Douglas Caroll in the Chicago Regional Transport Study (Chicago Area Transportation Study, 1962).

In the 60's and 70's the urbanization was gaining a bigger and bigger impetus.. For instance the population of San Francisco Bay metropolitan increased by more than 70% (ABAG, 1991) area between 1950 and 1970. This tendency reduced the applicability of traditional models, since those did not take into consideration the personal motivations of travellers and the temporal characteristics of their decisions. As a result of the demand for more effective estimation, the activity-based model of San Francisco Area was developed in the middle of seventies by Ruiter and Ben-Akiva (Ruiter & Ben-Akiva, 1978), where the econometric approach of transport planning was firstly applied in practice. Activity-based transport models are widely used by now, since those make it possible to forecast individual travel decisions more and more effectively.

3.2 Evolution of economic equilibrium models

Beside transport models, nowadays equilibrium models are also more prevalently applied in the field of urban modelling. However, their basic methodology did not contain any spatial representation until the end of the 20th century.

Most of the microeconomic models are prepared according general equilibrium conditions. The intellectual roots of the assumption of the self-regulating nature of markets – the formation of the economic equilibrium – can be originated from the so called "Invisible hand" theory of Adam Smith (Smith, 1776). The concept of general equilibrium – according to the apt statement of János Kornai – plays a similar role in the economics as the absolute-zero in the physics, which cannot be obtained in the real life, but is well defined in theory. It can be defined as an abstract point of reference. Real economic systems can be characterized by the distance they are away from the theoretical equilibrium point (Kornai, 1971).

The evolution of equilibrium theory continued in the second half of 19th century. In 1874, Léon Walras in the "Elements of Pure Political Economics" (Walras, 1871) laid down the foundations of equilibrium analysis. Walras assumes the economy to consist of households and firms. Both consumers and firms take prices as given, and market decisions depend on the equilibrium price. Each household owns resources and products applicable for final or intermediate consumption. Households gain income from the sale of resources.

After his retirement, in 1892, Walras was followed by Vilfredo Pareto in his chair as the head of the Department of Political Economics at Academy of Lausanne. Thanks to his work, general equilibrium theory's applicability in practice was greatly improved by "Pareto optimum" (Pareto, 1897). The "Pareto optimum" is not simply one of the possible market equilibriums, but the equilibrium, in case of which none of the actors can increase their utility, without declining the utility of another actor.

In the 20th century, among others for instance Paul Samuelson investigated the dynamic stability of the equilibrium models applying the methodology of mathematical statistics (Samuelson, 1941). Dixit and Stiglitz published their theory on "The "Monopolistic Competition and Optimum Product Diversity" in 1977, which extended the conventional equilibrium theory by formulating involving market imperfections and imbalance (Dixit & Stiglitz, 1977). And finally Paul Krugman has to be mentioned, who laid down the foundation of new economic geography by introducing spatial representation in equilibrium theory (Krugman, 1991). This approach with some modification made equilibrium methodology became applicable in urban environment as well.

3.3 Evolution of integrated models

Since the location of production and consumption mostly differs from each other, market processes generate transport demands both on the side of producers and consumers. Further, individual transport demands can be originated to the occurrence of individual needs. If the nature of these needs causes economic demand, then the individual economic decision will be affected by the transport decision. Hence the two kinds of decision made it necessary to integrate transport and economic models. In addition traditional urban planning approaches assume that mobility demand between zones is constant. However, congestions of zones connecting transportation corridors deeply affect the decisions of economic actors and so travel demand. Hence the assumption of constant demand does not prove to be realistic (Bokor, Török, 2011).

For example Bagwell mentions in "The transport revolution from 1770" that the number of passengers in England between 1770 and 1830 was multiplied by a factor of fifteen, while the travel time between the most important cities was cut by half (Bagwel, 1974). According to the data above, the conclusion was correct regarding the interaction between the economy and the transport network (reduction of travel time affected travel demand directly and indirectly through economic demand). So the internal transport links of an area affect the mobility structure of the area significantly.

The above introduced phenomenon draws the attention to another decision problem, which is also related the economy and the transport system. Transport system development processes happen on publicly owned land. Due to the limited-nature of public resources public decisions are required to be made. These problems can explain the next step of model development, which lead to the integration of land-use. The models focusing on the interaction between land-use and transport are mentioned as LUTI models in literature (Wegner, Fürst, 1999). The next generation of these models are mentioned as SETI (Spatial Economic and Transport Interaction) models (Russo, 2009).

4. Controlling urban development

Based on the connection of transport science and economics it is possible to develop models to describe system processes and interactions of the urban environment. Interventions of regional and transport development changes the life, the look and the structure of cities and influence people's everyday lives and the decisions of the economic actors.

System processes, estimated validity period assigned to the processes and regulating instrument system of the processes which are essential in transport and economical aspect are presented below to prepare the elaboration of the urban model which can describe processes differentiated in time.

Transport processes which taken place daily are defined by term of journey (e.g. work, consumption) as a short-term system process. Journey involves decisions related to transport processes which happen between an origin (e.g.: home) and a destination place (e.g.: office) which places can be assumed to be fix in short-term. The transport process (choice of mode and direction) is directly affected by the traveling person's social-economic characteristics (e.g. level of income), the transport modes which are available in the system and origin-destination relations. The process can be affected (controlled) by "regulation of land use" (e.g. establishment of pedestrian zone affecting route choices) and "traffic control" (e.g. reducing green-time). These regulations aim to maximize the efficiency of traffic flow and minimize the external costs (accidents, pollution, travel time). "Choice of the destination" is a decision made by the consumer. This process is mostly a medium-term system process related to people's mobility demands. Destination places can be assumed to be unchanged in short-term because workplaces or the daily visited supermarket can be related to formed habits. In medium-term, the choice of destination place depends on the traveling person's social-economic characteristics, the passenger's origin and possible destinations (e.g. service and production places in the system, workplaces, shops). The destination identification process can be regulated by the "influence of mobility demand" (e.g.: eco-friendly and socially-sensitive education, integrated transport pricing policy, preferring public transport, etc.). Producers' processes influencing the settlement structure can be defined as a "medium-term decision group" related to the "choice of production place". Firms can be characterized as rational decision makers since their market potentials, and just as well their decisions in reference to "production place" depend on current land prices and characteristic of the labour force. "Choice of residence" is the decision process which influences the settlement structure in long-term. This choice depends on the spatial and residential characteristic of the urban environment, the social-economic characteristics of the decision makers and the local capabilities of the urban transport network. Both the choices in reference to residence and production place can be affected by the control of "land use" (urban-planning and settlement development - e.g.: influencing the proportion of inhabited area- green area- transport infrastructure- production area).

Figure 1. presents the decision processes, their relationship and the estimated validity period of the decisions.

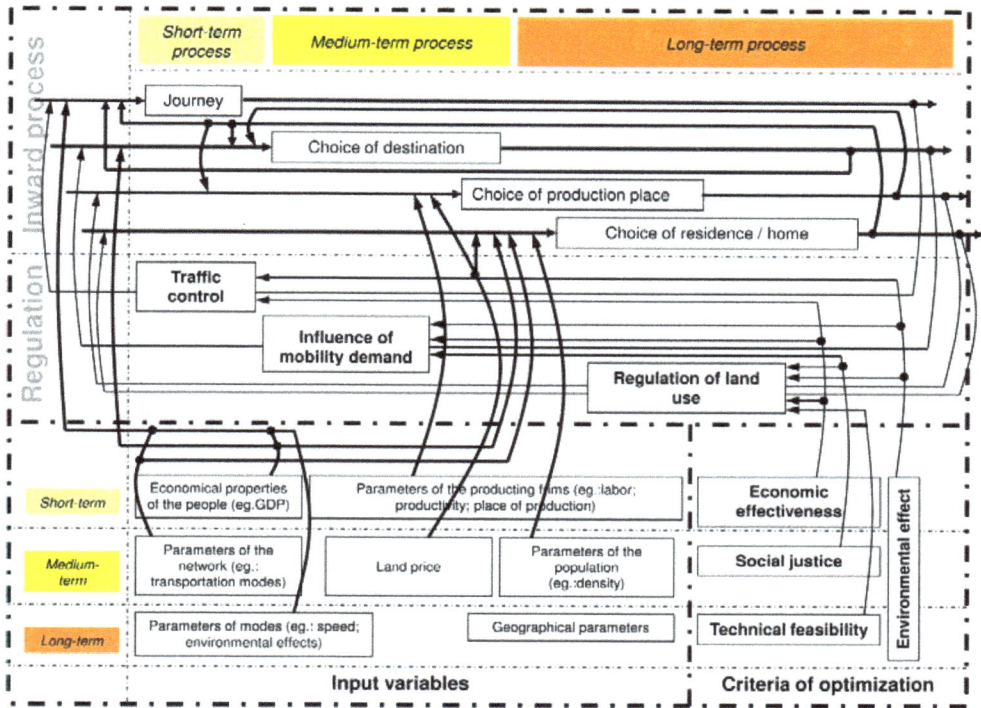

Fig. 1. Control model for the urban environment

Related to the possibly applicable measures and interventions influencing the operation efficiency of the urban environment, it has to be emphasized that the validity period of the decisions determines the complexity of the problem. For example, as it was examined the effects of a measure that is valid in short-term we can be assumed for the solution of the medium-term and the long-term decision problems to be fix parameters since e.g. an interim road maintenance work probably does not influence the choice of residence and production place.

However it has to be emphasized that according to the theory of "organic-urban environment" urban-planning should not focus on the solution of particular technological problems but on connecting segregated social groups and making urban-environment liveable (Jávor 2005). Hence every decision support system related to the urban environment should only be applied as an orientation tool for stakeholders.

5. The basic model

The applied basic model was developed by the authors; however it does not introduce any scientific innovation in the field of spatial general equilibrium models.

The urban environment is spatially divided into locations. These locations are linked with each other. Let us assume that each link aggregates all the possible routes between two locations (e.g. average distance). There is "i" amount of locations (zones) in the model. There are producers in every zone and all producers are specialized, hence there are "i" types of product. In case of special modelling it could be important to extend the model to order more producers to certain zones (see Anas et al., 2007). Firm makes no profit at any scale of operation (zero profit condition). "j" types of consumers are considered in the model to be able to describe different consumer groups.

There are different types of consumers in every zone and all types of consumers supply labour for one type of producers. Consumers travel from home to work and from home go shopping, generating income (1) and buying their demanded quantities of each differentiated good at each location. This pattern of shopping occurs because the consumer considers goods purchased from each location as essential commodity (this means also that each consumer will want to visit each). Land is involved in this simple model however it can be easily extended (see Anas et al., 2007). Furthermore it is necessary to be mentioned, that there is only one producer in the same location, but products are able to be substituted according to the Cobb-Douglas type utility (7) function hence pricing of commodities are quasi-competitive and vary among locations at equilibrium. The consumers take their location of employment and the shopping locations as given. Consumers are price takers in all markets and take as given all transport costs and travel times. The consumer chooses home locations and the shopping trip pattern. In the Cobb Douglas utility function, we assume that the taste coefficients $a_{11}..a_{ij}$ are different across consumers and that $\Sigma_i a_{ij}=1$ for every i product (homogeneity of degree one). Producers decide how much labour to demand. The equilibrium conditions involve product market and labour market (8). The equations presented below describing the defined model are derivable from the traditional consumer-producer constrained optimization problem (see Samuelson et al).

$$L_j * (W_j - T_{ij}) - Pi * \Sigma_i X_{ij} = 0 \tag{1}$$

$$(P_k + T_{kl}) * X_{kl} / c_{kl} - (P_i + T_{ij}) * X_{ij} / {}_{cij} = 0 \tag{2}$$

$$L_j - L_{sum} * U_j / \Sigma_i U_j = 0 \tag{3}$$

$$W_i - Pi * a_{ij} * \Sigma_j X_{ij} / L_i = 0 \tag{4}$$

$$P_i - (W_i{}^{awi}) / aw_i{}^{awi} = 0 \tag{5}$$

$$\Sigma_j X_{ij} - L_i {}^{awi} = 0 \tag{6}$$

$$U_j - \pi_i X_{ij}{}^{cij} = 0 \tag{7}$$

$$L_{sum} - \Sigma_j L_j = 0 \tag{8}$$

Where:

U_j: utility of consumers living in zone j (variable),

X_{ij}:   consumption of inhabitants living in zone j, visiting zone i hence choosing product or service type i (variable),

c_{ij}:   Cobb-Douglas taste coefficients of product or service type i consumed in zone j (variable),

a_{wj}:   Cobb-Douglas production parameter of labour used in zone j (parameter),
L_j: consumer living in zone j (variable),
P_i: price in zone i (variable),
W_i: wage in zone i (variable),
L_{sum}:   the amount of labour available in the urban area (parameter),
T_{ij}:   transport cost (parameter).

6. Defining the optimal measure toolkit of urban development

The below presented methodology was developed by the authors. With this new approach, it is possible to define optimal measure toolkit, which maximizes the positive effect of transport interventions on social welfare.

The optimal measure toolkit of urban development should lead to the maximum or minimum value of one or more system components' objective function. As it was mentioned above, the model contains two basic system components: consumers and producers. Since the developed control model is mostly focused on interventions related to the transport network, hence transport cost seems to be the adequate attribute to describe the effect of urban development measures. Accordingly, if we assume transport cost to be a system variable, we need to define i * j additional equations to keep the equation system solvable.

With the introduction of an additional constrained optimization problem the extended equation system can remain solvable. Since the aim of the extension is to define optimal public decisions (e.g. transport network development), it seems to be obvious that the chosen interventions should increase the objective function of a system component (consumers, producers). On the one hand consumers seem to be the better choice, because society consists of the individuals – and public decisions should primary support society. However, on the other hand deducing the form of social welfare function (corresponding to individual utility function) cannot be the scientific task of the economist according to Samuelson. Taking into account Samuelson's consideration, social welfare function can be carefully applied by paying attention to only being interpreted in a relative term (for comparison), which still satisfies our objectives.

First the commodities demanded by consumers have to be derived from equation (2) so that consumption is dependent on transport cost.

$$X_{ij} - _{cij} * (L_j * (W_j - T_{ij})) / (P_k + T_{kl}) = 0 \qquad (9)$$

This equation contains one or two variables depending on whether $T_{ij} = T_{kl}$ or not. If they are equal it means that individuals in the L_j consumer group go shopping to the same location as working.

Figure 2. presents the effect of transport cost on consumption, where T_{ij} is the cost of transport to work, T_{kl} is the cost of transport to the place of consumption and X_{ij} is the amount of consumed goods.

It can be observed that an increase in both the costs of travel to shopping and working induce a decrease of consumption. This phenomenon suits the behaviour of rational consumer, since if travel costs to work increase, then consumers' income will reduce. The same result can be observed in case of travel to the place of consumption getting more

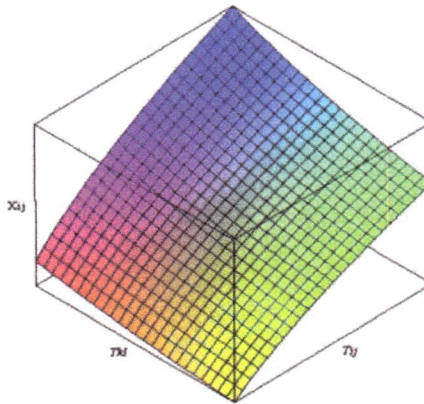

Fig. 2. The effect of transport on consumption

expensive, because then consumers can buy less commodity with the same income. Applying the new equations the social welfare function can be defined, which now can be the objective function of the constrained optimization problem.

$$U_j - \pi_i \, (_{cij} * (L_j * (W_j - T_{ij})) \, / \, (P_k + T_{kl}))^{\,c_{ij}} = 0 \qquad (10)$$

The objective function will be constrained by the available development budget due to the limited nature of public resources. To be able to derive the solution of the maximization problem, constraint has to contain the variables of the objective function, so the constraint has to be based on transport cost. The equation below represents a possible way of formulating the constraint condition of public decision.

$$D - (T_{ij} - N_{ij}) * UN = 0 \qquad (11)$$

Where:

D: available public resources (parameter),
T_{ij}:   cost of travel to consumption place between zone i and j after development (variable),
N_{ij}:   cost of travel to consumption place between zone i and j before development (variable),
UN: cost of a unit of transport development (parameter),

Now the Lagrangian can be derived from equation (11), (12) of constrained maximization problem.

$$Lagrangian_1 = \pi_j \, U_j(T_{ij}) - \lambda * (D - (T_{ij} - N_{ij}) * UN) = 0 \qquad (12)$$

The equation (12) presents the relation between transport cost from home to work and to the shops. The equation has to be derived for three cases. If the place of work and consumption are different there are two derived variables in the equation, and if the consumer works and consumes in the same zone there is only one variable in the equation. The above presented method is applicable to involve transport network in general equilibrium models as an internal variable, which allow us to define optimal development interventions maximizing social welfare function.

7. Optimal measure toolkit of urban development – Numerical method

In the previous section transport cost was involved in the general equilibrium model as a continuous, internal variable. An example is presented below (provided by the authors), although practical application of the methodology will in the next step of the research be demonstrated.

To present a simple example, which can fit to the above mentioned conditions, we applied the equation system of the basic model but extended it with a transport development module. The parameters of the basic model were set to queasy-symmetric. So every $c_{ij} = 0.33$, , $a_{wj} = 0.5$, L_{sum} = 10. The only asymmetric part of the model is the arbitrarily chosen relation matrix below, which defines the home-work pairs. The definition of the matrix was led by the objective of ensuring that only differing locations are ordered together as a home – work pair.

		Location of home		
		1	2	3
Location of workplace	1	0	1	0
	2	0	0	1
	3	1	0	0

Table 1. Relation matrix of home – work location pairs

Finally the equation system has to be solved. Change in T_{ij} represents the result of public resource allocation. To be able to evaluate the results the equation system is solved for four basic cases, which can provide us points of reference. First the equation system is solved with a basic transport network (no intrazonal travel time representation), the transport cost matrix of which is presented below.

		Location of home		
		1	2	3
Location of workplace	1	10^{-7}	$2*10^{-2}$	$2*10^{-2}$
	2	$2*10^{-2}$	10^{-7}	
	3	$2*10^{-2}$	$2*10^{-2}$	10^{-7}

Table 2. Basic transport cost matrix

Table 3. contains the results of the different solutions. There are twenty variables and five cases. The variables have already been introduced related to the basic model. The "A" case represents the "do nothing" case. The "B" case represents the effect of a transport cost reducing intervention between location 1 and to 2. In case "C" all transport costs have been reduced symmetrically. In case "D" the same transport cost reducing measure was concentrated on two links asymmetrically. The "E" case represents the optimal solution of resource allocation based on the conditions of the constraint function. Where the cost of a unit of transport development is one (UN=1) and available public resources are 0.0009 (D=0.0009).

Variable	A	B	C	D	E
L_1	3,33331	3,33334	3,33332	3,32991	3,333333
L_2	3,33332	3,33167	3,33332	3,33505	3,333333
L_3	3,33332	3,33499	3,33332	3,33501	3,333333
X_{11}	1,12615	1,11211	1,12386	1,12382	1,111873
X_{12}	1,10352	1,10967	1,10468	1,10273	1,11073
X_{13}	1,10352	1,11159	1,10468	1,10343	1,11073
X_{21}	1,10352	1,10971	1,10468	1,10247	1,11073
X_{22}	1,12671	1,11173	1,12427	1,12704	1,111874
X_{23}	1,10298	1,11030	1,10429	1,10529	1,110729
X_{31}	1,10352	1,11150	1,10468	1,10327	1,11073
X_{32}	1,10298	1,11017	1,10429	1,10533	1,110729
X_{33}	1,12671	1,11321	1,12427	1,12625	1,111874
P_1	1,00000	1,00010	1,00000	1,00061	1
P_2	1,00000	1,00024	1,00000	0,99954	1
P_3	1,00000	0,99966	1,00000	0,99984	1
W_1	1,03046	1,00275	1,02590	1,03015	1,001571
W_2	1,03046	1,00305	1,02590	1,02802	1,001571
W_3	1,03046	1,00184	1,02590	1,02864	1,001571
L_{sum}	9,99997	10,00000	9,99998	9,99998	10
U_{sum}	3,32953	3,32956	3,32961	3,32956	3,329751

Table 3. Results

Summing up the results, it is clear that the change in transport costs indirectly affect customers' utility. Investigating case "B" we can see that the asymmetric intervention (T_{ij} between location 1 and 2 became $1,8*10^{-2}$) enhanced social welfare, however a sensible shift in the equilibrium can be observed.

In case "C" the interventions were symmetric (T_{ij} between all locations were reduced to $1,8*10^{-2}$). It was expectable, that this more intensive "measure toolkit" enhanced social welfare function more and of course equilibrium remained symmetric.

In case "D" the interventions were asymmetric again. The total transport cost decline were the same as in case "C", however transport cost of link between location 1 and 2 was reduced with one unit, transport cost of link between location 2 and 3 was reduced with two units and transport cost of link between location 3 and 1 was reduced with three units. In this case the social welfare grew less than in the symmetric case.

In case "E" the optimal solution was defined. The optimal solution was symmetric and compared to other alternatives optimal solution generated the maximum value of social welfare function, how it expectable was.

8. Conclusion

In earlier times – as we have seen – travel time, speed and capacity problems already made people try to estimate mobility demand. OD matrices describing mobility demand between zones are traditionally assumed to be constant. However, when a network operates close to its capacity, then a realistic traffic assignment would modify the network's travel time matrix. That engenders changes in residence and production-place-choices and in long term this phenomenon would affect mobility demand structure and so the origin-destination matrix as well. Hence the assumption of constant demand does not prove to be realistic.

The continuous development of SCGE (spatial computable general equilibrium) models has made it possible to describe the behaviour of actors playing various roles in geographically closed economic space. Nowadays SCGE models can be applied at acceptable estimation efficiency to evaluate the expected spatial economic structure and the development of a given region.

Beside the traditionally produced output variables of general equilibrium models (e.g.: process, wages, rents) transportation cost or travel time can be involved in the equation system so as to be able to define the optimal measure toolkit leading to more efficient transport network . This approach makes it possible to enhance the efficiency of urban development processes, and beside this, it extends the traditional engineering approach by involving demand matrices in modelling process endogenously.

Thus a new module of equilibrium models has been introduced, which makes it possible to support the transport development. In this way it is possible to optimise our interventions on the investigated system.

In conclusion, the efficiency of urban transportation is getting more and more important because of the increase rate of mobility demand. To plan, control and organize urban transportation in the most efficient way, we also need to consider the aspects of land use. To handle both of the above mentioned urban planning areas, we shall develop models able to pay attention to all of their restrictive factors within temporal properties.

9. Acknowledgement

This work is connected to the scientific program of the " Development of quality-oriented and harmonized R+D+I strategy and functional model at BME" and "Modelling and multi-objective optimization based control of road traffic flow considering social and economic aspects" project. These projects are supported by the Szechenyi Development Plan (Project ID: TÁMOP-4.2.1/B-09/1/KMR-2010-0002) and by program CNK 78168 of OTKA.

10. References

ABAG Regional Data Center (1991). Association of Bay Area Governments: Final 1990 census and historical population data of the San Francisco Bay Area cities and counties. San Francisco, Association of Bay Area Governments.

Anas, A., Liu, Y. (2007). A regional economy, land use, and transportation model (RELU-TRAN): formulation, algorithm design and testing.

Anas, A. (2007). A unified theory of consumption, travel and trip chaining. Journal of Urban Economics. No. 62. pp. 162-186.

Auclair, C. (1997). "The UNCHS (Habitat) indicators program". Sustainability indicators — Report of the project on Indicators of Sustainable Development. Wiley, New York, Pp. 288–292.

Bagwell, P. S. (1974). The Transportation Revolution from 1770. New York, Barnes & Noble.

Bokor, Z., Török, Á. (2011). Introduction of a capacity sensitive OD matrix estimation process applying genetic algorithm based calibration of SCGE model. World Conference on Transport Research, Lisbon.

Burgess, R. (2000). The compact city debate: A global perspective. In: M. Jenks and R. Burgess, Editors, Compact cities: Sustainable urban forms for developing countries, E & FN Spon, pp. 9–24.

Carey, H. C. (1877). Priciples of Social Sciences. Philadelphia, J.B. Lippincott & Co.

Chicago Area Transportation Study, (April 1962). Chicago, Illinois, Harrison Lithographing, 1959-1962. (Study Findings (Volume I) December 1959., Data Projections (Volume II) July 1960., Transportation Plan (Volume III).

Dixit, A. K., Stiglitz, J. E. (1977). Monopolistic Competition and Optimum Product Diversity. The American Economic Rewiew. Pp. 297-308.

Gubbins, E. J. (2002). Managing Transport Operation. London, Kogan Page Ltd.

Hagerstrand, T. (1970). What about people in regional science? Regional Science Association Papers 24, 7–21.

Hunyady, Gy., Székely, M. (2003). Economic psychology. Osiris, Budapest.

Jávor, B. (2005). The living urban machine. Ökotrend, Budapest, 2, 20-21.

Juhász, J. (2003). Hungarian Dictionary. Budapest, Academy.

Kornai, J. (1971). Anti-Equilibrium. Budapest.

Krugman, P. (1991). Increasing returns and economic geography. Journal of Political Economy 99, 483–499.

Maslow, A. (1943). A Theory of Human Motivation. Psychological Review 50(4), 370-96.

Pareto, V. (1897). The New Theories of Economics. Journal of Political Economy 5, 485-502.

Plátön. The Republic (Politeia) http://etext.lib.virginia.edu/ebooks, Virginia: Copyright 2000, by the Rector and Visitors of the University of Virginia, 144. p.

Ruiter, E. R., Ben-Akiva, M. E. (1978). Disaggregate Travel Demand Models for the San Francisco Bay Area. Transportation Research Record 673, 121-128.

Russo, F., Musolino, G. (2009). Multiple equilibria in spatial economic transport interaction models. Association for European Transport and contributors.

Samuelson, P. (1941). The Stability of Equilibrium: Comparative statics and dynamics. Vol. 9., Nr.2.

Samuelson, P. (1948). Consumption theory in terms of revealed preference. Econometrica 15.: 243-253.

Samuelson, P. (1954). The Pure of Theory of Public Expenditure: The Review of Economy and Statistics. Vol. 36., Nr.4.

Smith, A. (1776.). An Inquiry into the Nature and Causes of the Wealth of Nations. London, Methuen & Co., Ltd.

Walras, L. (1876). Éléments d'économie politique pure.

Wegner, M., Fürst, F. (1999). Land-Use Transport Ineraction: State of the Art. TRANSLAND project. Universitat Dortmund – Fakultat Raumplanung.

WCED: World Commission on Environment and Development. (1987). The Brundtland Report, Our Common Future. Oxford: Oxford University Press.

2

Philosophical Urbanism and the Predilections of Urban Design

Abraham Akkerman

Department of Geography & Planning, and Department of Philosophy
University of Saskatchewan, Saskatoon, SK
Canada

1. Introduction

The scholarly thrust of urban planning is its use of, as well as its contribution to, social science. As a discipline bordering with civil engineering, furthermore, urban planning has always carried also an important cross-disciplinary message beyond social science. Yet the association of urban planning with *humanistic* disciplines and the fine arts has been undervalued or ignored altogether. At a time when the vast majority of humanity resides in cities, however, this association is implicit in the primary purpose of urban planning, as a constituent of contemporary social, scientific and technological progress, aimed at the advancement of both society as well as the individual human being.

Epitomizing the bond of urban planning with humanistic concerns and the liberal arts is urban design, sometimes considered a sub-discipline of urban planning, at other times viewed as an extension of architecture or landscape architecture. Also due to the long history of built form, urban design has a tradition of thousands of years. Whereas urban planning usually traces its origins to the nineteenth century, the deliberate design of built urban form goes millennia back, to Çatalhöyük, Mohenjo daro and Jericho.

The purpose of the present chapter is to explore urban design from the perspective of two of its historical, albeit overlooked, aspects: philosophy and psychoanalysis. Focusing on the historical perspective of urban design the present chapter aims precisely upon these two aspects as presently missing links. Philosophical concerns and psychoanalytic backdrop that throughout history have been instrumental in the built urban form have been largely ignored in discussions surrounding urban design. Yet urban design, and by extension, urban planning in general, ought to consider these two missing links in the construction and understanding of our built environments not only as historically significant, but also as guiding considerations in the planning and design of human habitat in the third millennium.

Some recent reflections upon the built environment have related urban design with what has been termed by Friedrich Nietzsche as the Dionysian and Apollonian dispositions of the arts, and with the philosophical urbanism of Walter Benjamin. The present chapter will explore the nature of this linkage suggesting a psychoanalytic discourse related to foundational gender aspects in the process of urban design. In the contemporary milieu of urban society, policy and politics, such a discourse has its own significance if only for its implications upon gender representation in the built environment.

Much of the discussion on gender in urban design had focused on operational significance and the underlying social, economic and political reasons for the historical lack of fair consideration of gender in the built environment. The gender discourse in urban design, however, has been lacking a fundamental, philosophical insight that would serve as a self-reflection for the designers themselves, rather than a mere operational guidance for the design process and its objectives, as has been the case so far.

The present chapter addresses the notion of philosophical urbanism as cognizance of a spatio-temporal progression whereby city-form and the mind are intensely intertwined. Philosophical urbanism focuses on this progression as a historical interaction projecting gender features upon the built environment, and in return, absorbing features of the existing built environment into mind's own thought-processes, of which urban design is only one facet. The recognition of gender-based myths is paramount in this context. Two such myths have been shaping the history of ideas as well as evolutionary change in city-form. The myths of the *garden* and the *citadel* are the intertwined agents of cerebral progression of minds and the transformation of built environments in a spatio-temporal interaction process that had commenced in prehistoric times and is still ongoing over our very own geographic space. It was during the early Greek antiquity that the myth of the garden had transformed into the Dionysian deity, and the myth of the citadel into representations of the god Apollo. While the feminine *garden* and the Dionysian have always represented nature, the masculine *citadel* and the Apollonian have been transmuting onto the myth of the ideal city. But whereas city-form has evolved mainly due to the myth of the ideal city, over historical times the myth of the garden has become subdued. Gradually throughout history of the built environment the myth of the garden has been replaced by the allegory of the Grand Designer as a companion myth to the ideal city.

Environmental allegories are universal imprints of the mind, and urban planners and designers ought to recognize the significance of such allegories and myths within their own consciousness as it projects itself upon the environments they plan and design on behalf of others. The underrepresentation of the garden myth in contemporary urban environments, in particular, is critical. In the spatio-cerebral amalgam, the mind-city composite, and in present-day urban civilization, the garden allegory ought to constitute a vital component. The functional aspect of recognizing the significance of the garden myth is not so much in the promotion of urban gardens and green spaces, but mainly in the endorsement of serendipity and surprise through safe walking opportunities in the city. At the profound level of self-reflection, not only designers but urban planners too may recognize their own place in the historical feedback between mind and city-form.

2. Origins of urban design and the human form

During the late Renaissance, inspired by the work of the Roman Marcus Vitruvius Polio, *De Architectura*, Leonardo da Vinci penciled a well-known male figure circumscribed by a square and a circle. Leonardo's drawing of the Vitruvian Man, itself a source of later inspiration in the fine arts, depicted proportions in the human body as a guide to both applied and aesthetic appeal in human-made artifacts. It is very likely that the impetus to

Leonardo's drawing was the discussion of human proportions by Leon Battista Alberti in his handbook on sculpture, *De statua*, published in mid-fifteenth century.[1] It is also of more than passing notice that Alberti's short treatise appears as well to have inspired the Tuscan painter and engineer Francesco di Giorgio Martini (1439 – 1501) who in his own treatise, *Trattati di architettura, ingegneria e arte militare*, showed the physical form of a human body as a standard for the optimal layout of an ideal city.

The Canadian art historian, Domenico Laurenza, has suggested that di Giorgio Martini and Leonardo had met in the late fifteenth century in northern Italy, where they would have also likely discussed the Vitruvian Man. The meeting with Leonardo took place at least several years following the publication of Francesco's treatise where his drawing, in Figure 1, appeared. [2] To whomever of the two Renaissance artists the antecedence in the Vitruvian-inspired drawing of a man is claimed, there could be little doubt that Francesco had been the first to advance an allegory of likeness between ideal urban features and the human body.

Resident at Florence, Francesco would have been familiar with, and very likely influenced by, the Florentine academy, led at the time by Marsilio Ficino (1433 – 1499). It would have been through adherence to the Neo-Platonic doctrine of the Florentine academy that Francesco's anthropomorphous city-form appears to have been also an alteration of the very first western concept of the ideal city – one by Plato in the 4th century BCE. In his ten volume philosophical treatise, *The Republic*,[3] Plato had detailed the social structure of his ideal city through an analogy with the makeup of the human soul. Plato had used the city-soul correspondence to advance his own vision of social stratification within the ideal city, but had said relatively little about the ideal city's physical structure. By extending Plato's city-soul analogy onto a city-*body* analogy Francesco, presaging Leonardo's pictorial notion of human body's outline corresponding to the circle and the square, had addressed the gap left by Plato:

> *I will describe the various parts of city areas and how they have the same structure and form as the human body. First, thinking of a human body stretched out on the ground, I will place a thread on the navel, and pull it in a circular motion around that body. Similarly, squared and angled the design shall be. Moreover, just as the body has all its parts and limbs with perfect measure and size, the same should be noted of those cities* [4]

In Francesco's urban planning proposal, and certainly in Plato's own philosophical doctrine, the Ideal City notion instills flair of universality to the fusion of minds with their built environment. Plato saw his ideal city both as a mirror of the human soul, as well as an impression of a cosmic prototype of the city found in heaven.[5] The fabled view that the city should reflect cosmic qualities was further developed by the Stoics who, a century later, extended Plato's view by conferring an organic character to the cosmic notion of the city. In the Stoic myth, advanced by Zeno of Citium (334 - 262 BCE) and Chrysippus of Soli (280 - 207 BCE), cosmopolis – the universe as a city – had been likened to an immense animal: The sun as its soul, the stars as godly creatures. [6]

The view that the terrestrial ideal community ought to mirror the cosmopolis was inherent in the teachings of the Stoic Cleanthes (c. 331-232 BCE), Zeno's successor.[7] Cleanthes furthered the myth depicting cosmos as a vast, rational animal, into an elaborate scheme where *pneuma* (fire or soul), which accounts for the structure of the universe and for the destiny of individual things, resides in the sun. This fabled view sees a measure of *pneuma* in each thing on earth, and the highest measure of it in humans. Creatures which most closely approximate the entirety of the universe, are rational life-forms – humans and gods, the latter being stars in heaven. [8]

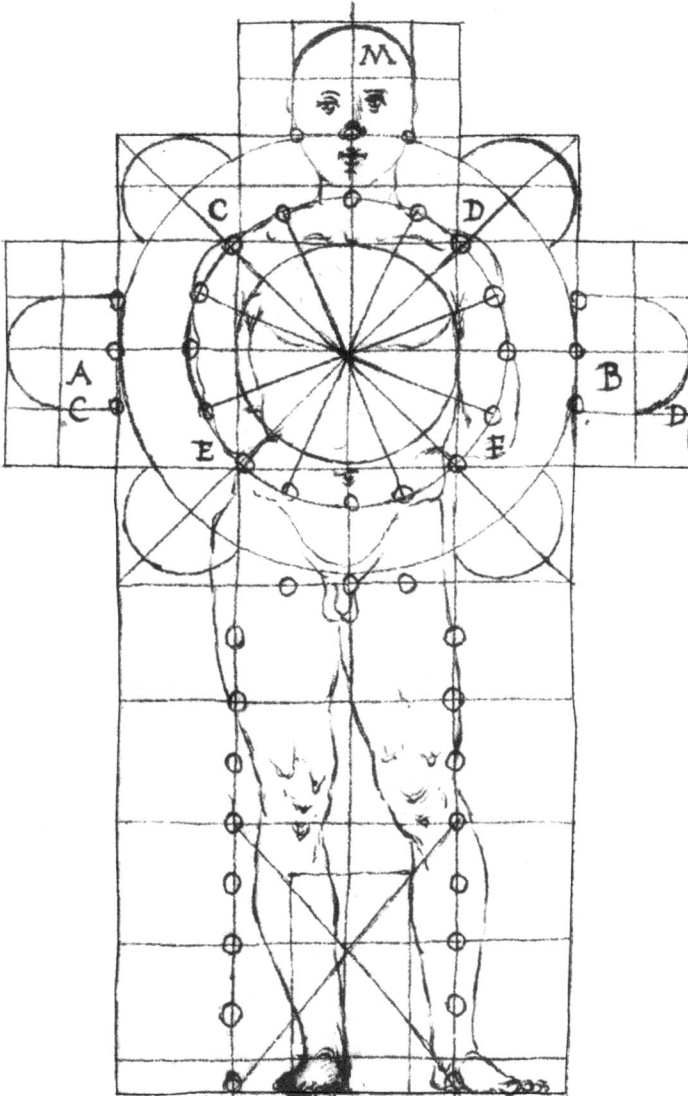

Fig. 1. Outline of an Ideal City. Francesco di Giorgio Martini (1482), *Trattati di architettura, ingegneria e arte militare*, Tomo I, Tomo II, transcrizione di Livia Maltese Degrassi (Milan: Ed. Il Polifilo, 1967).

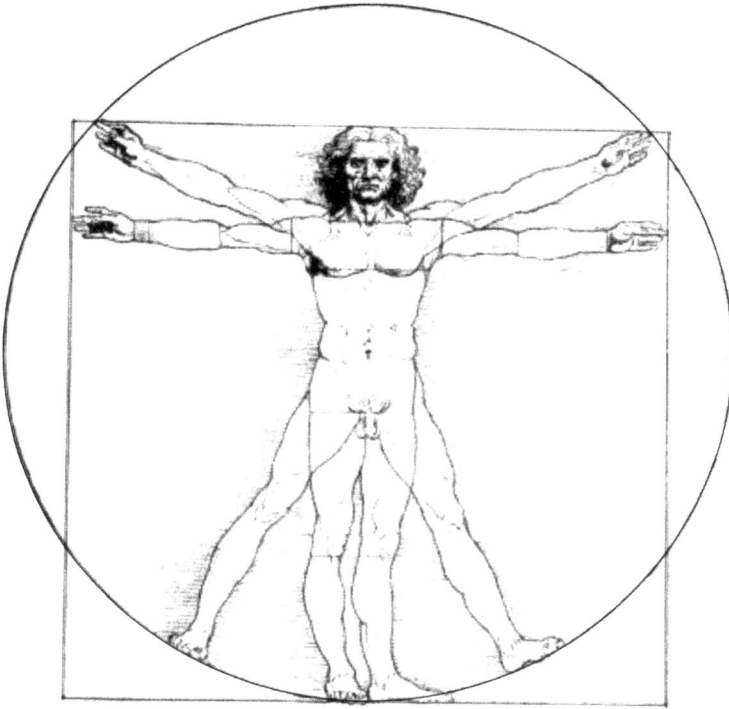

Fig. 2. The Vitruvian Man. Leonardo da Vinci (1490), "Vitruvian Man," pen, ink, and watercolour over metalpoint, 334x245mm (Venice, Italy: Gallerie dell'Accademia).

The organic penchant of the illusory cosmopolis may have seeped into Stoicism from a much earlier, Sumerian marriage myth of the Sky Father and the Earthmother. Both myths, it has been pointed out, are fundamental to the allegory of *hieros gamos*, central in Carl Jung's interpretation of the terminology of alchemy. [9] Francesco's urban concept, thus, seems to have embraced two mythical traditions of the ideal city: The one Platonic, in which he pictures the ideal city in an anthropomorphic analogy; the other, Sumerian-Stoic, in which the ideal city reflects an organic perfection of the cosmos itself.

3. The mind-city composite

The city-form of Siena, where Francesco was born and raised may have been conducive to his notion of an ideal city as a mechanistic analog to the human body. Said to exhibit to this day "planned organicism" [10] Siena, in spite of having been overtaken by neighboring Florence in 1484, had successfully and consistently retained urban features quite different from those of Florence. Striking among them, still today, is the absence of urban foliage in Siena, on the backdrop of the city's centuries-long close relationship with its agricultural hinterland. Siena's curved streets from the time of the Romanesque through to early Renaissance appear to extend the randomness of surrounding nature onto the built environment of the city. Siena's physical features, therefore, seem to have been instrumental in Francesco's own organic outlook of his ideal city, where ongoing interaction between the built environment (that of Siena) and an early concept (the Platonic and Stoic myths of ideal communities) evidently yield a new concept (Francesco's city-body analogy), which ultimately translates into a new urban landscape (or Francesco's proposal for one).

A feedback sequence pattern between city-form and thought, one that the example of Francesco's project seems to imply, has been alluded to throughout the twentieth century, but never pursued to the full extent. The immediate origin to inquiry on the mind-city interaction can be traced to Marcel Poëte[11] and Walter Benjamin [12] in the first half of the twentieth century, and such inquiry has been recently labelled, *philosophical urbanism*.

Poëte put forward the conjecture that the history of each city is reflected in its layout and projected upon its city-form. On this view, differences in current layouts between cities were said to inherently indicate their different histories, the innate memory of an early settlement being imprinted in its urban descendant centuries later. In a blend of mysticism and philosophy Poëte has been seen to bestow the flair of a living organism upon the city, implying a single hybrid of city-form and the minds within it. [13]

Similarly, but drawing on Carl Jung's notion of the collective subconscious, Benjamin observes "elements of ur-history" wedded in an epochal feedback pattern, producing in their contact with contemporaneous urban environment a newly shared ideal, or common elements in the visioning of a utopia. The shared elements of a community's image of a utopia stamp their mark upon contemporaneous configuration of urban objects, "from permanent buildings to ephemeral fashions." [14]

Arguably, the notion of a mind-city composite could be detected already in Plato's city-soul analogy, where the ideal city, as the Form of a city, is a universal paradigm shared by all mankind.[15] The broader context of inquiry could be seen reaching to the cultural geography of Carl O. Sauer in the first half of the twentieth century and to the psychosocial context of place offered recently by Charles Withers.

In "The Morphology of Landscape," Carl Sauer[16] advanced the notion of a cultural landscape as the imposition of culture upon nature. Culture, as defined by the shared myths, beliefs and behavioral standards, manifests itself in a cultural landscape, *i.e.*, in human intervention upon natural landscape. The urban landscape of Francesco's Siena, as an example of a cultural landscape, illustrates Sauer's insight on the ongoing feedback: human impact upon physical landscape yields a return sway in the reverse direction. Landscapes modified by human action, an aspect of culture, impact culture itself, whereby the feedback interaction between culture and landscape as a progression in time defines much of the history of civilization. Francesco's urban contemplation shows, as a case in point, how urban landscape feeds back onto Francesco's mind. Francesco's urban contemplations, then, are only a link in a feedback chain. On this view the mind-city composite is a spatiotemporal construct of which an urban landscape at a point in time is only a single "snapshot" instance of a cultural landscape.

On the other hand, too, philosophical urbanism comes to complement the contextualization of place as recently articulated by Charles Withers,[17] particularly in the foundational sense of location. To Withers the importance of place is within the context of two questions emanating from both myth and religion: *Who am I?* and, *How have I come to be?* The cornerstone of philosophical urbanism, both as a theoretical concept as well as an urban design issue, is addressed within the context of the question, *Where am I?* Withers posits his two questions as being of fundamental significance in the understanding of scientific revolutions, and in the interpretation of cultural change. The recognition that place is, in essence, an evolutionary-subjective entry in geographic space, rather than a posted static-objective structure, is at the heart of the dynamic interaction that emerges with the notion of the mind-city composite.

The notion of *urban* place as evolving through subjective experience is evoked by Michel de Certeau in his 1980 essay, "Walking in the City." An individual's memory forges a subjective urban milieu and transforms urban space into a mythical text.[18] De Certeau views streets as words, both of which involve "hollow places in which a past sleeps." Streets to him are non-verbal allegories which call upon our memory. Only through walking, the one mode of travel almost entirely overlooked by 20th century urban design and planning, can we, according to de Certeau, preserve the communal memory of urban places, thus conferring meaning to them and the streetscape enveloping them. This view has been echoed more recently by Rebecca Solnit.[19]

De Certeau's essay reiterates account given by the philosopher Henri Bergson some 60 years earlier. In his *Introduction to Metaphysics* Bergson gives an example of a continuum flow of human experience contrasted against discrete snapshots of the same experience.[20] The example Bergson chooses is a walk through a city, and the contrasting comparison is between the continuum of the experience of walking through a city against a series of separate photographs along the same walking route. The additional useful comparison Bergson had made was to match his city example with another example – the reading of a Homeric poem. The wholesome experience of the poem, contrasted with detached lines from it, is the analogy Bergson offers to his earlier example of a city walk against a series of discontinuous observations of the city.

4. The rise of environmental myths

The Bergsonian insight emerges from the Beaux-Arts architecture and urban design of the turn of the twentieth century, and the ensuing philosophical urbanism ought to be seen as rooted in the Europe of that time. Prominent within this context is Le Corbusier's urban scheme, The Radiant City (*Ville radieuse*). Joint with his cousin, Pierre Jeanneret, the 1931 project[21] is striking in its blueprint of zoning in parallel bands, from offices at the top end of the drawing, through housing in the centre, to industry at the bottom. In the words of Kenneth Frampton, the plan of the Radiant City is nothing but "a humanist, anthropomorphic metaphor [...] inserted into this model". [22] The resemblance to the allegoric project of Francesco's ideal city five centuries earlier is unambiguous. Permeating through the blueprint of Le Corbusier's Radiant City is an outline of a masculine super-creature, his entire plan thus standing for an anthropomorphic metaphor recalling the Platonic myth of the demiurge.

A mechanistic myth is evidently also behind the Modulor, Le Corbusier's own version of the Vitruvian Man, showing proportions within the human body as corresponding to the Golden Section ratio in two different series of anthropometric ratios. Both series, referred to as the Blue and the Red, constitute the guideline for Le Corbusier's design of furniture and urban dwellings.

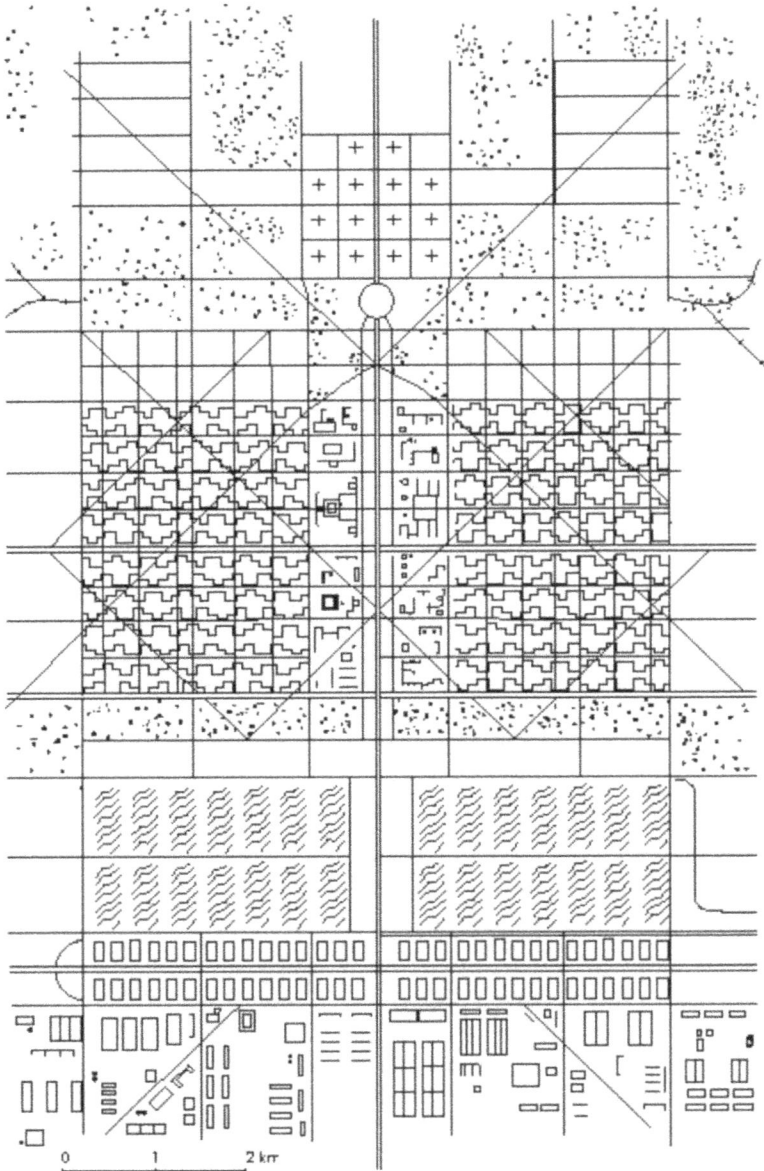

Fig. 3. *Ville radieuse*, an ideal city plan, showing offices at top, housing at middle, and industry at bottom. Le Corbusier (Charles Edouard Jeanneret) and Pierre Jeanneret (1933), *The Radiant City*, translated from the French by Pamela Knight (New York: Orion Press, 1967).

A symbiotic fusion of human minds and city-form, mutating through historic time and geographic space, the mind-city composite, as a pre-rational, spatio-temporal hybrid, gives rise to the notion of the ideal city.[23] This evolutionary outlook is premised upon the standpoint that the human mind possesses a prehistoric environmental imprint traceable to the Earthmother, a primordial feminine paradigm that Jung had identified in the 20th century. [24] On this view, *two* geographical and gender-related myths arose from the Earthmother. The feminine myth of the garden had evolved from the nurture-giving Earthmother symbolizing the female gatherers of fruits and vegetables. Contrasted with the comfort and nourishment of the Earthmother, the menace of nature's ever changing fortunes as well as male-made violence gave rise to the masculine myth of the citadel.[25] Through actual construction of citadels as man-made shelters and sanctuaries, both the myth and the stature of the citadel came to feed upon each other in an ongoing environ/*mental* progression.

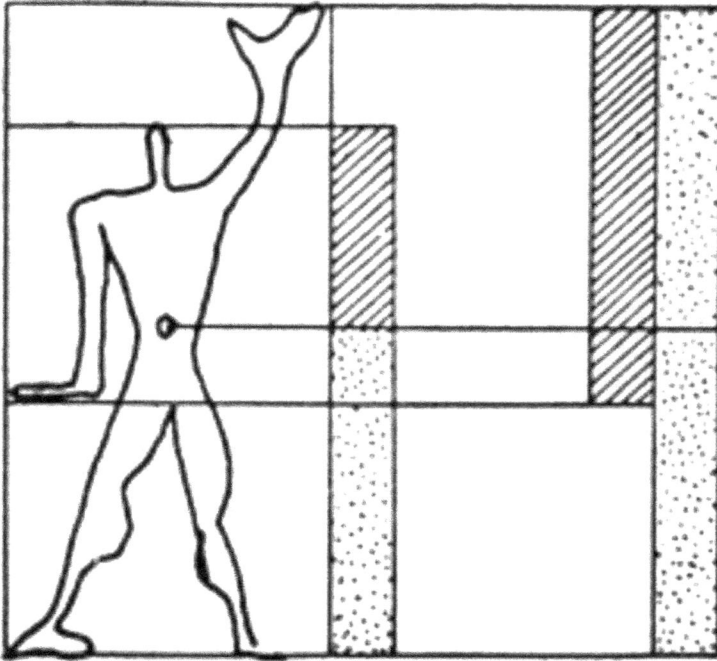

Fig. 4. Le Corbusier's Modulor (reproduction).

For the early man the source to the idea of construction, the creation of shelter in particular, was twofold. On the one hand, the flight for survival from the existential threat of scorching heat or violent storms, came to be attributed to the varying temper of the sky. On the other hand, equanimity and the feeling of safety had emerged from the observation of fixed stellar configurations in the sky, detectable through calm, clear nights. Emerging over millennia, verifiable observational tradition of the nightly sky inspired the notion of a conscious building scheme in early man. The observation of the pole star and its envisioning as a pivot round which sky vault revolves each calm night, or upon which the universe had been built, became the perceptual origin as well as inspiration to the idea of a plan and judicious construction.

Fig. 5. Apollo – the Greek god who personified masculinity, city-walls and colonies, prophecy, music, light, medicine and hunting.

Evidence to the link between the universe so envisaged and the myth of the ideal city can be thus sought in the pole star. While the emergence of early modern humans, some 120,000 years ago is believed to have originated in southern Africa between the Equator and the Tropic of Capricorn[26] the vast majority of urban civilizations had evolved in the northern hemisphere. As against the sky of the southern hemisphere a unique feature of the northern sky is that during the 26,000-year period of the precession of the equinoxes, northern direction of the earth axis points virtually continually to a pole star, a clear orientation beacon to migrating humans.The southern celestial pole, on the other hand, is almost never marked by a bright star, and when bright stars of the southern hemisphere, Canopus (Alpha Carinae) and Achernar (Alpha Eridani), come close to the southern pole (within about 10° c. 12000 BCE, and 8° c. 4000 BCE, respectively) they appear so close to the gleaming Milky Way of the southern hemisphere that their brightness contrast is lost to the naked eye.[27]

Northward street orientation has been ingrained in urban environments throughout history. The first sign of systematic urban design is an orthogonal grid plan of street blocks evidenced in the Indus Valley civilization of the third millennium BCE. At Mohenjo-daro, c.

2600 BCE, twelve orthogonal city blocks, each measuring 1200 x 800 feet were formed by three 30-foot wide avenues, forming a north-south axis, and two streets crossing them at right angles.[28] The careful observance of cardinal directions at Mohenjo-daro is evidenced in other early cities as well.

Excavations of cities as ancient as Jericho (9000 BCE) or Çatalhöyük (7500 BCE), seem to reinforce the notion that since the Mesolithic in the Near East humans have projected the myths of the garden and the citadel, or the ideal city, upon their environments, thus accelerating their own cerebral development. The built environment, along with much of the rest of material culture, has attained corresponding feminine and masculine facets. Ongoing attempts to ensure protection as well as strife towards envisaged perfection in newly built environments fed back onto the myths of the garden and the citadel with new observations and contemplations. Evolving into antiquity, during Greek classicism gender facets of myth and material culture came to be expressed, in Nietzsche's coinage, in the Dionysian and Apollonian dispositions of the arts.[29]

Fig. 6. Dionysus – the Greek god of fertility, wine and ecstasy.

5. The ideal city and the Grand Designer

With capricious skies overhead, and unpredictable nature at his surroundings, the prehistoric man's plight for survival was contrasted with the permanency of a pole star as cornerstone and axle of the nightly sky vault. Emerging from this contrast was the myth of the Sky Father. Through the Platonic myth of the demiurge and the Stoic allegory of the cosmopolis, almost two millennia before Francesco and Leonardo, the Sky Father had transmuted into a parable of the Grand Designer, an anthropomorphic creature of a cosmic scale.

In *De Opificio Mundi* (*On the Creation of the World*), written about the year 30 CE, Philo Judaeus (20 BCE – 50 CE) had reiterated a contention that came to be known as the Argument from Design, the professed proof for the existence of God. In his attestation Philo uses a line of reasoning that historically traces to Cicero's *De natura deorum* [30] and that came to be known in modern age as the evidence for "intelligent design" of the universe.[31] In his statement Philo says:

> (VI) ... It is manifest also, that the archetypal seal, which we call that world which is perceptible only to the intellect, must itself be the archetypal model, the idea of ideas, the Logos of God, already occupied in the creation of the world; for neither is a city, while only perceptible to the intellect, anything else but the reason of the architect, who is already designing to build one perceptible to the external senses, on the model of that which is so only to the intellect. [32]

On Philo's view the cosmos can be compared to a perfect urban design of an architect, and God is seen as the cosmoplast, a Grand Designer much in the image of the Platonic demiurge.

Leaving aside criticisms against the Argument from Design, notably those of David Hume [33] and Immanuel Kant [34] the Argument itself has a psychoanalytic aspect. The question as to whether or not the Argument from Design is a valid proof becomes secondary in significance, once a terrestrial craftsman is put forward as an analog to deity. The idea of God, as the designer of the universe, compared by a human being to the image of self or of another human being, is, arguably, a subliminal expression of an individual's megalomania.

Fig. 7. Statue of Alexander at 'The Mount Athos Colossus' in Johann Bernhard Fischer von Erlach, *Entwurf einer historischer Architektur* (*Sketch of Historical Architecture*), (Dresden: Sächsische Landesbibliothek - Staats- und Universitätsbibliothek Dresden, 1721).

It is edifying that Philo, himself known for humility and modesty, was hounded by persecution of the Jewish community of Alexandria, a city founded some 400 years earlier on a plan by Dinocrates of Rhodes, to serve the grandeur of Alexander the Great and his stature as divinity. Intended as a showpiece of Alexander's power, the meticulously planned metropolis was built instead of a hallucinatory proposal for gargantuan statue of Alexander that was to be sculpted in the flank of Mount Athos in northern Greece. According to Vitruvius, in Book II of his *Ten Books of Architecture*, Dinocrates bragged to Alexander that he had prepared a colossal design for Alexander's statue in whose left hand a most "spacious city" were to be fashioned. [35] It was also in Egypt where the oracle of Amun, the King of gods, declared Alexander as the new Master of the Universe.[36] Such particular settings from which Philo's version of the Argument from Design had emerged show the allegory of the Grand Designer as a veiled response, and place it as a companion myth to that of the ideal city.

The two myths, the ideal city and the Grand Designer, have been two evolutionary forces that have molded city-form – the configuration of spaces, edifices and infrastructure. City-form throughout the history of civilization, could be thus seen more an expression of an ongoing projection of allegories rather than the strict product of reason.[37] The lineage of mutual feedback between the myths of the ideal city and the Grand Designer ought to be seen, therefore, as genealogy of the mind-city composite reflected in the history of city form and in the history of ideas.

The myths of the ideal city and the Grand Designer have been essential in cerebral progression of humans, and in the evolution of city-form, to this very day. In the history of built environments intervention by a sovereign or by a community's representative, have pointed to aspiration for an ideal city, while the mythical image of the ideal city has undergone adaptation following the actual change in the built environment. This ongoing process has continually molded the myths of the ideal city and the Grand Designer, much as it has also continually transformed city-form.

There could be little doubt that Francesco, as well as, modern architects or planners such as Claude-Nicholas Ledoux, George-Eugène Haussmann, Robert Moses, Frank Lloyd Wright, or Le Corbusier, saw their own projects as urban ideals by a grand designer.[38] Architects or planners bending over their three-dimensional urban models, of course, do not necessarily point to a subliminal version of the Dinocratic delusion at Mt. Athos. But self-image and ostentation of the designer frequently elucidate the background of such projects from past times. It is for this reason too that superficial attempts in more contemporary urban design to counter mechanization and monumentalism by imitating medieval city-form have met with dismissal. Frampton has gone as far as dismissing New Urbanism as megalomaniac.[39] Impact of the existing city-form upon the forging of the myths of the ideal city and the Grand Designer, and the return feedback of the two myths upon the design of urban landscapes defines philosophical urbanism as a meeting point of philosophy, psychology and urban design.

6. The *garden* as a missing preconception in urban design

Uncovered by Jung as archetypes of femininity and masculinity, Anima and Animus are each inherent in the gender of its counterpart: All men carry a subconscious feminine component within their persona, and all females similarly carry a subconscious masculine ingredient.[40] Jung's finding had followed on the heels of Friedrich Nietzsche's classification of Greek art as having Dionysian, feminine, or Apollonian, masculine, attributes [41] The myths of the garden and the citadel can be viewed, in fact, as precursors of Dionysian and Apollonian dispositions not only in the arts, but as ecompassing much material culture. Inevitably, the suggestion that edifices and open spaces, intertwined within a city, represent masculine and feminine urban features is a reasoned extension. Furthermore, a grid, or any other predictable pattern of streets implies measurability or masculine imposition upon open space, while a labyrinthine street pattern paraphrases nature's surroundings of the city, implying retention of the city's feminine trait.

It was the practical and the mythical that had conjoined as a gender projection in ideal city-form from antiquity through to the pre-industrial age. Throughout urban history compass directions and grid layout of streets have addressed the need for navigation, for protection from sun or wind, or for other thermal comfort, and for measurement of land in ownership, all consistent with the desire for the geometrically regimented urban ideal mirroring the perfection of the cosmos. But countering such universalist intentions was topography, lack of advanced technology, incremental intervention by property owners, and deliberate curvature of streets against armed intruders. Throughout the ages these two opposing forces behind the construction of cities cast the ancient and medieval city-form into a fascinating maze of lanes and open spaces. The desire for regimentation and predictability in streetscapes of the ideal city simply gave way to joint dynamic impact of myriad mundane small changes continually occurring over centuries in cities and towns, such as Siena.

The myth of the ideal city can be thus viewed as emerging from the *citadel* in an interplay with the myth of the Grand Designer, the latter having evolved from Jung's archetype of the wise old man, possibly through end of the Neolithic.[42] The myth of the citadel, progressing into the ideal city sometime during the late Bronze or early Iron Ages across the Fertile Crescent and the Indus Valley, had remained the prime constituent in the organization, growth, development and defence of human communities throughout the ages. This is how primal city-form had arisen during the late Bronze Age and early antiquity. Giambattista Vico's contention that the myth of the citadel arose through the outlook of warfare [43] further illuminates masculine features in city-form throughout history. The consolidation of masculinity in city-form in the west appears to have, indeed, continued intact from antiquity through to the post-industrial age and modernity.

It was the archetype of the Earthmother that gave rise to later allegories of the garden and the citadel. Whereas the masculine *citadel*, the initial consort of the *garden*, had evolved into the ideal city to become the driving force in the history of urban design, the feminine myth of the garden had turned sometime in the late prehistory into a secondary premise, and has become increasingly and gradually subdued ever since. By late Renaissance the myth of the garden had been all but ousted from city-form by the *citadel*'s emerging escort, the Grand Designer, a latter day version of the Sky Father and the archetype of the wise old man. Throughout much of history city-form has been thus moulded by visions of the Ideal City and the acumen of the Grand Designer. Through the two myths, both interacting and continually transmuting into new allegoric imprints, city-form, has been subject to corresponding historic transformations. Architectural and streetscape styles, historic or current, are therefore best looked upon as facets of the mind-city composite, rather than being simply viewed as observable features of an objective, physical reality of edifices, spaces and infrastructure. [44]

Epitomizing the garden allegory, green spaces within cities until early modernity have usually not been within the public domain. Since antiquity there has been only a sporadic record of *public* gardens, parks or green commons within cities. During the French revolution the royal Garden of the Tuileries was opened to the public, after the mob stormed it in 1792. In Britain successful park design in early 19th century's London by the architect John Nash, preserving nature within the city, may have inspired Patrick Geddes into becoming one of the most influential urban thinkers for whom natural wilderness was a key ingredient of the urban environment.[45] But in spite of the recognition Geddes has received in the annals of modern urban planning, the allegoric *garden*, as a public space, has continued to be only a secondary component in modern urban design. Best exemplified by Ebenezer Howard's Garden City concept from the turn of the 20th century, this ideal city was a forerunner of the North-American suburb intended to provide acceptable living conditions to the English working class.[46] But the brilliantly simple idea to inject nature into a small-size city almost immediately became a commercial success, and the two garden cities in England, Welwyn Garden City and Letchworth, had become exclusionary pieces of real estate space sought out by anyone but the working class.[47]

In contemporary city urban green spaces designed or designated for public use have usually fulfilled attendant function to the mechanized and automated city-form.[48] Even in cities where deliberately carved green spaces constitute their centre-piece, such as New York's Central Park, Vancouver's Stanley Park or Sydney's Botanical Gardens, safety and security concerns deem them anything but 'open.' The Apollonian overrun the Dionysian, Nietzsche might have said of twentieth century's city form.

And yet, even as a subdued and discarded parable, unassumingly throughout history, inefficiently and submissively, the Dionysian myth of the garden has been perched, as if in a slumber, against the Apollonian myths of the ideal city and the Grand Designer. As a spontaneous, authentic trait the *garden* could hardly be sought out in urban spaces purposely devised and assigned, but as spatial allegory the garden has never been entirely expunged from city-form; not even from the contemporary city. In fact, irrationality, spontaneity and authenticity, ought to be seen as fundamental ingredients of the *garden*, and they emerge in forsaken and overlooked places in every city still today.

Furthermore, it is not due to a defined space that the *garden* in the city endures, but rather due to *time* that is unstructured, immeasurable and entirely subjective, void of clock or calendar, marked by accidental events and sporadic surprise in urban spaces of abandon. It is, in truth, precisely this irrational aspect of city-form that is being assailed to be expunged from planned, "rational" city-form still today. Spontaneously emerging small urban spaces, neglected by planners, have become places of abandon throughout cities, while twentieth century urban design had focused on an Ideal City of the superscale: Brasilia of Oscar Niemayer; Chicago of Daniel Burnham; Canberra of Walter Griffin, to cite only a few examples.

Towards the end of the twentieth century the failure of the Ideal City had become evident through the increasingly frequent ad-hoc attempts to maintain mass transit between homes and workplaces, safety and public health in streets, to ensure the physical and economic survival of urban dwellers. Within the process of urban design, from a mythical Ideal City onto a city-form aimed at maintenance and survival, alienation arose turning the city into a community of reclusive strangers. And yet, it appears to be precisely alienation, urban dysfunction and the urban place of abandon, along with the immeasurable temporality of encounter with such places or events, that becomes the new fertile ground of the myth of the garden in city-form.

7. Self-reflection in planning and design

The pre-modern, pre-industrial city, as exemplified by Siena, is in stark contrast to the surprise-free, controlled and solemn Ideal City, or its modern version, the Rational City. The streets of Siena, much as those of extant Romanesque, Gothic or early Renaissance old-town sections throughout many European cities, exude organic ambience, and an ongoing opportunity for a fortuitous encounter. Security, along with transportation efficiency and safety, are the overriding concerns in the industrial and post-industrial city, where surprise has become tantamount to danger.

As facets of amorphous time and space medieval streetscapes offer anticipation and surprise as their very essence. The sense of authenticity in the streets and squares of urban places, such as Siena, appears inspired or absorbed from surrounding nature. Siena's maze of streets has no foliage or green spaces. Instead of an omniscient or benevolent "grand designer" behind its city-form, Siena's urban spaces were carved each with a startling originality through the profusion of edifices by individual builders of the past. Centuries-long accumulation of ongoing small changes in the built environment of Siena have introduced humane measure of disorder and unexpectedness into its streetscapes, but also genuine civic places of spontaneous encounter, as against solemnity and control of the ideal, envisaged in the Rational City of the twentieth century.

On the contrasting backdrop of the example of Siena, heroic slogans such as "Make No Small Plans" (Daniel Burnham) or "A City Made for Speed Is a City made for Success" (Le Corbusier), point to the state of mind prevailing during the twentieth century. The ultimate goal of the individual has been to optimize returns from participation in the urban community. But whereas the overarching goal of the twentieth century designer-planner has been to control or streamline the urban crowd, the goal of urban dwellers has been to extrude themselves from imposed crowd control and the corresponding rigidity of city-form. The urban public comprising such individuals has created, in turn, a runaway discordance: Traffic congestion, line-ups, systemic failures, inadvertent breakdowns or wanton vandalism are the overt symptoms of the disparity between minds and their built environment. From mere nuisance in the city of the mid 20th century, these manifestations have turned into defining traits of city-form at the present time.

In the escalating cycle of cacophony, individuals focus on ever-new means at extrusion from the rest of the crowd and at insolence against city-form. In this *super*conscious state individuals' extrusion from the crowd continually amplifies the conflict between the crowd and a city-form intended to streamline it – only to make incoherence and malfunction increasingly acute. The bewildering trait of the urban superconscious is that all attempts to escape it through extrusion only ascertain its very perpetuation and intensification. Focused on urban superstructures, automation and mechanization, twentieth century urban planning had paid its own tribute to the ghost of the superconscious.[49]

Within the post-modern urban context of spiraling incoherence, it is the self-reflection of the designer, the recognition of his or her own humanity and the attendant, unavoidable exposure to myth and existing city-form alike, that fashions thought – and within it, urban-design ideas too. It is within this discernment that one may recognize existing city-form as the culmination of the Apollonian attempt to rationally structure and configure urban *objects*, rather than focusing on the *voids* between them. The juxtaposition of a structure against a void, quite analogous to the contrast between the masculine and the feminine, between the Apollonian and the Dionysian, is also the contrast between the Citadel and the Garden, between noise and silence.

From the depth of one's own consciousness emerges the premonition that urban void ought to become the new sanctuary in the postmodern city. The focus of urban design in the 21st century ought to shift from urban superstructures to the understanding of urban voids. The acceptance of urban voids by favoring, first and foremost, their mindful preservation could constitute a much needed paradigm shift in urban design – a shift from the failings of the Apollonian to the ascent of the Dionysian.

The Apollonian-Dionysian split within city-form has, indeed, not only spatial but also a temporal manifestation. In the industrial and post-industrial city amorphous, unstructured time has become an impediment to the construction of urban environments, and to survival within them. The fashioning of modern city-form could not have occurred without the structuring of time through the ability to measure small temporal sequences within the diurnal cycle. As an ultimate affront to the *garden*, the industrial and post-industrial city-form has turned temporality in the metropolis into compartmentalized or – in the words of Henri Bergson – mechanized and spatialized sequence.[50]

As opposed to the amorphous time, as a medium to events in the medieval streetscape, time in the metropolis came to be considered an adversarial entity, objective and measurable, that could be confronted only through scheduling, optimization or another human scheme.

Yet against the measured, segmented and wholly objective time of clocks and machines in the planned city-form, there is the authenticity of subjective, emotive experiences of an individual. Hardly noticeable through an objective, measurable observation, seldom shared with or communicated to others, unstructured temporality, referred to as *durée* by Bergson,[51] is a lasting and unique quality of any living individual. In its organic vitality the subjective, internal and unstructured temporality supervenes on feelings, perceptual episodes, or individual experiences and provides a vital contrast to the mechanized form of the contemporary city. The challenge to urban design is to address traits that constitute our very humanity: To reintroduce amorphous time into city-form, to bring authenticity and spontaneity back into the city's streetscapes.

8. Myth of the Rational City

A hybrid evolving in geographical space and in historical time, the mind-city composite has given rise to the myth of the Rational City, and with it, to twentieth century city-form of the superscale featuring the urban superconscious as its own psychosocial trait. This latest transmutation of the myth of the ideal city can be viewed as constituting a transition stage of urban environments that were built to perform through mechanically structured time. This transition stage, expressed in Apollonian adherence to an ideal-city concept of structured temporality has led the contemporary city-form to the verge of urban dysfunction. Failure in attempts at planning and control of the city and its multitudes of people has been increasingly manifest in conflict and contradiction due to, exactly, attempts – necessary as they are – to structure time round clocks and rules of safety and security. One of the important traits of the modernist myth of the Rational City becomes thus the occasion to juxtapose the fable of the well-functioning city against the reality of the *mal*functioning city.

It is the pandemonium itself, along the disarray and incomprehension of present-day metropolis that seems to be turning into fertile ground to the rise of the Dionysian garden paradigm, inherent in the city-form of medieval communities such as Siena. The words of the Czech-French writer Milan Kundera resonate of a commonplace yet deep, shared and intense experience epitomizing the contradiction of the garden and the citadel, and its impact upon the human individual in the metropolis:

> She said to herself: when the onslaught of ugliness became completely unbearable, she would go to the florist and buy a forget-me-not, a single forget-me-not, a slender stalk with miniature blue flowers. She would go out into the streets holding the flower before her eyes, staring at it tenaciously so as to see only that single beautiful blue point, to see it as the last thing she wanted to preserve for herself from a world she had ceased to love [...] She said to herself: the world has arrived to the frontier of something disastrous; if it crosses it, everything will turn to madness: the people will wander through the streets with forget-me-nots in their hands or will kill each other on sight. It will take very little, the drop of water that overflows the glass: just one car, person or decibel more.[52]

It is the attempt to eradicate amorphous time from the midst of structured city-form that has brought the metropolis to the verge of collapse. This observation leads Kundera to make ominous distinction between human authenticity implicit in natural and spontaneous walk, and human automatism ensuing from mechanized and automated urban transportation of contemporary city-form:

> Before roads and paths disappeared from the landscape, they had disappeared from the human soul: man stopped wanting to walk, to walk on his own feet and to enjoy it. What's more, he no longer saw his own life as a road, but as a highway: a line that led from one point to another, from the rank of captain to the rank of general, from the role of wife to the role of widow. Time became a mere obstacle to life, an obstacle that had to be overcome by ever greater speed.[53]

Kundera's is a call to recognize time and change, not as a problem to be measured, structured or confronted, but as the golden rule of our world. Most of twentieth century urban design, including Ebenezer Howard's Garden City adaptations,[54] has never been attuned to such a stance, solidifying instead the Platonic myth of the ideal city and the Neo-Platonic notion of a solemn perfection. As another aspect of this disregard, mainstream urban planning has seldom acknowledged the failure of planned public parks and gardens, even as the very notion of twentieth century's *private* suburban garden, into which Howard's concept had evolved, has become mainly parlance of possession, division and control.

During the 3rd century BCE in Athens, Epicurus had established an arboretum known as the Garden, where women and slaves would be welcome to join in debates. About the same time, near a public park where the statue of Apollo-Lyceus, the effeminate Apollo, was standing, Aristotle had founded his *lyceum*. Aristotle's was a *peripatetic* philosophy, taking its very name from the boardwalk in his *lyceum* where walking was the physical complement of thinking and discoursing. A generation after Aristotle, and frequently at odds with his teachings, the Stoics had emerged in Athens fostering their own thought by simultaneously strolling *and* debating. On the north side of the city's agora the Painted *Stoa* stood, and it was there where Zeno of Citium used to discourse pacing up and down the painted colonnade: "[...] Hither, then, people came henceforth to hear Zeno, and this is why they were known as men of the Stoa."[55]

A million years ago, bipedalism detached breathing from striding in the early humans, thus facilitating the advent of speech, and the progress towards orderly observation and thought.[56] Throughout millennia, walking, perception and thought have become intertwined and integrated. The automotive and automated city-form of twentieth century's urban superscale had minimized, sometimes effectively eliminated, walking in the city. The automated superscale brought with it not only the frenzy of the urban superconscious, but it also perturbed an important kinetic-cognitive link in humans. Mechanized, automotive and automated city-form of the superscale leads to automatism in human behavior, promoting a course of severance of the cognitive process from kinetic propensity. Whether such development could lead, in the longer run, to weakening in the cerebral cortex of humans throughout metropolitan areas of the industrialized world, is a disturbing question that is yet to be answered.

In his *Laws* Plato envisions people in the ideal polis as if made out of waxwork.[57] The array of absolutist doctrines later in history, inspired by Plato's designed society as pointed out by Karl Popper,[58] could conceivably include some portions of twentieth century's urban planning. The lone voices of Jane Jacobs and Camillo Sitte, [59] inclined more to the dynamics of the community than to the wisdom of the planner, have been always politely received, but seldom acted upon. From the dawn of civilization onward the faith in a being (or beings) behind the design of the universe found, a matching, self-reflective notion in the priest or the overlord on their own role as the Grand Designer – or his faithful intermediary. The myth among the folk was reinforced by priests through cultic rituals that in time became automatic, uniform and universal. Mimicking a technological project and borrowing from the image and spirit of inventiveness twentieth century's myth of the Rational City has been an attempt at a surprise-free city – automated, uniform and universal, at times also celebrating the "grand designer" behind it.

Provided that, first and foremost, one sees a genuine human predicament in present-day city-form, the breaking away from automation and uniformity in city-form could hardly be found in yet another set of objects. If urban edifices are a masculine trait of city-form, while urban voids are its feminine quality, one cannot but recognize that a structure evokes association of a sedentary function, while urban void suggests movement, calling to mind a processual, rather than sedentary, function. Siena is an excellent example of such a processual attitude in city-form, where feminine essence has been preserved, as if the myth of the garden had prevailed in it.

Five hundred years after Francesco's treatise urban planning and design ought to be able to assimilate also the wisdom, knowledge and reflection that humankind has gained, uncovered or discerned since. It is to a large extent due to the spiralling dysfunction of the post-industrial city-form that urban design has begun to address human scale, and walking in particular. That myth has been behind the failure of the contemporary city-form is admission only seldom made; and unheard of entirely would be the suggestion that another myth could come to the rescue. Yet a comprehensive stance recognizing the presence of myth in city-form, and recognition as well as acceptance of its measured impact, much as acknowledgment of the mutual impact between mind and city-form as an integrated, ongoing and vital dynamics, can no longer be ignored in urban planning and design.

9. Conclusion: Recognizing the urban subconscious

City-form and human cerebral development, in their mutual and ongoing impact, confer an ambience of affirmative, hybrid interaction between minds and their built environment. This, in turn, puts a heavy responsibility upon the planners and designers of urban places.

Such responsibility starts with the recognition that informal, veiled urban landmarks in old dilapidated neighborhoods, never subject to deliberate plan, have *always* been the keepsakes of communal memory. Elisabeth de Bievre has called such communal memory the urban subconscious, the "sum of physical circumstances [...] and historical events, experienced collectively by a group of people living for several generations in the same environment." [60] City-form within which shrines of communal memory are preserved, doesn't resolve urban dysfunction, but provides a respite from one. Through perpetuation of authentic urban places, human authenticity too can be addressed. The entire scope of human disposition within city-form, from authenticity to alienation, can be in fact viewed in correspondence with the full range of urban infrastructure, edifices and voids, seen as conduits to human experience.

The photographer Ryuji Miyamoto, in his *Architectural Apocalypse* (1986), has called urban subconscious the decay of architecture disintegrating into ruins – either by way of deliberate destruction, through planned urban growth and modernization, or as a consequence of a natural disaster. Similar flair emerges from places for which the architect Ignasi de Solá-Morales coined the term, 'terrains vagues,' landscapes of contempt: large, empty, disused, and abandoned spaces of the city's fringes – monuments of the urban subconscious. [61] Bursts of entreaty that sometime emanate from such places of contempt, are due to disruption in control and rule inherent in mainstream city-form, but also from the silence these places imbue.

To Helen *Armstrong* the bizarre, unsightly voids, the various instances of urban decay manifest a peculiar kind of the subconscious, as if reiterating "Freud's argument that the power of the uncanny place of dereliction is that it ought to remain hidden; [yet it] keeps coming to light." [62] The dereliction of urban decay to Armstrong expresses passage of time – not as an object frozen in time, but as a living, organic continuum of change. Urban decay, seen as the urban subconscious is the city's forgotten feature, the *other*, often unacknowledged face of the city, the contrasting facet of urban perfection, or in the words of Jean-Paul Sartre, the French philosopher, the city's *reverse* side. Armstrong considers landscapes of contempt, places of urban decay, as the most foundational, albeit overlooked, urban voids in the contemporary metropolis. The urban subconscious, in its various interpretations, and human authenticity are tied in a bond that does not seem to be subject to a clear definition. Yet recognition of this nebulous link is perhaps the first step to a genuinely successful urban design, not so much through projection of yet another ideal-city concept, but through self-reflection of the planner and the designer.

A major dictum of twentieth century's urban planning was the fight against urban decay and its various manifestations in the city. One common such manifestation, undoubtedly, has been graffiti and the constant fight for its removal. Pointedly, in his documentary satire, *The Subconscious Art of Graffiti Removal* [63] the filmmaker Matt McCormick lampoons the masking or erasing of graffiti as an expression to the suppressed artistic desires of city planners. Indeed, city planners are human too. And it is precisely for this reason that self-reflection and recognition of the alienation automated city-form carries onto one's own psyche is perhaps the very first step that urban designers and city planners ought to undertake in the road towards successful city-form in the twenty-first century. In his novel, *Nausea*, Sartre expressed thus the feeling of recognizing one's own alienation in the automated city against the authenticity of a landscape of contempt:

> *I am on the curb of the Rue Paradis, beside the last lamp-post. The asphalt ribbon breaks off sharply. Darkness and mud are on the other side of the street. I cross the Rue Paradis. I put my right foot in a puddle of water, my sock is soaked through; my walk begins. [...] I am cold, my*

ears hurt; they must be all red. But I no longer feel myself; I am won by the purity surrounding me; nothing is alive, the wind whistles, the straight lines flee in the night. The Boulevard Noir does not have the indecent look of bourgeois streets, offering their regrets to the passers-by. No-one has bothered to adorn it: it is simply the reverse side. [...] [64]

In his urban-design classic, *Learning from Las Vegas,* [65] Robert Venturi had taught city planners the significance of the urban ugly, not as a quality to be subjected to acceptance or rejection, but as an urban feature from which one can learn to produce a *successful* city-form. Venturi's observations of Las Vegas are mainly those of an automobile driver. In similar vein Sartre's *Nausea* should be seen as a secular guide to walking. The authenticity of human experience, described in the short excerpt from *Nausea,* could never have taken place other than through the simple act of walking.

A city walk is the quintessence of *human* disposition in urban void as it encompasses change, variety and diversity. Not solely a mode of occasional urban recreation, but again as the main means of access in cities, walking could bring back human spontaneity, creativity and serendipity. Recognition that city-form is a symbolic, gender-based landscape deriving from primordial allegories could be an important *step*, indeed, towards forging a city-form conducive to favorable development of human communities. Permeating urban voids, the myth of the garden, subdued and irrational, can be brought back to the fold of urban communities and their built environments, in a city-form emerging anew from the old. Walking, a biosocial expression of change, is as profoundly inherent a human component of *any* garden, as it is a forceful, cumulative process against urban uniformity. The greatest challenge to urban design in the 21st century is to fashion urban voids to strolling rather than to arterial transportation flows, and crafting quiet voids for listening and tranquillity rather than devising noise barriers against urban decibels. Self-reflection might be the utmost skill required by designers and planners if they are to achieve the goal of building livable, and living, cities.

10. Acknowledgment

I am indebted to Joseph Garcea for translating the quoted section of the Italian original in Francesco's *Trattati* into English.

11. References

[1] Jane Andrews Aiken, "Leon Battista Alberti's System of Human Proportions," *Journal of the Warburg and Courtauld Institutes* 43 (1980): 68-96.

[2] See Domenico Laurenza, "The Vitruvian Man by Leonardo: Image and Text," *Quaderni d'italianistica* 27(2006): 37-56.

[3] *Republic* II, 368d-369a; IV, 434d-435c.

[4] Francesco in *Trattati* I, tav 8-9, quoted in David Friedman, *Florentine New Towns: Urban Design in Late Middle Ages.* (Cambridge, Mass. and London: MIT Press, 1988), p. 252, note 4.

[5] See also, *Republic* IX, 592.

[6] Alan Scott, *Origen and the Life of Stars: A History of an Idea* (Oxford: Clarendon Paperbacks, 1991), 39-52.

[7] Vernon Arnold, *Roman Stoicism: Lectures on the History of Stoic Philosophy* (London: Routledge & Kegan Paul, 1958), 204-205.

[8] Alexander S. Kohanski, *The Greek Mode of Thought in Western Philosophy*. (Rutherfors, N.J.: Fairleigh Dickinson University Press, 1983), 76-82.

[9] See Steven F. Walker, *Jung and the Jungians on Myth: An Introduction* (New York and London: Routledge, 2002), p. 78. For Jung's reference to the Earth Mother see his very last essay "Approaching the Unconscious" in *Man and His Symbols*, ed. Carl G. Jung, Marie-Louise von Franz, Joseph L. Henderson, Jolande Jacobi, Aniela Jaffe New York: Doubleday, 1964), 1-94.

[10] Thomas Harvey, "Siena and Sustainability: City and Country in Tuscany," *Terrain.org: A Journal of the Built and Natural Environments* 20 (Summer/Fall 2007), http://www.terrain.org/articles/20/harvey.htm

[11] Marcel Poëte, *Introduction a l'urbanisme* (Paris: Edition Anthropos, 1929).

[12] Walter Benjamin (1934), *Das Passagen-Werk*, ed. Rolf Tiedemann and Herrmann Schweppenhäuser, with the collaboration of Theodor Adorno and Gershom Scholem (Frankfurt am Main: Suhrkamp Verlag, 1972).

[13] Donatella Calabi, "Marcel Poëte: Pioneer of 'l'urbanisme' and Defender of 'l'histoire des villes,'" *Planning Perspectives* 11 (1996): 413 – 436; Diana Periton, "Generative History: Marcel Poëte and the City as Urban Organism," *Journal of Architecture* 11(2006): 425-439.

[14] For quoted text in translation see Susan Buck-Morss, *The Dialectics of Seeing: Walter Benjamin and the Arcades Project* (Cambridge, MA: MIT Press, 1990), p. 114. See also, Susan Buck-Morrs, "Benjamin's *Passagen-Werk*: Redeeming Mass Culture for the Revolution," *New German Critique* 29 (1983): 211-240.

[15] *Republic* IX, 592.

[16] Carl O. Sauer, "The Morphology of Landscape," *University of California Publications in Geography* 2 (1925):19-53.

[17] Charles W.J. Withers, "Place and the 'Spatial Turn' in Geography and in History," *Journal of the History of Ideas* 70(2009): 637-658.

[18] Michel de Certeau, "Walking in the City," in *The Practice of Everyday Life*, translated by Steven Rendall (Los Angeles: University of California Press, 1984), 91-110.

[19] Rebecca Solnit, *Wanderlust: A History of Walking* (New York: Viking, 2000), p. 5.

[20] Henri Bergson (1903), *Introduction to Metaphysics*. (Totowa, N.J.: Littlefield Adams, 1970), 159-162.

[21] Le Corbusier (Charles Edouard Jeanneret) and Pierre Jeanneret (1933), *The Radiant City: Elements of a Doctrine of Urbanism to be Used as the Basis of Our Machine-Age Civilization*, translated from the French by Pamela Knight (New York: Orion Press, 1967).

[22] Kenneth Frampton, *Modern Architecture: A Critical History* (London: Thames and Hudson, 1992), p. 180.

[23] Abraham Akkerman, "Femininity and Masculinity in City-Form: Philosophical Urbanism as a History of Consciousness," *Human Studies* 29(2006): 229-256; "Urban Void and the Deconstruction of Neo-Platonic City-form," *Ethics, Place and Environment* 12(2009): 205-218.

[24] Carl Gustav Jung, *The Archetypes and the Collective Unconscious: Collected Works of C.G. Jung*, Volume 9, translated from the German by Richard F.C. Hull (New York: Pantheon, 1959), 81-84.

[25] Giambattista Vico (1744), *Scienza Nuova* in *The New Science of Giambattista Vico*, translated and edited by Thomas G. Bergin and Max H. Fisch (Ithaca, N.Y.: Cornell University Press, 1968), Book IV, 982.

[26] Alan J. Almquist and John E. Cronin, "Fact, Fancy, and Myth on Human Evolution," *Current Anthropology* 29 (1988): 520-522.

[27] Bart J. Bok and Priscilla F. Bok, *The Milky Way* (Cambridge, MA: Harvard University Press, 1974), p. 8; George Lovi and Wil Tirion, *Men, Monsters, and the Modern Universe* (Richmond, Va.: Willmann-Bell, 1989), p. 65.

[28] Jacquetta H. Hawkes, *The First Great Civilizations: Life in Mesopotamia, The Indus Valley and Egypt* (London: Hutchinson, 1973), 271.

[29] Friedrich Nietzsche (1871), *The Birth of Tragedy and the Genealogy of Morals*, translated from the German by Francis Golffing. New York: Doubleday, 1956).

[30] Marcus Tulius Cicero, *De natura deorum* (ii 34)

[31] William Paley (1809), *Natural Theology: or, Evidences of the Existence and Attributes of the Deity* (London: Oxford University Press, 2006).

[32] Charles D. Yonge, *De opificio mundi: The Woks of Philo*, Vol 1, translated and edited by Charles D. Yonge. (Peabody, MA: Hendrickson Publishers, 1993), 1-52.

[33] Part XXII in David Hume, Dialogues Concerning Natural Religion (Dublin: John Exshaw, 1782).

[34] Immanuel Kant, *Critique of Pure Reason*, translated by Norman Kemp Smith (New York: St. Martin's Press, 1929), A:625-629; B:653-657.

[35] Marcus Vitruvius Polio, *Ten Books of Architecture*, translated by Morris Hicky Morgan (New York: Dover, 1960), Book II, Introduction, § 2.

[36] See Nicolas-Christophe Grimal, *A History of Ancient Egypt*, translated from the French by Ian Shaw (Oxford: Blackwell, 1992), p. 382. In this regard, and on the linkage between Dinocrates and Francesco Di Giorgio Martini, see Veronica Della Dora, "Alexander the Great's Mountain," *Geographical Review* 95 (2005): 489-516.

[37] *Cf.* M. Christine Boyer, *Dreaming the Rational City : The Myth of American City Planning* (Cambridge, Mass.: MIT Press,1983), 214-215, 286-287.

[38] See Roger Friedland and Harold Zellman, *The Fellowship: The Untold Story of Frank Lloyd Wright and the Taliesin Fellowship* (New York: Regan Books, 2006), 251-340, 543-561. On self-centeredness and despotism of Baron George-Eugène Haussmann and Le Corbusier, see Norma Evenson, *Paris: A Century of Change 1878-1978* (New Haven, CT: Yale University Press, 1979), 199-219; 232-38).

[39] See Kenneth Frampton, "The Work of Architecture in the Age of Commodification," *Harvard Design Magazine* 23 (Fall 2005/Winter 2006).
http://www.gsd.harvard.edu/research/publications/hdm//

[40] Carl Gustav Jung (1912), *Symbols of Transformation: Collected Works of C.G. Jung*, Volume 5, translated from the German by Richard F.C. Hull, edited by Herbert Read, Michael Fordham, Gerhard Adler and William McGuire (Princeton, N.J.: Princeton University Press, 1956); also, Carl Gustav Jung (1945), *Two Essays on Analytical Psychology: Collected Works of C.G. Jung, Volume 7*, translated by Richard F.C. Hull and Gerhard Adler, and edited by Herbert Read, Michael Fordham and Gerhard Adler (New York: Pantheon Books, 1953).

[41] Friedrich Nietzsche, *ibid.*, p. 102.

[42] Carl Gustav Jung and James L. Jarrett (1936), *Nietzsche's Zarathustra: Notes on the Seminar Given in 1934-1939* (Princeton, N.J.: Princeton University Press, 1988), 153-163

[43] Giambattista Vico, *ibid.*

[44] Abraham Akkerman, "The City as Humanity's Evolutionary Link: Walking and Thinking in Urban Design," *The Structurist* 47/48: 28-33.

[45] Patrick Geddes, *Cities in Evolution: An Introduction to the Town Planning Movement and to the Study of Civics* (London: Williams & Norgate, 1915). On Nash's successful design

of Regent's Park from 1811 onward to 1835 see John Summerson, *The Life and Work of John Nash* (Cambridge, MA: MIT Press, 1980), 114-129.

[46] Ebenezer Howard, *Garden Cities of To-morrow* (London: S. Sonnenschein & Co., 1902). See also Jane Jacobs, *The Death and Life of Great American Cities* (New York: Vintage Books, 1961), 62-74.

[47] Jane Jacobs, *ibid*, 17-25.

[48] Abraham Akkerman and Ariela F. Cornfeld, "Greening as an Urban Design Metaphor: Looking for the City's Soul in Leftover Spaces," *The Structurist* 49/50 (2010): 30-35.

[49] Abraham Akkerman, "Urban Superconscious and the Return of the Garden Myth," *The Structurist* 45/46 (2006): 62-68.

[50] See, for example, Henri Bergson (1896), *Matter and Memory*, translated by N. M. Paul and W. S. Palmer (New York: Zone Books, 1991), p. 65; Henri Bergson, *Creative Evolution*, translated by A. Mitchell (New York: Henry Holt, 1911), 118-163). On this aspect of Bergson's influence on Poëte's urban ideas see Marcel Poëte, *Les idées Bergsoniennes et l'urbanisme*, Melanges Paul Negulesco (Bucharest: Imprimerie nationale, 1935); also, Kermit C. Parsons and David Schuyler, *From Garden City to Green City: The Legacy of Ebenezer Howard*. (Baltimore: Johns Hopkins University Press, 2003); and Volker M. Welter, *Biopolis: Patrick Geddes and the City of Life* (Cambridge, MA.: MIT Press, 2003).

[51] Henri Bergson (1896), *Matter and Memory*, translated by N. M. Paul and W. S. Palmer (New York: Zone Books, 1991), 191-192.

[52] Milan Kundera, *Immortality*, translated from the Czech by Peter Kussi. (New York: Grove Weidenfeld, 1991), pp. 21-22.

[53] Kundera, *passim*, p. 223.

[54] Howard, *passim*.

[55] Robert Drew Hicks, *Diogenes Laërtius: Lives of Prominent Philosophers* VII, 5, (Cambridge, Mass., and London: Harvard University Press, 1972).

[56] Craig Stanford, *Upright: The Evolutionary Key to Becoming Human*. (Boston: Houghton Mifflin, 2003), 38-60.

[57] *Laws* (V, 746).

[58] Karl R. Popper, *The Open Society and Its Enemies: Vol I, The Spell of Plato* (London: Routledge & Kegan Paul, 1969), p. 51.

[59] Camillo Sitte (1889), *City Planning According to Artistic Principles*, translated from the German by George R. Collins and Christianne Crassemann Collins (New York: Random House, 1965).

[60] Elisabeth de Bievre, "The Urban Subconscious: The Art of Delft and Leiden," *Art History* 18 (1995): 222–252.

[61] Ignasi de Solá-Morales, 'Terrain Vague,' in Cynthia C. Davidson (ed.), *Anyplace*. (Cambridge: MIT Press, 1995), p. 119.

[62] Helen Armstrong, "Time, dereliction and beauty: An argument for 'Landscapes of Contempt'," *International Federation of Landscape Architects – Eastern Region*, Conference 2006 (Sydney, Australia: Darling Harbour, 25-27 May, 2006).

[63] Matt McCormick, *The Subconscious Art of Graffiti Removal*. USA (2001), 16mm, 16 min.

[64] Jean-Paul Sartre, *Nausea*, translated from the French by Lloyd Alexander (New York: New Directions, 1959), pp. 24-26.

[65] Robert Venturi, *Learning from Las Vegas*, with Denise Scott Brown and Steven Izenour, (Cambridge: MIT Press, 1972).

3

Fluid Planning: A Meaningless Concept or a Rational Response to Uncertainty in Urban Planning?

Torill Nyseth
Department of Sociology, Political Science and Community Planning
University of Tromsø
Norway

1. Introduction

Many contemporary writers have developed and elaborated upon various fluid metaphors to capture aspects of contemporary social life (Urry, 2000). Flows and fluidity are some of the catchwords in social, cultural, and urban thinking that is used as building blocks in theorizing contemporary trends with a focus on process, connectivity, and mobility at the expense of the previous focus on boundedness, hierarchy, and form (Simonsen, 2004: 1333). Fluidity and flows could even be seen as a whole new paradigm (Shield,1997). In the social science literature these consepts are primarily used as metaphors to describe turbulence and instability. Also in the planning field, fluidity is a concept that first of all is used as a metaphor to describe conditions of uncertainty. Today's turbulent conditions represent particular challenges for planning and policy making, including new spaces of politics, radical uncertainty, awareness of interdependence, the importance of "difference," and dynamics of trust and identity (Hajer and Wagenaar, 2003). Strategic spatial planners are faced with a world of potentialities, possibilities and uncertainties that are mostly beyond their control (Hillier et al., 2011). Traditional strategic spatial planning practices are failing to cope with contingency and uncertainty. In the planning literature, fluidity is however, more than a metaphor. The conditions of radical uncertainty also call for new forms of planning that in a sense are fluid too. Situations of uncertainty require a form of planning that is more exploratory and open to change; planning could be more equivalent to a "voyage of discovery" rather than a "road map" (Balducci, 2011: 2).

The collaborative planning tradition is the most straightforward form of planning that addresses some of these challenges. Among the main characteristics of this tradition are openness, transparency, dialog and consensus building (Innes and Booher, 2003; Healey, 1997; Forester, 1999). Collaborative practices may represent a new sort of institution emerging that can take many shapes and forms but also have shared characteristics: they are fluid, evolving, networked and involves dialogs and distributed intelligence (Healey et al., 2000). A more radical answer, very distinct from collaborative planning theory can be found within the emerging post-structural, multi-planar theory of planning, developed by Jean Hillier (2007). In her book "Stretching beyond the horizon", Hillier argues that planning has to be open to what may come: "We need to re-invent planning as a strategic future-oriented activity, taking into account the unknown, open up for new possibilities, towards a planning as becoming instead of planning as fixing" (Hillier, 2007: 17). Hillier suggest that spatial planning practice requires redefinition and a new theoretical foundation if it is to be relevant to the dynamic complexities and contingencies of the modern world (Hillier, 2008:259). According to Hillier, the task is to move from "what is" to "what if" (Hillier, 2007: 17). A momentum of experimentation could then be seen as a kind of "virtual planning" that focuses especially on "the unknown" (Hillier, 2007: 232) in a time of "contingent openness" (Hillier, 2007: 224). This kind of planning is largely concerned with "possibilities" and "what ought to be" (Albrechts, 2005: 265–266), and may be seen as an argument for a more fluid form of planning.

What is the basis for these ideas about fluid planning? Can fluid planning be anything more than a vague idea or, at best, influence architectural projects by bringing new and exciting ideas about urban design towards realization? Is it possible to identify the borders between mainstream planning and forms of fluid, experimental planning? In what planning situations is a fluid "approach" relevant and is it possible to imagine fluid planning as a practice? These questions are tricky, with an ambition that is far from mainstream planning rationalities. There will be tensions between the experimental and the regular, the fluid and the fixed. The idea in this chapter is to address the above questions through a review of the literature on fluid, open, and contingent planning. The aim is to elaborate further on the concept of fluidity in a planning context, asking whether it is useful and where it takes us. The chapter will search for forms of evidence of fluid practices in the planning field. The chapter is divided into six sections. The next section addresses the concept of fluidity in the social sciences, seeking to identify its roots and how it has been conceptualized. The third section attempts to distinguish between different dimensions of fluidity, as can be extracted from the planning literature. Four different forms of fluidity are discussed: fluidity as a particular form of uncontrolled space; fluidity as a planning condition of radical uncertainty; fluidity as a norm; and fluidity as potentiality and chance, as momentum from which planning can be re-invented. The fourth section addresses fluidity in urban planning. The fifth section illustrates aspects of fluid planning through a case study of an urban planning project that has been called "The Tromsø Experiment." The final section sums up and discusses some critical aspects of the concept of fluidity in planning.

2. Setting the scene: The notion of fluidness

The metaphors of "flow" and "liquidity" have recently captured the attention of social theorists concerned with emergent social processes in a world perceived as being increasingly disorganized and complex (Sheller, 2004; Bauman, 2000; Castells, 1996). Castells speak of "space of flows," and Urry of "global fluids." Zygmunt Bauman talk about liquid modernity and suggests that there are reasons to consider "fluidity" or "liquidity" as fitting metaphors when attempting to grasp the nature of the present phase of the history of modernity (Bauman, 2000: 2). The quotes below are examples of the elaboration of these concepts from some of the most widely profiled social scientists.

> "The 'spatial concepts' – networks, flows, and fluids – are used as building blocks of a new orthodoxy of the theorization of social life, a theorization that is argued to favor a focus on process, connectivity, and mobility at the expense of an alleged former focus on boundedness, hierarchy, and form." (Simonsen, 2004: 1333)

"Flows have direction but no purpose. They are intentional but not purposeful or teleological. Similarly, flow is related to its own sense; it has no transcendental meaning or direction. It is not flowing to any specific place. Analytically, the differential of flow is a temporary, mathematical reduction. For example, a curve, mathematically differentiated yields a degree of change of direction. The flow metaphor is used to signal the qualities of motion, materiality, and viscosity." (Shields, 1997: 3)

"... The new lightness and fluidity of the increasingly, mobile, slippery, shifty, evasive and fugitive power." (Bauman, 2000:3)

"The network concept involves flows of people, information, and money within and across national borders. Flows and networks – defined as 'sets of interconnected nodes' are then conceived of as universal organizational principles, be it of infrastructure, companies, finance, information, or media." (Castells, 1996)

"Governance is no longer only about government but now involves fluid action and power distributed widely in society." (Innes and Booher, 2004:11)

"Places can be loosely understood as multiplex, as a set of spaces where ranges of relational networks and flows coalesce, interconnect and fragment. Any such place can be viewed as the particular nexus between, on the one hand, propinquity characterized by intensely thick co-present interaction, and on the other hand, fast flowing webs and networks stretched corporeally, virtually and imaginatively across distances." (Urry, 2000, p.140)

"Relational networks of connected elements are inherently unstable and fluid. Society performs by recording, channelling and regulating the flow of energies through such networks." (Thrift, 1996: 285)

"The idea of a gel of vicious liquid implies fluidity, slipperiness, instability, movement and transformation a form which nevertheless has the capacity for momentary stabilisation." (Hillier, 2007: 58)

"... smooth space is the fluid space of light and becoming, and striated space is controlled." (Hillier, 2007:65)

Fluidity and flows in the social science literature are primarily used as metaphors. John Urry, for instance, argue that in order to understand the new mobilities, we need metaphors that "view social and material life as being like the waves of a river." Such fluid notions are necessary to capture the multiple transformations of collective representations in which "collective relations are no longer societal and structural" (Urry, 2003: 59). Another body of literature is that of Appadurai and others who argue for the metaphors of flow, uncertainty, and chaos (1996). Deleuxe and Guattari use the term "bodies in a vortex" (1986). White characterize the social world as being constituted of disorderly and sticky "gels and goos" (1992). Mol and Law, representing Actor-Network theory, generally elaborate a fluid spatiality (1994). Social movements can be described with similar metaphors; "they flow along various channels but may overflow or ebb away or transformed into powerful waves" (Urry, 2003: 71). This is illustrated in a contemporary setting on the Internet, for instance through the Arabic Revolution in Cairo in January 2011. Fluids are subject to mixtures and gradients with no necessarily clear boundaries (Urry, 2003: 43), always moving and changing as they go. Fluids are not solid or stable and relates to turbulence and rupture.

Particularly influential is the network metaphor about society. Fluidity is often associated with the concept of network. The "rise of the network society," by Manuel Castells (1996), grasps some of the transformations that have taken place in society with consequences for planning and policy making. Castells write about spaces of flows when describing the network society; the increasingly mobile, technologically mediated spatial form that dominates contemporary capitalist societies. Networks are not stable structures; instability is one of their basic characteristics. "The network society should be conceived of as made up of open or unstable structures that expand, readjust, shift and evaporate" (Hajer and Wagenaar 2003: 5). Therefore, the shift in language from institutions to networks implies a change from stability (institutions) to fluidity (networks). Networks are seen as synonymous with flows. Another link can be made to the theories of network governance, where the blurring of borders between organizations is increasing, a consequence of which is increased interdependence (Rhodes, 1996; Sørensen and Torfing, 2007). Stein and Harper (2005) claim that complex, contested, and somewhat fluid boundaries between public and private allow for creativity and innovation in the public realm.

Fluid conditions could then be related to society, to social relationships, and to governance relations. Fluidity is used to illustrate instability, movement, uncertainty, complexity, and something uncontrollable; this is, in contrast to stability, which is static, fixed, ordered, and controlled.

From the planning field, fluidity characterizes what Miraftab (2009), for example, called "insurgent citizenship" practices – those radical planning practices that respond to neoliberalism: "through the entanglement of inclusion and resistance they move across the invited and the invented spaces of citizenship" (p. 35) (see also Sandercock, 1998). Graham Houghton and Philip Allmenninger also elaborate on fluidity in their discussion of "soft spaces" in planning. They make a distinction between hard and soft spaces representing two different approaches in planning; "Hard spaces are the formal, visible arenas and processes, often statutory and open to democratic processes and local political influence. Soft spaces are the fluid areas between such formal processes where implementation through bargaining, flexibility, discretion and interpretation dominate" (2007: 306).

In the tradition of pragmatism, Patsy Healey (2009) and John Forester (1989) developed planning ideas based on a focus on concrete problems in specific situations, joint development of shared understanding of problems across multiple rationalities, and of actions or policies upon which there is agreement. People learn by experience and by interacting with each other, experimenting with ideas in real cases. Healey stress the power of agency and the "unique situatedness of particular instances of practice" (Healey, 2009: 444), as well as how agents and institutionalised structural forces interrelate in complex networks. Pragmatists celebrated the experimental, encouraging creative exploration and discovery (Healey, 2008). Forester saw planners as "reflective practitioners" who, in their practice, "learn about the fluid and conflictual, deeply political and always surprising world they are in" (Forester, 1999: 26).

Post-structuralist approaches to planning theory and practice, which Jean Hillier deal with in particular, open up considerations of profoundly important questions about strategic spatial planning in the uncertainty that has increasingly been identified over the last decade or so in scholarly publications. Hillier see planning and planners as experiments that are "enmeshed in a series of modulated networked relationships"(Hillier quotes Charles Laundry who suggests, "we need to look into the sun, think at the edge, and cross boundaries."(Landry 2006, cited in Hillier, 2008:25). Hillier explores the potential of the concept of "becoming" as creative experimentation, "where problems are not 'solved' once and for all but which, over the 'lifetime' of a strategic plan, are constantly recast by changing actors, situations and preferences, to be reformulated in new perspectives" (Hillier, 2008a: 26). In a special issue of *Town Planning Review* (Vol. 82/5 2011) on "Strategic Spatial Planning In Uncertainty," Hillier et al. raise questions such as "how might spatial planners seek to affect and "manage" environments in undecidable situations? Can we develop theories and practices which rely less on closure and more on openness to possibilities and opportunities? How might we plan in situations of fluidity and complexity?" (Hillier et al., 2011: 4). To Hillier, plans are moments of stability, a temporary fixity, and spatial planning an experimental practice. Hillier argue for foresighting, speculation, and experiments because these methods entail thinking about futures that we may not be able to recognize directly, futures that do not simply extend our current needs and wants but may actively transform them in ways we may not understand or control (Grosz, 2008: 260, cited in Hillier, 2008: 34). These futures recognize the possibilities and potentials of a particular space. Planners are understood as navigators positioned as "helmsman steering the city" (McLoughlin, 1969: 86, in Wilkinson, 2011: 10). Such ideas have a lot in common with post-structural theories of planning and geography, where the internationalization of flux and instability is identified as a key navigational strategy available to negotiate the "in-between" spaces (Murdoch, 2006: 97). The main notion of relational geography is the performance of social practices, and the performance of space goes hand in hand; they are both entangled in the heterogeneous spatial processes of becoming (Murdoch, 2006: 18).

The table below summarizes and lists some of these theoretical inspirations to the concept of fluidity.

Theoretical inspirations	Relational geography	Post-structural philosoph	Social theory	Actor-network theory	Multi-Planner theory	Neo-pragmatism
References	Massey, Amin and Thrift, Murdoc	Deleuze and Guattari.	Beck, Castells, Urry	Law, Mol, Latour	Hillier	Healey, Forester
Concepts	Spaces of flow, the in-between spaces	Rhizome, Trajec-tories, Lines of flight	Liquid moder-nity, Network Society	Hetero-geneous networks	Fluid planning, planning as experi-mental practices	Creating explo-ration and discovery

Table 1. Some theoretical inspirations to the concept of fluidity in urban planning.

Fluidity then is more than a metaphor. In summing up the conceptualization of fluidity in the social sciences it might be useful to distinguish between fluidity as, 1) an ontology, 2), as epistemology, and 3) as a certain form of practice. Fluidity in its ontological status is related to a post-structural view of a world of flows, where the reality is fluid, temporary and always in the making. In this fluid ontology the world (and the city) is conceived through an ontology of process and potential, through the work of networks of enrolment, fluid-like flows, and multiple encounters (Amin and Thrift, 2002). As an epistemology, fluidity refers to the radical uncertainty of a world that has become too complex, but still something that can be and has to be managed. Planning means in this setting to control uncertainties and fluidities (Abbott, 2005). Fluidity as a form of practice could in a planning context be understood as a more experimental planning practice that is more open and transparent towards future possibilities (Hillier, 2007). In the next section, these aspects of fluidity will be elaborated. Fluidity is also much related to what has been labeled "fuzzy planning" in some of the literature on urban and regional planning (De Roo and Porter, 2007).

3. Forms of fluidity

What does fluidity actually mean in a planning context? Planning may become fluid when there are no solutions to a problem, when the problem itself is complex, fuzzy or "wicked". In this section, we will distinguish between four forms of fluidity. Fluidity may be understood first, as a particular form of *space*, second, as a planning *condition*. A third understanding is when fluidity becomes a *norm*, something that might be encouraged. The most common understanding of fluidity, however, is when it is related to planning experiments, as a potentiality, a chance, a *momentum* from which planning can be re-invented.

3.1 Fluidity as a particular space

The most obvious form of fluid spaces in a planning context is exemplified by Kim Dovey in his book "The Fluid City" (2005). The fluid space is here represented by the Melbourne Docklands Waterfront, which focused on the transformation of this industrial space into "fun-space"; the waterfront constructions, perhaps the most commodified spaces in any modern city. The Dockland case is also an example of a fixed, limited place with a clear identity that becomes something else through the re-facing of the city to the water (Dovey, 2005:2). More interesting, however, is fluid space understood as non-regulated space, free-zones, liminal space, and border-zones. Border-zones (the area on both sides of a national border) are spaces in which cultural identities are blurred. Some borders are fluid, multiple, intersecting, and not fixed (Aure, 2011: 174), while others are highly controlled and regulated. In contrast to regional space, which is defined by drawing boundaries around clusters of objects, practices, or people; and networked space, which is defined by the form of relation between entities; fluid space is defined by boundaries that come and go, allow leakage, or disappear altogether, while relations transform themselves without practice (Mol and Law, 1994: 643). Fluid space is defined by a lack of clear boundaries, being defined by liquid continuity and gradients rather that binary identities.

Another form of fluid space are 'temporary urban spaces', spaces that for different reasons have avoided regulation and not given a particular function and therefore open to more creative use, open spaces for political action, leisure, or other purposes, for instance cultural expressions (Haydn et al., 2006). Temporary spaces provide an experimental opportunity for an urban platform for democratic action and human expression. Haydn encourages planners to look beyond the city's fixed boundaries so that citizens can participate in the creation of temporary spaces, rather than being automatons in fixed spaces that planners negotiate with private development (Mayo, 2007). Temporary spaces cannot be planned, which lead the authors to conclude that these spaces are politically freer than interim uses, which are easily appropriated uses that sustain the political economy.

A similar concept is what Groth and Corijn (2005: 503) call "indeteriminate spaces," that is "spaces left out of 'time and place' with regard to their urban surroundings, mainly as a consequence of rampant deindustrialization processes." Interminate spaces have an unclear status, representing a sort of "no-man's-land, which may allow for the emergence of a non-planned, spontaneous urbanity. Groth and Corijn relate indeterminate space to Lefebre's concept of "differential space," which is "created and dominated by its users from the basis of its given conditions. It remains largely unspecified as to its functional and economic rationality, allowing for a wide spectrum of use which is capable of integrating a high degree of diversity, and stays open for change" (Groth and Corijn ,2005: 521). The qualities of the "indeterminate" spaces (Groth and Corijn, 2005), which Sandercock saw as a form of "insurgent urbanism", "embraces uncertainty as potential space of radical openness which nourishes the vision of a more experimental culture, a more tolerant and multifocal one" (Sandercock, 1998: 120).

According to Groth and Corijn, in such places that are not coded by marked-led urban development, distinct possibilities for practices of innovation and playful intervention arise. Such indeterminate spaces cannot be completely planned, because if they are, they lose their fluid status, and thereby also their creative potential. To survive they depend on the investment of informal actors that occupy these spaces.

A more metaphoric use of space is the concept of "smooth space" elaborated by Deleuze and Guattari (1987), who distinguish it from "striated" space. Striated space is strict and stringent space, stable points of order, while smooth spaces are spaces without boundaries, the spaces of "becoming." Smoothness implies slipperiness and movement, where one slides seamlessly from one site to another. These are not to be understood as real spaces, but tools for thinking about space; every real space will be a combination of smooth and striated space. The urban waterfront in transition is an example of a smooth space, a boundary between the stable and striated space of the city, and the smooth flows of the water (Dovey, 2005: 24).

3.2 Fluidity as a condition

The issue here is how to plan in situations of fluidity and complexity, when planning is like "walking in the mist." Today's planning problems are often quite open-ended, which raises the question of what kind of knowledge is relevant in a society that is in "a state of flux". Several planning theorists have grappled with new ways of thinking about strategic spatial planning in connection with coping with issues such as the unknown (Abbott, 2005), fluidity, and dynamic diversity (Healey, 2007). The fluid is related to the unexpected, uncertainty, contingencies. Fluidity and uncertainty go hand in hand, and uncertainty is seen as a "danger" to planning (Sandercock, 2003) as well as to planning politics (Flyvbjerg, 1991). While flexibility may be an advantage, it also means a lack of certainty, such as for investors, changing the rules of the game through the process, etc. Uncertainty may lead to manipulation, holding back vital information, and distrust; this situation characterizes many contemporary urban development processes.

However, planning has always been related to handling and reducing uncertainty (Abbott, 2005). Fluidness, on the other hand, is more than just handling uncertainty in the sense of a lack of knowledge about the future or not being able to control the tools needed to manage the situation. Radical changes in the way humans interact with their social, financial, and natural environments have led planning theorists and practitioners to increasingly discussing spatial planning in conditions of radical uncertainty (Christensen, 1985, 1999; Hillier, 2005).

The world in which strategic spatial planners attempt to plan is littered with potentialities, possibilities, and uncertainties, most of which are beyond their control. It is this "radical uncertainty" that paves the way for discussions about fluid planning. Such planning work involves "taking risks, the consequences of which can be thought about, but cannot be known" (Healey, 2008: 28). Traditional ideas of an orderly and hierarchical planning system that mobilizes resources according to planned or projected events hold little conviction in an age of simultaneity and juxtaposition, the contiguous and the fragmented, the anticipated and the unpredictable. Many problems are simply too complicated, too contested, and too unstable for schematic, centralized regulation (Forester 1999). Politics and policymaking are made in new spaces, which operate in an institutional void; there are no pre-given rules that determine who is responsible, who has authority over whom, or what sort of accountability is to be expected (Hajer and Wagenaar, 2003: 9).

3.3 Fluidity as a norm

In planning, fluidity is also treated as a norm that is related to open-ended processes, continuous transformation, and planning as speculation and becoming. A situation of fluidity may destabilize established discursive frames and routines and open up new connectivities and opportunities. Destabilizing releases energy and focuses on the dynamic forces present in every urban planning context, rather that the stable, fixed aspects. According to Dovey, planning should include "a proactive context," whereas "flexibility is built into the planning schemes" (Dovey, 2005: 134). Dynamics and fluidities are emphasized in contrast to the static geography of modernist strategic plans. Dovey also call for more fluidity. Proposals and decisions are expected to be "accompanied by affect-laden discourses" because cities "produce new desires and identities through the planning process itself" (Dovey, 2005: 211). According to Louis Albrechts (2006), the need for strategic plans to produce a competitive city necessitates a more fluid, generalized spatial structure to allow for the insertion of major private sector initiatives.

From the experiences of collaborative dialogues, Innes and Booher (2003) claim that a new sort of institution is emerging that is fluid, networked, and involves dialogue and distributed intelligence. These institutions "are more like the standing wave that keeps its shape while millions of molecules flow through it" (Innes and Booher, 2003: 57). These institutional forms reward experimentation, risk-taking, and new ideas. Healey et al.'s concept of "institutional capacity," which also relates to the concept of fluidity, is better understood as a complex, fluid, and evolving infrastructure that flows at deeper levels. "New elements and relations coexist, combine, and shatter as they encounter older ideas and ways of going on" (Healey et al., 2003: 86).

3.4 Fluidity as momentum

A fourth conceptualization of fluidity is to see it as a momentum: a situation in which what has been taken for granted about "good" planning politically is destabilized, which unsettles the apparent fixity of formal planning processes and calls for alternative strategies.

In its place, ideas of planning as a deliberately open and fluid process may be introduced as a response to the complex and plural field of stakeholders and interests involved in public planning and urban development, and where a condition for mobilizing new energies and setting new directions into motion occur. A momentum can be understood as what Buitelaar et al (2007) refer to as rupture in an institutional path: "During rupture there is scope for path-breaking and path-creating forms of action, when there is sufficient pressure, whether internally or externally driven, a "critical moment" for change arrives (Buitelaar et al, 896). A critical moment may turn into a critical juncture encompassing a break with past patterns, inducing the overhaul of discursive hegemonies through which institutional transformations may occur" (Burch et al, 2003).

Here, spatial planning becomes a field of experimentation, where tools are frequently based on communication and the involvement of actors rather than the top-down imposition of goals and policies. Hillier's theory explores the potential of the concept of "becoming" as creative experimentation. Following Gilles Deleuze and Felix Guattari (Hillier, 2007: 76), Hillier claims that experiments do not seek solutions, but instead ask the question "what comes next?" after the "contingent encounter" of an experiment. To Hillier, an experiment is "a transgression of boundaries" that works with "doubt and uncertainty" and pays attention to "the aleatory and chance" (Hillier, 2007: 230). A momentum of experimentation could then be seen as a kind of "virtual planning" that pays special "attention to the unknown" (Hillier, 2007: 232) in a time of "contingent openness" (Hillier, 2007: 224). As a particular situated human action, experiments are performative practices, searching into an openness without knowing where it ends, but aware of the possibility of the openness and fluidity of the situation, experimenting with and experiencing "how it works" (Hillier, 2008: 4). According to Hillier, the task should be "to test out ... how different innovations may perform in different spatio-temporal circumstances" (Hillier, 2007: 250). This is not in order to "fix" things, but rather to "test out" how to work with uncertain, temporary and open circumstances, because there are "always too many unknowns to give certainty" (Hillier, 2007: 250). To Hillier, experimentation involves "connection, interaction and duration – lines of flight that might involve new experimental discourses and new understandings of place" (Hillier, 2007: 281). A momentum, then, is a situation of instability, of unstable forces that can be shattered by a lack of decisive and ruling power, a lack of hegemonic discourses,

Characteristics/ Dimensions of fluidity	Particular space	Condition	Norm	Momentum
Examples	Indetermined space, Free-zones, Border-zones, Soft spaces, In-between space, Smooth space, Temporary spaces	Contingencies, Complexities, Dynamic diversity, governance	New connectivities and energies	Critical moment, Windows of opportunity, Planning as experimentation,

Table 2. Four dimensions of fluidity.

because they intervene in what Dovey saw as a future vision. Or, as Laws and Rein put it, "These moments of doubt are precisely the moments when systems are open to new insights, ideas and behaviour" (Laws and Rein, 2003:175). Any floating situation and fluid planning process shapes experiments, and experiments have the potential to influence the direction of progress. The strategic incentives to use such openings as opportunities to gain control combined with the cognitive tendencies to remove the irritation of doubt to make scarce those moments in which doubt is available and something new is really possible. Such moments create a "liminal space" (Shields, 1991; Hetherington, 1997) which open the way for reflection and reframing.

The forms of fluidity could even be extended beyond these four. For instance, fluidity could be related to the context of planning. Philip Allmenninger and Graham Haughton (2007) pointed at the fluid scales and scope of UK spatial planning, referring to the contested and fluid nature of both regions and the scalar complexity of the roles of the planning authorities. Gerd De Roo and Geoff Porter discuss the fluidness and fuzziness in planning. Planning concepts and doctrines, such as sustainability, participation, urbanism, and the compact city, are essentially fuzzy, fluid, or illusive themselves, according to Roo and Porter (2007); they are concepts that have multiple meanings. Fluidity could also be understood in the form of dynamics: flows are spatial and temporal, but above all, material. They have tempo and rhythm as well as direction (Shields, 1997: 2–3).

4. Urban planning as a fluid planning field

Cities, in particular, are spaces of flows, dynamics, and multiple relations. They are increasingly structured around flows of people, images, information, and money moving within and across national borders (Amin and Thrift, 2003: 51). Amin and Thrift (2003) maintain that circulation is one of the main characteristics of a city, saying "cities exist as means of movement, as means to engineer encounters through collection, transport and collation" (p. 81). Of course, cities are also ordered, but according to Amin and Thrift, this ordering is "often exacted through the design of flows as a set of serial encounters which construct particular spaces over time" (p. 83). Amin and Thrift use fluid ontologies when practising an urban theory based on "the transhuman rather than the human, the distanciated rather that the reflexive" (2003: 5). The aim is to conceive of the world (and the city) through an ontology of "process" and "potential", through the work of networks and enrolment, fluid-like flows, and multiple encounters. Cities are seen as fields of movements and moments of encounter between spatially stretched and distant connections. Some even talk about "fluid cities" in the sense that they are dealing with "a confluence of flows of different forces" (Dovey, 2005: 2). In what Healey defined as "the multiplex city" (Healey, 2000), she emphasized the diversity of the relations that transect urban areas, and the complexity and unevenness of their inter-relations (Graham and Healey, 1999). The "networked" urbanism discussed by Graham and Marvin (2001) is one articulation of the fluid social dynamics of cities. Space itself, particularly urban space, is considered more complex, fluid, and fragmented. They describe this fragmentation by referring to the "liquefaction of the urban structure" and of the "splintered city" (p. 115) producing unstable fluid structures. Throgmorten see cities and their planning-related organizations as nodes in a global-scale web, "a web that consists of a highly fluid and constantly changing set of relationships" (Throgmorten, 2003: 130). Even the process of planning becomes fluid

(Dovey, 2005). Cities as such cannot be comprehensively understood and planned for, because their dynamics are too complex (Healey, 2007).

So how can one conceptualize the complexities of urban dynamics, their openness to chance and their potential to become otherwise? According to Boolens (2006), urban planning as a distinctive area that uses a cautious approach to come up with proposals for the use of urban space on the basis of well-defined and far-reaching view over time, is outmoded. Contemporary urban planning is situated within this challenge: seeking to control fluidity through either spatial plans and political decisions seeking to tame critical voices and discourses by binding them to binary hierarchies (for instance: reason-unreason, good-bad, real-unreal etc) as well as to the discourse of spatial "answers" to the political need for a 'comprehensible plan'. To Healey, the work of strategy formation becomes an effort to create a nodal force in the ongoing flow of relational complexity (Healey, 2007: 228).

To Dovey, the fluid city is both a metaphor and a material reality. The material meaning has to do with the city facing the water, as illustrated by the construction of the urban waterfront in Melbourne's Docklands. The metaphor of a fluid city is related to a city becoming "unsettled"; an understanding of urban change as a confluence of flows of different forces, both local and global. Dovey is inspired by Appadurai's ideas about the various global flows, which he term "scapes": the "ethnoscapes" (flows of tourists, refugees, and immigrants), "mediascapes" (flows of information and images), "technoscapes" (flows of technology), "finanscapes" (flows of capital) and "ideoscapes" (flows of ideas, values and ideologies) (Appadurai, 1996).

Situated within discursive spaces and the diverse forces of change cause planning to oscillate between fluidity and ground. In facing these challenges, cities can try to ignore change and tame criticism or, according to Hillier, they can try to make analytical grounded flexible strategies. Hillier argue for post-structural urban planning that focuses on "change, transgressions, contingency, temporality, fluidity, immanence and emergence," giving "an open-endedness of social contexts" (Hillier, 2008: 25) that makes urban space to an aleatory field of meaning and action. On the other hand, fluid planning could also mean "anything goes," as was the case in Melbourne's Docklands, where the focus was entirely on flows of capital and not at all on what it was actually becoming; in other words, everything was fine as long as it sold and someone was willing to consume (Wood, 2009). The problem with the Docklands planning process was not that it was too fluid or ungrounded, but, according to Stephen Wood, "that it was not ungrounded enough." Its openness stopped with the capital, while the other positive and productive forces and desires did not find their way through.

In many cities, the failures of classical, modernist, comprehensive and rational planning, and top-down governing schemes have opened the door for a new social awareness, or rather uncertainty, regarding the best way to develop and govern the city. In recent years, it has become increasingly important for cities to be "open" to their multiple ways of living, diverse interests and ethnic difference and to open up the planning process to "experiments" that involve the public and stakeholders in new ways. Patsy Healey sees the challenge as having two components: understanding the contingencies that make it appropriate "to challenge fixities in one context and seeking to stabilize fluidities in another" (2007: 15).

5. Practicing fluidity: The Tromsø Experiment[1]

In order to illustrate how fluidity way work in practice in an urban planning context, the chapter now presents a case study of the "Tromsø Experiment" or, more precisely, "The City Development Year" (CDY), a planning experiment that took place in this relatively small city in northern Norway in 2005-2006. The formal planning process related to a city centre plan was put on hold for one year. In its place, planning as a deliberately open and fluid process was introduced as an idea and as an answer to the complex and plural field of stakeholders and interests that were involved. Openness often means losing control; indeed, the situation becomes unpredictable and, in fact, in this case, no one was in complete control of what happened the following year. This made a public space for mobilizing new energies and setting new directions into motion. A fixed planning process had, overnight, become a fluid one. Cities rarely have the courage to make a "new beginning" and open up to the unexpected via multilateral cooperation between city planning authorities, citizens, local businesses, production, civil society, and professionals. Collective efforts risk ending in "low politics" rather than competitive strategies. The Tromsø Experiment (the CDY) was a year in which to experiment and to develop alternative ideas and methods for a reformulated city center plan. Table 1 below summarizes the key events and the fluidities of the experiment.

The experiment allowed for new becomings by allowing the aleatory or chance to occur, and stimulated the unexpected through new methods of participation, mapping, discovering and sensing the city. With the citizens, the city was investigated neighborhood by neighborhood, looking at how the form, meaning, and social significance of space and place are dependent on the space's past, present, and imagined future. Contextual conjunctures were analyzed rather than facts. The city was analyzed by highlighting the type of dynamics and driving forces that were working in particular areas, and the rhythms of change to which the areas were exposed. In some areas, for instance, there had been an extensive "appartmentification" or gentrification. Questions were raised in each quarter, such as "What is the history of this space?" and "What is its future?" Other questions included, "What narratives have been played out here?" "What emotions and stories are buried here?" "What is the relationship to the surrounding streets and quarters?" and "What are the threats and what are the possibilities?" People were invited to consider strategies for formulating regeneration and transformation of the neighborhood. Each of these socio-spatial analyses was then put together, linked to maps and visualized. In the end, all of the focus areas were presented as a "City Reader" that provided citizens with a new way of reading about the city, or to give them an opportunity to discover new aspects of the city through other concepts and perspectives.

---

[1] The case study is based on the authors' observation of urban planning in Tromsø for a period of more than 10 years, through a combination of two different perspectives. The first of these was the insiders' perspective, in participating in some of the activities that led to the experiment. The study also builds on focus group interviews with 40 city stakeholders, representing different businesses, creative and cultural industries, developers, and research and higher education institutions. Document studies, particularly of the CDY report, and its publications, including chronicle articles in the local newspaper published throughout the year, have also been important sources. The section builds on two other papers by the author (Nyseth et al. ,2010; Nyseth, 2011).

Period	Forms of fluidity	Event
1994	Experiments with new forms of spatial planning	The concept of Tromsø as an architectural "experimental zone" is introduced and the dialogic planning model, "The Tromsø Game," is developed
1999	Ordinary planning process	The planning administration starts the process for a new town plan
2000–2005	Public debate - rupture	The City Forum arranges a number of public meetings on issues related to urban planning and development
February 2005	Formal planning process are stopped	The proposal for the new town plan is due for its final decision-making by the planning committee
March 2005	A fluid situation occurs	A last public meeting on the proposal is held in the city and the "time out" is introduced
March 2005 – March 2006	Experiments with different forms of public participation, dialogs, planning discourses, analyses,	The planning administration is removed from finishing the town plan and City Development Year takes over the process, introducing a range of activities, such as; Tromsø X-Files (Expedition) City walks, Feature articles Public meetings and seminars University Conference Focus group interviews Interactive web-page
April 2006	CDY project ends	The City Development Year submits its report to the planning committee and ends its activities
June–August 2006	International publicity	The experiment is presented at the Venice Biennale
to 2007	Ordinary planning process	A new town plan is finally completed

Table 3. Overview of key events of the fluid planning situation and the Tromsø experiment.

The Reader was primarily an opening up to the ordinary citizens of the "black boxes" of planning, through new ways of analyzing the city using a different rhetoric than formal plans normally do. Terms like "appearing and disappearing city landscapes" and "curing" were used to illustrate the degree of transformation and damage that the different areas had undergone. Political and planning intentions and considerations are often hidden in maps and texts, so the Reader also provided citizens with insight into how planning works. Such methods can also make planners more open about their intentions and their use of rhetoric. Although the experiment departed from the traditional focus on the qualities of place and governance, the planning discourse was, through the CDY, extended from a formerly narrow focus on planning, architectural programs, and urban form, towards an understanding of the city as a complex embodiment of everyday life processes made of subjects and practices. This made it possible for the participants to acknowledge that places are made of flows of becoming, and to realize the significance of forces of the imaginary and desire as well as capitalism and politics. The CDY committee and the way it performed was, in itself, an example of networks, flows, and contingencies between formal and informal arenas. The construction of the group with professionals from outside the planning authorities makes this an example of a governance network (Nyseth, 2008), in that it involved relatively stable, horizontal articulations of actors that were interdependent but operationally autonomous (Sørensen and Torfing, 2007). The committee's performance and activities also constructed new networks and a connection between actors that did not normally have contact with each other. This new connectivity can be exemplified by the idea suggested by a narration of a marine researcher, of a "marine fish market" at the harbor. The story involved a visit from some Japanese colleagues who commented that it was peculiar in a city that has fish as its basic industry that "it was not possible to eat fish anywhere in the city at lunchtime." This story led to a project about the construction of a coastal food and adventure centre at the harbor. In a rich fishing region, such a project is not exceptional, but it had not been realized before, and it did occur through the opening and new connectivities that had been created through the CDY. The fluid situation had created new spatial and urban life connectivities. The publication of ideas from the public also mobilized normative statements about, for instance, the need for more collaboration in the city. Through a number of seminars and conferences, webs of relations were established with other cities facing similar problems and to the central level of government, which connected planners, sectors, and other central participants in urban governance networks.

Through an exhibition called Tromsø X-files and city walks, the city became an arena for potential ongoing explorations by the citizens. Its ambition was to give citizens a unique opportunity to get to know their city in a new way; that is, to take them on a sensory journey through the city's past, present, and future architectural and physical landscape. Opening up planning in this way challenged the established knowledge and practices of planning. It became more legitimate to ask questions about what was going on, and especially the potential effects on planning itself and future city development. This exhibition put almost all the new, but not yet realized buildings and development projects in the city on display. Even architects' drawings and sketches of their ideas for new projects were used. This enabled anyone to see not only one project at a time, but every project arranged within its context; the surrounding buildings and quarters. The exhibition made it possible to see what it would mean for the city as a whole if all these projects were realized. A concrete model was produced of the existing city, where some of the new projects were added, so everyone could judge their potential effects on the city. Guided tours of the exhibition by a member of the network or the planning office were arranged on a daily basis. This exhibition represented the essence of the experiment, which one of the members of the network expressed as: "To exhibit is to open up!" This slogan became a brand for the ideology the CDY wanted to represent. In this context, "opening up" refers to the number of new and unknown projects that were in different phases of realization, including sketches on drawing-boards, and also to the process and methods used to uncover the city's "hidden future." "Opening up" also referred to critical dialogues between the project and the citizens, which was one of the goals of the project.

So, from a situation where plans had been more or less closed to the public, the new openness produced new possible "lines of flight" concerning conflicts and impasses. It was an experiment that moved contingency and fluid planning into a political situation that could question the city's hegemonic planning discourse, not least because in charge of this experiment was a network of professionals and "bricoleurs" independent of local government and planning authorities (Nyseth et al., 2010).

6. Conclusion: Does the concept of fluidity take us anywhere?

The metaphor of fluidity makes it possible to move from the focus on fixed, ordered, and regulated landscapes and planning as the tool to achieve this, towards exploring processes, flows, movements, open boundaries, informal relations, etc. Fluidity is flexibility and change; it is flows of money and desire; it is the formation of new identities of both people and places (Dovey, 2005: 243). But the fluid is also fragile; it can be there one moment and vanish in the next. To follow these lines of thought, one needs to have a certain taste for the unknown.

It is rare for any city to look for a truly fluid planning practice as part of its ordinary everyday planning; instead it will seek a "temporary resting" (Healey, 2006) in which it tries to regain an order of development, for instance, by looking for inspiration by promoting debate on an "open situation" in urban planning and development. Cities do not necessarily look for scenarios, rather for ideas about contemporary forces of development to be dealt with and how to cope with them. If cities were to have truly fluid planning, they could include, as in the case of Melbourne, Australia: (1) a "future vision," (2) "scribbles indicating possible functional zones," (3) a "collage creating composite pictures," for instance, by only working from (4) "a diagram presented as fluid blobs," that have (5) "no content" because the "fluid city" is only discursive (Dovey, 2005: 133–134). Alternatively, as in the Tromsø case, this could be achieved through openness, through inviting "bricoleurs" or "outsiders" into the planning process, and through new discourses about urban planning.

In summing up this chapter, it is necessary to raise some concern about the concept of fluidity. Fluidity may have some advantages related to flexibility, openness and the production of new ideas, but it also means destabilization, which could lead to a situation where no one is in control. Dovey, for instance, expressed a deliberate ambivalence towards the fluid conditions of urban development, saying: "there are values in both 'going with the flow' and in resisting its place-destructive tendencies" (Dovey, 2005: 5). The flows of desire for a better future are the very basis of urban place-making, yet unregulated desires are also the source of urban destruction. Fluidity also has connotations of uncertainty, difficulty, and ambiguity. Too much fluidity, or fluidity going "wild," would mean not only losing control, but also, in a sense, giving up the ambition of steering, which would certainly give other forces more room to maneuver. Questions about power must be raised. Who gains and who loses in situations of fluidity? What forms of power dynamics are played out when a planning process is "opened up to the unknown"? Fluid conditions may marginalize civil society, giving too much power to private investors. Therefore, there must be limits to fluidity. Decisions have to be made and plans have to be adopted, which means stabilization and fixity. On the other hand, in order for urban planning and democracy to become alive, processes need to be open for the unexpected.

Fluidity, therefore, is a condition to which all cities must face up to. Like its opposite – "stability" – "fluidity in urban development is both good and bad" (Dovey, 2005). Fluidity and stability must be understood as a continuum; there is never complete fluidity or complete stability. There will be temporary resting, and at the same time moments, situations and spaces of temporariness which call for a new approach. The challenge seems to be how fluidity can be managed without losing control? There needs to be a form of institutional capacity that can translate the fluid moment into a strategy. The case study in this chapter has illustrated one form of control that was organized within a project with a strict time limit and accountability placed in the planning council. This might be a solution that worked in this case, but there may be many other models to develop. The role of public planning is perhaps not to control but to manage fluidity, to stand against the destructive forces of the marked as a mediator of public interests. But the "public interest" is also fluid as it consists of a multiplicity of interests that is never stable. Managing this ambivalence is perhaps the most difficult task that urban planning will face in the years to come.

7. Acknowledgements

Thanks to Ronny Kristiansen for comments and suggestions that improved the chapter considerably and also the PLANNORD network at the Aarhus meeting august 2011.7.

8. References

Abbott J. (2005). Understanding and managing the unknown. *Journal of Planning Education and Research*, Vol. 24, No 3, pp.237–251, ISSN 0739-456x

Albrechts, L. (2005). Creativity as a drive for change. *Planning Theory*, Vol 4, No 3, pp.247–267, ISSN 1473-0952

Amin. A. and Trift, N. (2003). *Cities. Reimagining the urban*. Cambridge: Polity Press, ISBN 0-7456-2414-6

Appadurai, A. (1996). *Modernity at large: Cultural dimensions of globalization*. Minneapolis: University of Minnesota Press, ISBN 0-8166-2792-4

Aure, M. (2011). Borders of understanding: Re-making Frontiers in the Russian-Norwegian Contact Zone. *Ethnopolitics*, Vol 10, No 2, pp.171–186, ISSN 1744-9065

Balducci, A. (2011). Strategic planning as exploration. *Town Planning Review*. Vol 82, No 5, pp. 529–546, ISSN 0251-3625

Balducci, A., Boolens, L., Hillier, J., Nyseth, T. and Wilkinson, C. (2011). Introduction: Strategic Spatial Planning in uncertainty. Theory and exploratory practice. *Town Planning Review*, Vol 82, No 5, pp. 481-501, ISSN 0251-3625

Bauman, Z. (2000). *Liquid Modernity*. Cambridge: Polity, ISBN 0-7456-2410-3

Boolens, L. (2006). Beyond the Plan; Towards a New kind of Planning. *disP* 167.4/2006, ISSN 0251-3625

Buitelaar, E., Lagendijk, A. and Jacobs, W. (2007). A theory of institutional change: illustrated by Dutch city-provinces and Dutch land policy. *Environment and Planning A*, Vol 39, pp.891–908, ISSN 0308-518x

Burch, M., Hogwood, P., Bulmer, S., Carter, C., Gomez, R. and Scott, A. (2003). *Charting Routine and radical change: A discussionpaper.* Manchester Papers in Politics: Devolution and European Policy Series, 2, 2003.

Castells, M. (1996). *The Rice of the Network Society.* Oxford: Blackwell, ISBN 1-55786-616-3

Coaffee, J. and Healey, P. (2003). "My voice: My Place": Tracking Transformations in UrbanGovernance. *Urban Studies*, Vol 40, No 10, pp.1979–1999, ISSN 0042-0980

Christensen, K. (1985). Coping with uncertainties in Planning. *Journal of the American Planners Association*, Vol 51, pp. 63–73, ISSN 1939-0130

Dahl, K.E. and K. Uhre. (2004). *Framtida i nord: Kommende og forsvinnende landskaper.* (Northern futures; Coming and disappearing landscapes) http//www.dahl&uhre.no

Deleuxe, G. and Guattari, F. (1986). City state. *Zone*, Vol 1–2, 194–199, ISSN 0162-1904

Dikec, M. (2007). *Badlands of the Republic. Space, Politics, and Urban Policy*, Oxford, Blackwell, ISBN 9781-4051-5631-8

Doel, M. (1999). *Poststructuralist Geographies. The Diabolic Art of Spatial Science*, Edinburgh, Edinburgh University Press, ISBN 0-7486-1243-2

Doucet, I. (2008). Centrality and/or Centrality: a matter of placing boundaries. In: Maciocco, G. (Ed): *Urban Landcspae Perspectives.* Springer, New York: pp. 93–123, ISBN 9783540767992

Dovey, K. (2005). *Fluid city.* London: Routledge, ISBN 0-415-35923-6

Engwickt, D (1992). *Towards an Eco-City; Calming the Traffic.* Sydney: Envirobook, ISBN 0-85881-062-x

Flyvbjerg, B. (1991). *Rationalitet og magt I. Rationality and power.* Copenhagen: Akademisk Forlag, ISBN 87-500-2994-0

Forester, J. (1999). *The Deliberative Practitioner. Encouraging Participatory Planning Processes.* Cambridge: MIT Press, ISBN 0-262-06207-0

Graham, S. and P. Healey (1999). Relational concepts in time and space: issues for planning theory and practice. *Europeanplanning studies*, Vol 7, No 5, pp. 623–646, ISSN 0965-4313

Graham, S. and Marvin, S. (2001). *Splintering urbanism: Networked Infrastructure, Technological Mobilities and the Urban Condition.* London: Routledge, ISBN 0-415-18964-0

Grosz, E. (2004). *Chaos, Territory, Art; Deleuze and the framing of the earth.* New York: Columbia University Press, ISBN 9780231145183

Groth, J. and Corijn, E. (2005). Reclaiming Urbanity: Indeterminate Spaces, Informal Actors and urban Agenda Setting. *UrbanStudies*, Vol 42, No 3,pp. 503–526, ISSN 0042-0980

Hajer, M. (2003). A frame in the fields: Policy making and the reinvention of politics. In: M. Hajer and H. Wagenaar (eds): *Deliberative Policy Analyses.* Cambridge: Cambridge University Press, ISBN 0-521-82366-8

Hajer, M. and H. Wagenaar (eds.) (2003). *Deliberative Policy Analyses.* Cambridge: Cambridge University Press, ISBN 0-521-82366-8

Haughton, G. and Allmendinger, P. (2007). 'Soft spaces' in planning. *Town and country planning*, September 306–308, ISSN 0040-9960

Haydn, F. and Temel, R., and Arlt, P. (2006). *Temporary Urban Spaces. Concepts for the Use of City Spaces.* Basel, Birkhaüser, ISBN 3-7643-7460-8

Healey, P. (1997). *Collaborative Planning: Shaping places in Fragmented Societies.* London: Macmillian Press, ISBN 0-333-49574-8

Healey, P. (2006). *Urban complexity and spatial strategies. Towards a relational planning for our times.* London: Routledge, ISBN 978-0-415-38035-5

Healey, P. (2007). Re-thinking key dimensions of strategic spatial planning: Sustainability and complexity. In: Gert De Roo and Geoff Porter (eds): *Fuzzy Planning. The role of actors in a fuzzy governance environment.* London: Ashgate, ISBN 978-0-7546-4962-5

Hetherington, K. (1997). *The Badlands of Modernity,* London, Routledge, ISBN 041511469-1

Hillier, J. (2005). Straddling the post-structuralist abyss: Between transcendence and immanence? *Planning Theory,* Vol 4, No 3,pp. 271– 299, ISSN 1473-0952

Hillier, J. (2007). *Stretching beyond the horizon: A multiplanar theory of spatial planning and governance.* London: Ashgate, ISBN 978-0-7546-4749-2

Hillier, J. (2008). Plan(e) speaking: A multiplanar theory of spatial planning. *Planning Theory,* Vol 7, No 1, pp. 24–50, ISSN 1473-0952

Huxley, M. (2006). Spatial rationalities: Order, environment, evolution and government. *Social and Cultural Geography,* Vol 7, No 5,pp. 771–787, ISSN 1464-9365

Innes, J. and J. Booher (2003). Collaborative policymaking: governance through dialogue. In: Hajer, M. and H. Wagenaar (eds): *Deliberative Policy Analyses.* Cambridge: Cambridge University Press, ISBN 0-521-82366-8

Innes, J. and J. Booher (2004). Reframing Public Participation Stratgies for the 21st Century. *Planning Theory and Practice,* Vol 5, No 4, pp. 419–436, ISSN 1464-9357

Jacobs, Jane (1961). *The Death and Life of Great American Cities.* New York, A Vintage Book, ISBN 0-679-74195-x

Jensen, O.B. (2008). Culture stories: Understanding cultural urban branding. *Planning Theory,* 6(3): pp. 211–236, ISSN 1473-0952

Kollbotn, K. Dahl, K.E. and E.F. Johannessen (2006). BY05: *The end Report for the City Development Year.* Tromsø: Tromsø Municipality.

Laws, D. and Rein, M. (2003). "Reframing Practice". In: Hajer, M. and H. Wagenaar (Eds): *Deliberative Policy Analyses.* Cambridge: Cambridge University Press, ISBN 0-521-82366-8

MacCallum, D. (2008). Participatory Planning and Means-Ends Rationality: A Translation Problem. *Planning Theory and Practice,* Vol 9, No 3, pp. 325–343, ISSN 1464-9357

Mayo, J.M. (2007): Comments on Haydn, F. and R. Temel (Eds): Temporary Urban Spaces: Concepts for the Use of City Spaces. Canadian Journal of Urban Research, Winter 2007, ISSN 1188-3774

Miraftab, F. (2005). Insurgency and spaces of active citizenship. *Journal of Planning Education and Research,* Vol 25, No 2, pp. 200–217, ISSN 0739-456x

Mol. A. and Law, J. (1994). Regions, networks and fluids: anaemia and social typology. *Social Studies of Science,* Vol 24, pp. 641–671, ISSN 0306-3127

Murdoch, Jonathan (2006). *Postmodern Geographies,* London, Sage, ISBN 0-7619-7423-7

Netto, V. (2008). Practice, space, and the duality of meaning, *Environment & Planning D: Society & Space,* Vol 26, pp. 359–379, ISSN 1472-3433

Nyseth, T. (2008). Network governance in contested urban landscapes. *Planning Theory and Practice,* Vol 9, No 4, pp. 497–514. London: Routledge, ISSN 1464-9357

Nyseth, T, Pløger, J. and Holm, T. (2010). Planning Beyond the Horizon; The Tromsø Experiment. *Planning Theory,* Vol 9, No 3, pp. 223–248, ISSN 1473-0952

Nyseth, T. (2011). The Tromsø Experiment: Opening up to the unknown. *Town Planning Review,* Vol 82, No 5, pp. 573–594, ISSN 0251-3625

Osborne, T. and Rose, N. (1999). Governing cities: notes on the spatialisation of virtue. *Environment and Planning D: Society andSpace,* Vol 17, pp. 737–760, ISSN 1472-3433

Osborne, T. and Rose, N. (2004). Spatial phenomenotics: making space with Charles Booth and Patrick Geddes. *Environment and Planning D: Society and Space,* Vol 22, pp. 209–228, ISSN 1472-3433

Pløger, J. (2004). Strife: Urban Planning and agonism. *Planning Theory,* Vol 3, No 1,pp. 71–92, ISSN 1473-0952

Pløger, J. (2009). Contested Urbanism – Struggles over representation, *Space and Polity,* December 2009, ISSN 1356-2576

Rabinow, P. (1989). *French modern.* Norms and Forms of the Social Environment. Chicago: Chicago University Press, ISBN 0-226-70174-3

Rhodes, R.W. (1997). *Understanding governance: Policy networks, governance, reflexivity and accountability.* Buckingham: Open University Press, ISBN 0-335-19727-2

Roo, G.D. and G. Porter (eds.) (2007). *Fuzzy Planning. The role of actors in a fuzzy governance environment.* London: Ashgate, ISBN 978-0-7546-4962-5

Sandercock, L. (2003). *Cosmopolis II: Mongrel cities in the 21st century.* London: Continuum, ISBN 0-8264-6463-7

Sennett, R. (1991). *The conscience of the eye.* London: Faber and Faber, ISBN 0-571-16192-8

Sheller, M. (2004). Mobile publics: beyond the network perspective. *Environment and Planning D: Society and Space,* Vol 22, pp. 39–52, ISSN 1472-3433

Shields, R. (1997). Flow as a New Paradigm. *Space and Culture,* Vol 1, pp. 1–7, ISSN 1552-8308

Simonsen, K. (2004). Networks, flows, and fluids – reimagining spatial analyses? *Environment and Planning A,* Vol 36, pp. 1334–1340, ISSN 0308-518x

Stein, S. M. and Harper, T. L (2005). Rawl's justice as fairness: A moral basis for contemporary planning theory. *Planning theory,* Vol 4, No 2, pp. 147–172, ISSN 1473-0952

Sørensen, E. and J. Torfing (eds.) (2007). *Theories of Democratic Network Governance.* Basingstoke: Plagrave Macmillan.

Throgmorten, J.A. (2003). Planning as persuasive storytelling In a global-scale web of relationships. *Planning Theory,* Vol 2, No 2, pp. 125–151, ISSN 1473-0952

Thrift, N. (1996). *Spatial Formations*: London: Sage, ISBN 0-8039-8545-2

Urry, J. (2000). *Sociology beyond societies: Mobilities for the Twenty-First Century.* London: Routledge, ISBN 0-415-19088-6

Urry, J. (2003). *Global complexity.* Cambridge: Polity Press, ISBN 978-0-7456-2818-9

Wilkinson, C. (2011). Strategic Navigation: in search of an adaptable mode of strategic spatial planning practice. *Town Planning Review,* Vol 82, No 5, pp 595–613, ISSN 0251-3625

Wood, S. (2009). Desiring Docklands: Deleuze and Urban Planning Discource. *Planning Theory,* Vol 8, No 2, pp. 191–216, ISSN 1473-0952

4

A Model of Urban Infrastructural Planning in a Traditional African City: A Case Study of Ilorin, Nigeria

Olutoyin Moses Adedokun[1] and Dickson Dare Ajayi[2]
[1]Department Of Geography, Federal College Of Education, Zaria,
[2]Department Of Geography, University Of Ibadan, Ibadan
Nigeria

1. Introduction

A model is defined as a simplified structure of reality, an organized principle, a synthesis of idea or a mental construct in which an actual life situation may fit in (Timmermans, 2000). Analysis of urban areas, structure of cities, urban activities and their linkages at the level of individual city has witnessed, over the years a number of improved models (Carter, 1975, Ayeni, 1979; Wegner 2001). These are the ecological/economic competition model, spatial interaction models, urban activity pattern and activity system models, spatial behavioural and microbehavioural models. The paper is viewed within the urban activity pattern and activity system models, spatial behavioural model, and microbehavioural model. These models were reviewed as a precursor to the proposed model - the model of Urban Neighbourhood Activity Centers.

2. Models of urban analysis

2.1 Models of urban activity pattern and activity system

Most theoretical work in the area of activity analysis stems from geography and urban planning. This is largely based on two important publications in the 1970s first by Haggerstrand (1969) and Chapin (1974). The formulation of the Activity Approach in the 1970's was a reaction against certain weaknesses of the four stage modelling approach such as: appropriateness of the model structure, poor data quality, poor quality of the model calibration and on poor quality of forecasts made with it. (Atkins, 1987, Axhausen, 1990). Until the end of the 1980s, these questions were not thought to be serious enough to challenge the practical use of the models. The propositions of Activity Approach have been explained by Clarke (1985), Eurotopp (1989) , Axhausen (1990) and Kwan (2005) as follows: (i) activities are primary, travel is secondary (ii) activities and travel take place in time-space (iii) behaviour is influenced by space-time and other constraints (iv) household activity patterns are related to household structure (v) interactions within households change over time (vi) activity and travel behaviour is constrained by the experience of the traveler.

The 'Activity Approach' stands apart from the other traditions of modeling identified by Pas (1988). The concern of the 'Activity Approach' has not been the derivation of mathematical descriptions of human behaviour as was the concern of social physics, econometric and psychometric but to derive a modeling system to explain human spatial behaviour. One important concept that emerged from the models of urban activity pattern and activity system is the concept of "Action Space" which is the collection of all urban locations about which the individual has information and the subjective utility or preference associated with these locations (Horton and Reynolds, 1971).

The core of an individual's action space is termed the "activity space" which is the subset of action space, that is, all locations or areas that an individual has ever visited or had direct contact with as a result of day-to-day activities. In terms of activity system, persons and firms are regarded as behavioural units, which perform some activities in some locales. The activity space is the most structured by individuals because it comprises locations, which have actually been visited. A location will change from being part of the action space to being part of the activity space once a trip has been made. The difference between activity space and action space is that although both are part of time-space environment, action space is a part of time space environment for which an individual has positive information through media, relatives or friends while activity space is a part of time- space environment frequently used.

The pattern displayed by individuals as they carry out their activities is the "activity pattern" of individuals. An Activity Pattern (AP) is an ordered sequence of activities of an individual that takes place within a space-time continuum, the activities being linked by travel. An activity pattern is, therefore, an activity programme with its schedule. An activity programme, on the other hand, is the collection of all the activity undertaken during a certain period of time irrespective of the order of their occurrence. The activity pattern is determined by individual's propensity and opportunity to engage in particular activities (Chapin. 1974), based on "predisposition" factors (such as role obligations, traits, etc.) Associated with activity pattern are certain features, which characterized its inherent utility to the individual. These features include (i) the total number and types of activities and activity types included in activity pattern, (ii) time of the day each activity was performed, (iii) duration of each activity, (iv) activity sequence, (v) location of each activity, (vi) total distance travel, and (vii) amount of flexibility or space-time autonomy.

The advantages of the activity system approach is that it is broad: its values - behavioural - patterns framework permits the introduction of the widest range of operating factors without limiting assumptions and deductive arguments which characterized the neo-classical economic/ecology models that ignore all the competing forces and identifying only one link - that with the centre of the city.

2.2 Spatial behavioral models

The weakness of economic and ecological models gave rise to spatial behavioural models. Lloyd (1976) attempted an examination of the linkages among *cognition, preferences* and *migration behaviour* in different states in the United States of America and found a strong structural linkage among them. He concluded that, based on the three components, it is possible to predict the direction of movement in that preferred locations of cognitive space would receive larger percentages of migrants and vice-versa. Hanson (1977) revealed that when the spatial form and extent of information levels are compared with the travel patterns that emerged from the travel diary of the urban residents, it is evident that the set of locations actually contacted is but a small sub-set of the cognitive opportunity. Cadwallader (1978) used information and preference surfaces in explaining individual's cognitive space. Preference surface reflects the varying attractiveness as a place to live in, which people attach to different locations. It was, however, discovered that the underlying structures of information and preference surfaces are more difficult to disentangle, especially in the case of information surface, and there is no evidence that these structures are strongly related.

Information about the objective environment as stored subjectively in the individual's brain is called the cognitive constructs of the urban environment. Cognitive image is, however, different from locational schematic or mental map. The term "cognitive" is used to indicate the non-locational character of most images, and to suggest that thinking and verbal behaviour form a stronger component of cognitive images than concrete visual imagery (Wong, 1979). Cognitive affective maps are mental orderings of the environment that involve preference ratings that guide residential choice, but personal and financial considerations often preclude selection of the preferred location (Preston, 1982). Behavioural models provide psychological oriented accounts of the destination choice process in repetitive urban spatial behaviour (Pipkin, 1981).

Behavioural approach is increasingly finding a place in urban and transportation planning in an attempt to improve the policy contents of plan proposals (Chorkor, 1986a; 1987, 1988). The earlier approach of urban/transportation planners has leaned heavily towards the environmentalists' viewpoint which, according to Herz (1982) maintains that spatial behaviour, including personal and social constraints is essentially a function of the material environment. Thus, what appears to be close to environmental deterministic view of behaviour has been jettisoned and planners are currently adopting behavioural approach, which sees the spatial structure, material environment and man as mutually interacting elements of the urban space.

Man is considered as having the ability to evaluate, interpret and react to his environment as observed and perceived. However, Lundqvist (1978) observes that existing planning models are not capable of integrating structure and behaviour in a theoretically sound way and at a level of detail that is useful for planning purposes. He suggests that it may be necessary to work with less ambitious approaches built on iterative use of structural and behavioural models. This problems is still evident despite the fact that sophisticated techniques are being developed and employed in planning to analyse spatial behaviour (Burnett, 1977: Arad and Berechman, 1978; Smith, *et al*, 1982). Timmermans and Veldhuisen (1982) observe that behavioural models offer a potentially more valuable approach to predict the likely effects of physical planning schemes as compared with the gravity-type approaches. They, however, noted that behavioural models share with the gravity type models the problem of whether equations could be developed solely on the basis of policy sensitive factors or whether other influential factors should be included.

2.3 The micro-behavioral approach

The micro-behavioral approach, otherwise known as systems approach to urban activity network, seeks to explain urban spatial structure, not in a mechanistic term as done by the neo-classical economists. Rather it explains urban spatial structure, especially the interactions in an urban setting, as an interaction between population (people), activities, locations and time. This approach is different from activity system model, in the sense that it incorporates the time element and individual instance of behaviour and activity, obtained in the form of disaggregated data. The important features of this approach are (i) incorporating time as a commodity, (ii) giving new meaning to land (multiplicity of land use), (iii) perceiving people's behaviour as linked to the ways in which they perceive space, and (iv) providing a holistic approach to urban structure. This approach of integrating individual spatial behaviour overtime was pioneered by Hãggerstrand (1969). He used a simple diagram to illustrate his concept of space-time dimensions. Hãggertrand postulated the geographers' two-dimensional space on the surface of the earth or on the surface of a map. A line on this surface indicated movement in space but not in time. He suggested a third dimension to signify time.

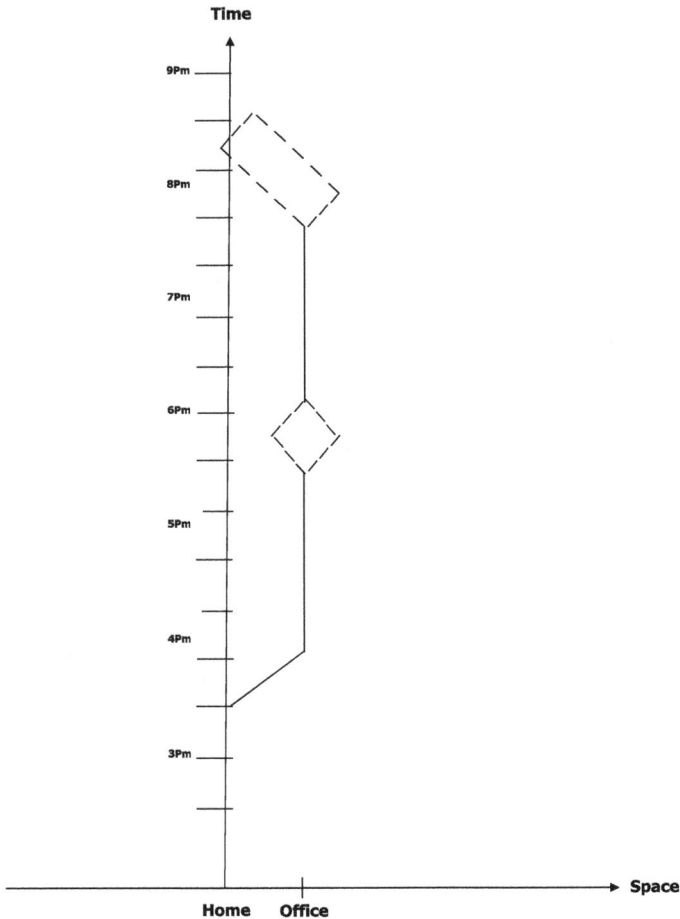

Fig. 1. Man's Daily Space-time Dimensions. Source: *Adapted from Haggerstrand* (1969).

Figure 1 represents a very simple working day. Solid lines represent the path of all obligatory activities and dotted lines the prism or feasible regions of movement in periods for which there are no fixed activities. The worker is assumed to be effectively fixed at home until 7:30am to 8:00am where he can conveniently sleep and take breakfast. He must then take a direct route to work, where he is obliged to stay until lunchtime. During the lunch hour he has a certain amount of freedom; he must be back in the office exactly an hour. From 2pm until about 5pm he is again expected to stay at work. But after 5pm he has no need to be home until 7pm for supper. In this period, 5-7pm, he can stay on at work or he can go somewhere near or stop off on the way back for a drink or visit. The main feature implicit in this model of daily behaviour is the idea that certain activities are fixed in both space and time. This model (of the individual's activity -time -space decision) is based on a set of assumptions Six of these assumptions were identified by Cullen and Godson (1975). They are: organised behaviour, the Action Space, Priorities, Constraint, Flexibility of Activities and Scheduling.

2.4 Spatial interaction function model

The Spatial Interaction Function has made it possible to recognize the interaction between people (Population), activities, locations (of the activities) and times (of the day) when the activities took place. This is not in the conventional mechanistic method, but rather in terms of attractiveness of places. Also through this modeling, it is possible to predict at any point the location and quantity of new population through what can be described as Inter-Element Linkages as shown in Figure 2.

Population	Social contact (1)		
Activities	Activity pattern (2)	Activity Transition (3)	
Locations	Individual spatial behaviour (4)	Activity Location (5)	Time (6)
Time	Individual Temporal behaviour (7)	Activity Timing (8)	Use Density (9)
	Population	Activities	Locations

Fig. 2. Inter-element Linkages. Source: *Procos and Harvey, 1977*

Figure 2 shows the linkages among the time and the space in which the activities of a given population take place. Linkages between population sub-groups would produce social contacts (cell 1). On the other hand, the linkages between locations and population, locations and activities and locations and locations produce individual spatial behaviour, activity location and time of day when activities take place (cells 4, 5 and 6). Finally the linkages between time and population, time and activities and time and locations produce individual temporal behaviour, activity timing and use density of land respectively (cells 7, 8 and 9).This modeling technique is thus unique and different from the conventional method which only explains linkages through origin destination using the gravity modeling technique (Abler *et al*, 1992, Ayeni, 1979; 1992).

2.5 The functioning model

The Function Model unlike the conventional modeling technique did not consider the attractivity of a space in a mechanistic or unidirectional way rather it recognizes attractivity in terms of interaction between population (people), location activities and times. The most important aspect of this modeling technique is its multidirectionality as shown in the **Figure 4.**

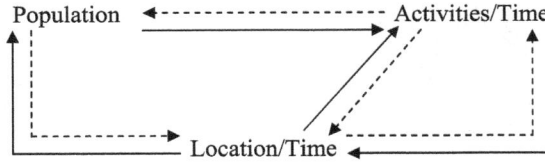

Fig. 3. Cycle of Inter-Element Interaction. Source: *Procos and Harvey, 1977*

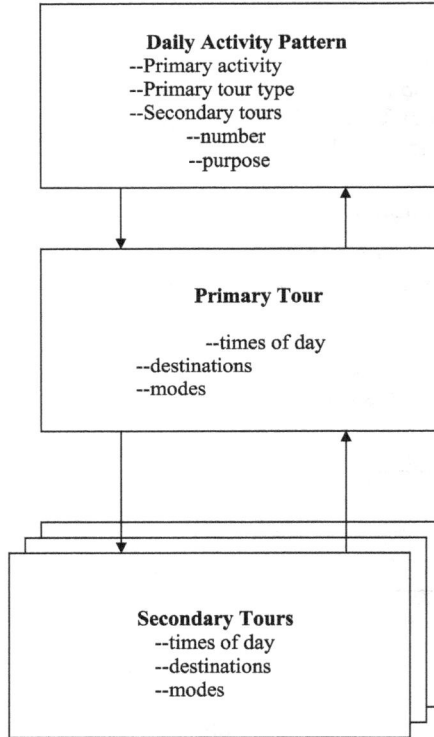

Fig. 4. The daily activity schedule. Source: *Adapted from Timmermans* (2000).

Recent works on time-space activity unlike the earlier ones, have been largely concentrated on modeling time-space activity pattern (Zahavi, 1978; Lantrip, 1990; Axhausen and Garling 1991; Garling, Kwan and Golledge, 1993; Vander Knaap 1996; Golledge, Kwan and Garlin 1994; Bhat, 1997; Garling, Romanes and Vilhelmson, 1996; Timpf, 2002), computational process and the application of Geographical Information System (GIS) in modeling time-space activity (Hirsh Prashkea and Akiva, 1988; Kitamira, 1988; Axhausen, 1990; Kwan, 2000; Root and Recker 1990; Timmermams 2000; Golob and Mcklally, 1995; Ben-Akiva *et al* 1996; Kwan, 2003, 2004, 2005; LLoyd and Shuttle 2005). For instance Ben-Akiva *et al* (1996) applied the daily activity schedule model to data from the Boston metropolitan area, including a 24-hour household travel diary survey collected in 1991, as well as zonal and transportation system attributes from the same time period to produce the daily activity schedule model.

The model groups the elemental decisions of the daily activity schedule into five tiers or hierarch: (a) daily activity pattern, (b) primary tour time of day, (c) primary destination and mode, (d) secondary tour time of day, and (e) secondary tour destination and mode (Fig.5).

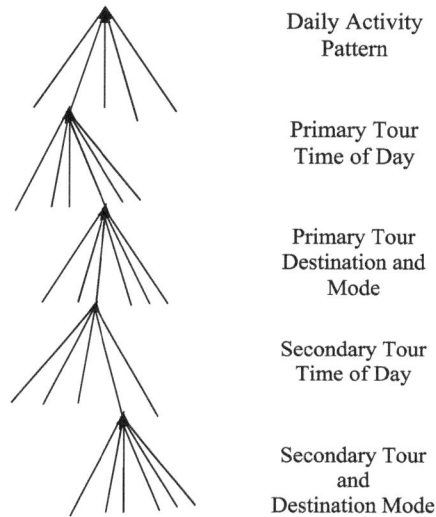

Daily Activity
Pattern

Primary Tour
Time of Day

Primary Tour
Destination and
Mode

Secondary Tour
Time of Day

Secondary Tour
and
Destination Mode

Fig. 5. The daily activity schedule. Source: *Adapted from Timmermans (2000)*

The choice of daily activity pattern determines the number of secondary tours in the daily activity schedule. The choice of secondary tour time, destination and mode are conditioned upon the choice of a daily activity pattern. For daily activity patterns with 2 or more secondary tours, the conditional choice probabilities of the secondary tours are mutually independent, calculated from the same models. This approach, though represent a state-of-the art in modeling, it is, however, deficient in the sense that it ignores time and space constraints across secondary tours, but simplifies the model structure. (Timmermans, 2000)

Most of the works in developed countries especially in the United States have been in the field of transportation geography especially in the area of urban travel activity pattern and telecommuting in the city. A few of the most recent time-use studies have examined the impact of new Information Technology (IT) on people's everyday lives. For instance, Robinson *et al* (1997) compared the mass media use and social life of internet users, based on a 1998 survey of 3,993 adults in the U.S. They found no significant or consistent evidence of time displacement by mass media use of social activities as a result of internet or computer use. Another study by University of California, Los Angeles Center for Communication Policy (UCLA – CCP, 2000) involving 2096 households in the U.S found that internet use has considerable impact on a person's time use pattern and shopping behavior. The literature, although revealing about how people spend their time, have shown that studies do not explicitly incorporate the spatial dimension in their data collection procedure and analysis. The time displacement perspective therefore needs to be expanded to a space-time displacement perspective to be useful for understanding people's activity pattern in space - time and the urban economy. The space- time displacement perspective suggests that there are distinctive geographical consequences associated with such time displacement Kwan

(2000) explored and implemented the extensibility diagram as an analytical tool within a three dimensional (3D) geographical information system (GIS). Using the navigational history of a person's web browser and data collected through personal interview, she developed a method that takes into account the multiple spatial scales and temporal complexities (simultaneity and disjuncture) involved in individual activities both in the physical and virtual worlds. Kwan (2002, 2003) also proposed a conceptual frame work of human cyberspatial behaviour based on time - geographic perspective, she argued that space-time constraint are still important even in the information age.

Although all these recent time geographical studies observed that the telecommuting did not induce a significant increase in non-work trips and activities, they also suggested that significant insights can be gained through examining the impact of new IT on people's space time constraint, and indicated the kind of data and methods useful for geographical studies of everyday life and the urban economy (Mokhtarian, 1990; Mokhtarian et al 1995, 1999; Kwan, 2001). But as noted by Kwan (2003, 2004 and 2005); Kwan, Murray, O'Kelly and Tiefelsdorf (2003), Weber and Kwan (2002, 2003) this kind of research (IT - Cyber spatial - Time geographic research) is still uncommon in geography to date in both advanced and developing countries. This is due to the fact that the required data are complex, costly and time consuming to collect, and there are a few established methods for analyzing these complex data.

In developing countries in general and Nigeria in particular, works on urban activity linkages through time and space are rare or non-existing. The few existing works in Nigeria like those of the American and European scholars focus mainly on interaction or movement within urban areas. For instance, Fadare (1986, 1989) and Ojo (1990) examined the factors affecting household trip generation in Ibadan while Aderamo (1990, 2000, 2004) examined the impact of road development on urban expansion in Ilorin, the spatial pattern of Intra-Urban trips and the role of transport factor in the structure and growth of Ilorin. Adeniji-Soji (1995) also analyzed the pattern of telecommuting and intra-city travel in Ibadan. Others such as Adagbasa (1995) and Bello (1995) studied intra-pattern urban movement in Benin City. One thing that is common to all these works is that they are all carried out within the framework of transportation geography without necessarily linking movement in space with the activity of the people. Thus a study was conducted to explore the missing link between people, their activities, locations of these activities and the times of the day when these activities take place with the aim of providing information that could be used to develop a model applicable to activity location and land use planning and management in a traditional African city in transition (Adedokun, 2008, 2009). From the data collected and analyzed in the study area, about 90 percent of the respondents had their activities fixed in space especially at home/place of work and about 89 percent also had their respondents are engaged in informal sector and are self-employed. These population characteristics and the activity pattern (fixed in space and time) generated mainly home and office/work place based activities which the people have adjusted and accustomed to. The implication of this is that certain spaces would be in greater (or lesser) demand at a particular time. This forms the basis for the proposed model for this study.

3. A model of urban neighbourhood activity centres

Based on the foregoing, we would like to construct a generalized model of land use planning and facility location in a traditional medium size urban center using Ilorin as a case study (see Figure6).

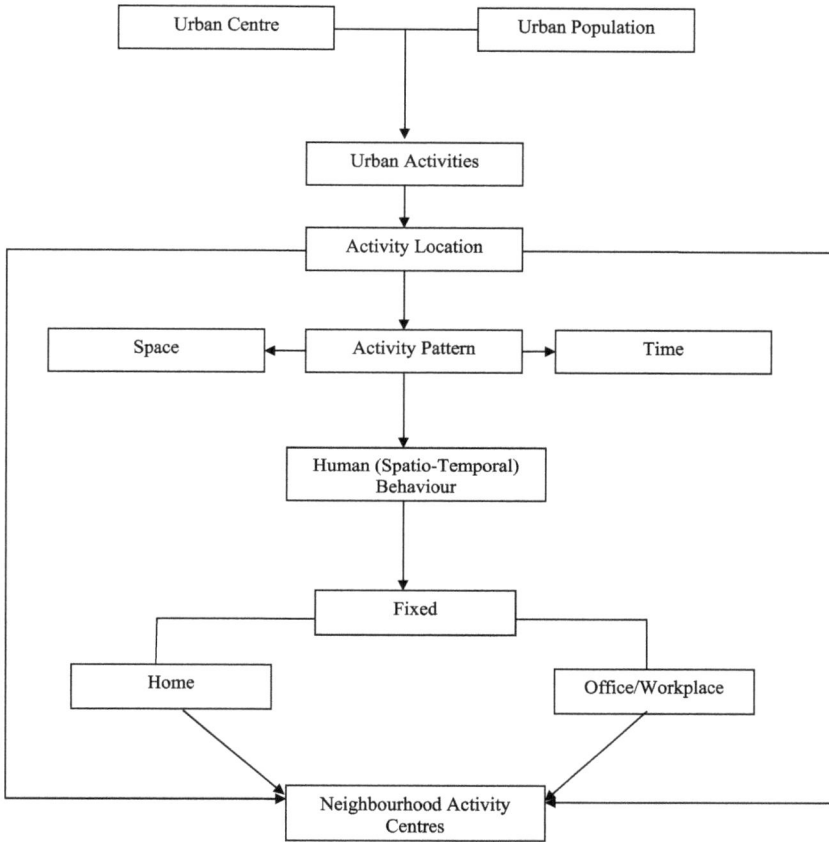

Fig. 6. A Model of Urban Neighbourhood Activity Centers. Source: *O.M. Adedokun*, 2009.

Figure 6 shows the linkages among population, time and space and urban activities. Given a medium size urban centre with its population, there would emerge various types of urban activity located in different parts of the city. The location of these activities would in turn generate activity pattern with space and time (or spatio-temporal) dimensions. The activity pattern itself would generate human spatio-temporal behavior. In this paper, it is shown that human behaviour in space and time is fixed. The fixity in human spatial behaviour is shared between home and office/work place. The urban neighbourhood activity centre model is of the view, therefore, that in planning for a medium size urban centre in developing world, there is the need to adopt a strategy that would incorporate the behaviour of the people. Instead of strict land use zonation approach; facilities may be located closely to or around neighborhoods where people are fixed. In this case and as demonstrated, facilities and infrastructures should be located between homes and work places. Obviously, if there is a demonstrable linkage between two activities in time, it makes sense to locate the facilities housing them in the same space so as to eliminate time and energy consuming travel.

4. Conclusion

Over the years, studies on urban land use have been dominated by increasing mechanistic approach. These studies have not elucidated on a better understanding on dynamic model of urban infrastructural planning. We are not oblivious of the fact that oftentimes, models showcase the peculiarity of the context within which they are produced. Also, most of the earlier models were based on experiences of European and American cities that have little or direct bearing with orthogenetic or traditional African cities in transition. There is, therefore, the need for an active involvement of those who are directly impinged upon (the people) in issues relating to land use planning and activity location. Here lies the importance of Urban Neighbourhood Activity Centers Model. This model if implemented, especially in medium sized traditional or orthogenetic cities, where modern day town planning theory is alien and unenforceable, will help to achieve efficiency and optimality in land use planning and activity location. It will also help to achieve and ensure spatial and social justice, equity and equal opportunity for all environmental (urban land users) consumers.

Although this research used disaggregated data to determine house hold activities which were latter classified into three category of activities i.e. home based, office/work place and outdoor based activities respectively and the linkages between the populations these activities, their location in time and space. Future research frontiers could pick on any of the three groups (home, workplace and outdoor activities) and further disaggregate them to determine their locations, time spent on each and the level of linkages between the activities. Furthermore, future research could focus on larger cities that are not necessarily state capitals (even two adjacent or contiguous cities (Conurbations) and compare their spatio-temporal activity linkages for the purpose of achieving optimality in urban land use planning and management.

5. References

Abler, R; Adams, J and Gould P.R. (1972) *Spatial Organisation: The Geographers view of the World*. New York, Prentice Hall.

Adagbasa, J.O. (1995), *Spatial Analysis of Intra-Urban Patterns of Movements in Benin-City, Nigeria*. Unpublished Ph.D Thesis, University of Benin, Nigeria, pp 25-75

Adedokun, O. M. (2008), "A Spatio-Temporal Analysis of Activity Linkages in Ilorin, Kwara- State", *Savannah*, Volume 21, Number 2, pp.41-52

Adedokun, O. M. (2009), *A Spatio-Temporal Analysis of Urban Activity Linkages: A Case Study of Ilorin, Nigeria*, Unpublished Ph.D Thesis, Department of Geography, Ahmadu Bello University, Zaria, pp 17-37

Aderamo, A.J. (1990), *"Road Development and Urban Expansion: The case of Ilorin"*, Unpublished Ph.D Thesis, University of Ilorin, Nigeria pp 10-25

Aderamo, A.J. (2000), 'Spatial Pattern of Intra – Urban Trips in Ilorin, Nigerian' *Geo-Studies Forum* 1, 1& 2 pp. 46-57.

Aderamo, A.J. (2004), "Transport Factor in the Structure and Growth of a Traditional Settlement: A Case study of Ilorin", Nigeria' *Geo-Studies Forum*, Volume 2 (1), pp. 145-156.

Adeniji-Soji, (1995), *Patterns of telecommuting and Intra-City Travel: A case study of Ibadan Metropolis; Nigeria*, Unpublished Ph.D Thesis, University of Ibadan.

Arad, R.W. and J. Berechman (1978), "A Design Model for Allocating Inter-related Land Use Activities in Discrete Space", *Environment and Planning A*. Volume 10, pp. 1319 – 1332.

Atikins, S. T. (1987), "The Crisis of Transportation Planning Model" *Transport Reviews,* Volume 7, pp. 307 – 325.

Axhausen, K.W. (1990), *An Introduction to the "Activity – Approach".* Transport Studies Unit (TSU) Working Paper 533. Oxford University, London.

Axhausen, K.W. and T. Gashing (1991), "Activity Based Approaches to Travel Analysis: Conceptual Frameworks, Models and Research Problem", *Transport Review,* pp. 1-25.

Ayeni, B. (1992), *A Place for everything and everything in its Place.* An Inaugural Lecture, University of Ibadan, .

Ayeni, M.A.O. (1979), *Concepts and Techniques in Urban Analysis.* London, Croom Helm Books.

Bello, A. S. (1995), *Urban Public Transportation in a Growing city: The case of Ilorin, Nigeria,* Unpublished Ph.D Thesis, University of Ilorin.

Ben-Aktiva, M.E., S.R. Lerman and J. Swait Jr. (1996), *'Discrete Choice Analysis - Theory and Application to Travel Demand',* MIT Press Cambridge, Massachusset.

Bhat, C. (1997), *Recent Methodological Advances Relevant to Activity and Travel Behaviour Analysis.* Resource Paper for the International Association for Travel Behaviour Research. Austin Texas.

Burnett, P. (1977), "Perceived Environmental Utility Under the Influence of Alternative Transportation Systems", *Environment and Planning A,* Volume 11, pp. 393 – 401.

Cadwallader, M. (1979), "Neighbourhood Evaluation of Residential Mobility", *Environment and Planning A,* Volume 11, pp. 393 – 401.

Carter, H. (1975), *The Study of Urban Geography,* London, Edward Arnolds

Chapin, E.S. (1974), *Human Activity Pattern in the city.* New York, John Wiley and Sons.

Chokor, B.A (1986), "Developments in Environment-Behaviour-Design Research: a Critical Assessment in the Context of Geography and Planning", *Environment and Planning A,* Volume 18, pp.5-26

Chokor, B.A (1986b) 'City profile: Ibadan' *Cities,* 3, 2, pp.106-116

Chokor, B.A (1987), "Environmental-Behaviour-design-research Techniques: An appraisal and Review of Literature with Special Reference to Environment and Planning in Third World", *Environment and Planning A,* Volume 19, pp.7-32

Chokor, B.A (1988), "Environment-Behaviour-Design Research: An Agenda for the Third World", *Environment and Planning A,* Volume 20, pp. 425-434

Clarke, M. I. (1985*), "Activity Modelling – a Research Tool or Practical Planning Technique?",* Paper presented at International conference on Travel Behaviour, Nordwijk.

Cullen, I. and V. Godson (1972), *"The Structure of Activity Patterns: a bibliography",* Research Paper, No 2, Joint Unit for planning research.

Eurotopp (1989), *'The Potential for Using Activity Based Approaches Report to Drive'* Transport Studies Unit, Oxford University, U. K.

Fadare, S.O. (1986), *Intra – Urban Travel Characteristics. A study of socio-economic Attributes in Ibadan.* Unpublished Ph.D Thesis, Department of Town and Regional Planning, University of Sheffield.

Fadare, S. O. (1989), 'Analysis of Factors Affecting Household Trip Generation in the Residential Areas of Ibadan' *Ife Research Publications in Geography*, Number 3, 34 – 48.

Garling, T., T. Kalen; J. Romanus, M. Selart and B. Vilhelmson (1996), "Computer Simulation of a Theory of House hold Activity Scheduling", *Environment and Planning A*, 1 – 37.

Garling T., J. Saisa, and R. Linderberg (1986), "The Spatio-Temporal Sequencing of Everyday Activities in the Large Scale Environment", *Journal of Environmental Psychology*, 6, 261-268.

Golledge, R.G; Mei-Po Kwan and T. Garling (1994), "Computational Process Modelling of House-hold Travel Decision using a Geographical Information System", *Papers in Regional Science*, Number 72 (2), pp. 99 – 117.

Golob, F.G. and M.G. McNally, (1995), *A Model of Household Interactions in Activity Participations and the Derived demand for Travel*, Paper Presented at the Conference of Activity Patterns, Endhoven University of Technology, the Netherlands.

Haggerstand, T. (1969), "What about People in Regional Science", *Regional Science Association*, 24, pp. 5-35

Hanson, S (1977) "Measuring the Cognitive Levels of Urban Residents". *Geographical Annaler* 58(2) Pp 57 – 81.

Hert, R. (1982) 'The Influence of Environmental Factors on Daily Behavior' *Environment and Planning A* Vol. 14 Pp 1175 – 1193.

Hirsh, M; J.N. Prashkea and M.B. Akwa (1986) 'Dynamic Model of Weekly Activity Pattern' *Transport Science*, Volume 20 (1), pp. 24 – 36.

Horton, F.E. and Reynold, D.R. (1971) "Effects of Urban Spatial Structure on individual Behaviour", *Economic Geography*, Volume 47 (1), pp. 10-25

Joh, G.H., T.A. Arentze and H.J.P. Timmermans (2005) 'A Utility-based analysis of activity time allocation decisions underlying segmented daily activity-travel patterns' *Environment and Planning A*, Number 37 (11), pp.105-125.

Kitamora, R. (1988) "An Evaluation of Activity-Based Travel Analysis", *Transportation Research*, Number 15, pp. 9 – 34.

Kwan, Mei-Po, (2002) 'Time, Information Technologies and the Geographies of Everyday Life' *Urban Geography*, Number 25 (5), pp. 471-482.

Kwan, Mei-Po, (2003a) *New Information Technologies, Human Behaviour In Space Time and the Urban Economy* Paper Presented at the 82nd Annual Meeting of Transportation Research Board (TRB) Washington D.C. Jan 12-16.

Kwan, Mei-Po, (2003b) 'Geovisualization of Activity Travel Patterns Using 3D Geographical Information Systems' Paper Presented at 10th International Conference on Travel Behaviour Research. Luierne. Aug. 10-15.

Kwan, Mei-Po, (2004) 'GIS Methods in Time Geographic Research: Geo-computational and Geovisualization of Human Activity Patterns' *Geografiska Annaler B*, Volume 354, pp. 341 – 353.

Kwan, Mei – Po (2005) 'Measuring Activity and Action Space/time' in Martins E.H.,Lee-Gosselin and T. D. Seen (eds) *Integrated Land-Use and Transportation Models: Behavioural Foundaations*. Oxford: Pergamon-Elsevier, pp. 101-132.

Kwan, Mei-Po, (2002) 'Time, Information Technologies and the Geographies of Everyday Life' *Urban Geography*, Number 25 (5), pp. 471-482.

Kwan, Mei-Po, (2003a) *New Information Technologies, Human Behaviour In Space Time and the Urban Economy* Paper Presented at the 82nd Annual Meeting of Transportation Research Board (TRB) Washington D.C. Jan 12-16

Kwan, Mei-Po, (2003b) 'Geovisualization of Activity Travel Patterns Using 3D Geographical Information Systems' Paper Presented at 10th International Conference on Travel Behaviour Research. Luierne. Aug. 10-15.

Kwan, Mei-Po, (2004) 'GIS Methods in Time Geographic Research: Geo-computational and Geovisualization of Human Activity Patterns' *Geografiska Annaler B* 354, pp. 341 – 353.

Kwan, Mei – Po (2005) "Measuring Activity and Action Space/time", In: Martins E.H., Lee-Gosselin and T. D. Seen (eds) *Integrated Land-Use and Transportation Models: Behavioural Foundaations*. Oxford: Pergamon-Elsevier, pp. 101-132

Lantrip, D. (1990) 'The Built Environment as a constraint to Human Freedom: An Activity Based Stochastic Model' *Environment Design Research Association* (EDRA), Number 21, pp. 52 – 63.

Lloyd, R.E. (1976) "Cognition, Preference and Behaviour in Space : An Examination of the Structural Linkages" *Economic Geography*, Number 52, pp. 241 – 253.

Lloyd C and I. Shottle worth (2005) 'Analysing Commuting using local regression techniques: Scale, Sensitivity and geographical patterning' *Environment and Planning A*, Number 37 (1), pp. 81 – 103.

Londquist, L. (1998) "Urban Planning of Locational Structures with due regard to user Behaviour' *Environment and Planning A*, Number 10, pp. 1413-1429.

Ojo, O. E. (1990) "Urban Travel Activity Pattern: The example of Ibadan", *Unpublished Ph.D Thesis*, University of Ibadan, Nigeria.

Pas, E. (1988) Weekly travel – activity Behaviour. *Transportation Research*, 15 B, pp. 89-107.

Pipkin, J.S. (1981) "The Concept of Choice and Cognitive Explanations of Spatial Behaviour", Economic *Geography*, Volume 57(4), pp. 315-361.

Root, G.S. and Recker W. W. (N.S.) "Towards a Dynamic Model of Individual Activity Pattern", *Formulation*, 1, 371 – 382.

Smith, T.R., Pellegrino, J.W. and Golledge R.G. (1982) "Computational Process Modeling of Spatial Cognition and Behaviour", *Geographical Analysis*, Volume 4(4), pp. 305-325.

Stinson, L.L. (1999) 'Measuring How People Spend Their Time : A Time use Survey Design' *Monthly Labour Review*, Number 122, pp. 12-19.

Timmermans, H.J.P. (2000) "Theories and Models of Activity Patterns", *Albataros*, pp. 6 – 70.

Timmersmans, H.J.P. and Veldhuisen, K.J. (1982) 'Behavioural Models and Spatial Planning: Some Methodological Considerations and Emperical Tests' *Environment and Planning A* Volume 13, pp. 1485 – 1498.

Timpf, S. (2002) "Geographic Activity Models", In: Goodchild, M; M. Duckham and M.F. Worbays (eds) *Perspectives in Geographic Information System (GIS)* New York, Taylor and Francis.

Van-der-Knap, W.G.M. (1995), *Analysis of Time-Space Activity Patterns in Tourist Recreation Complexes*. Working paper, centre for Recreation and Tourism studies Wagenirogen Agricultural University, The Netherlands.

Wegener, M. (2001) "New Spatial Planning Models", *International Journal of Applied Earth Observation and Geo-Information*, Volume 3(3), pp 15-25

Wong, K.Y. (1979) "Maps in Minds: An Empirical Study" *Environment and Planning A* Volume 11, pp. 1289-1304.

Zahavi, Y. (1978) *"The Measurement of Travel Demand and Mobility"*, Paper presented at the Joint International Meeting on the Integration of Traffic and Transportation Engineering in Urban Areas. Tel-Aviv Israel.

Part 2

Quantitative and Computer Spatial Planning Methods

5

A Program Management Information System for Managing Urban Renewals

Hyeon-Jeong Choi and Ju-Hyung Kim
Hanyang University
Republic of Korea

1. Introduction

The urban renewal project is an activity to regenerate existing city areas that have become run-down. In South Korea, urban renewal is led by the government, which perceived the importance of reducing environmental burden and enhancing the efficiency of use of existing urban infrastructure (Bae et al., 2008). Managing these projects is challenging for the government because of its insufficient experience with projects of a such massive scale and input resources (Lee, 2000). Furthermore, since the locations designated to be renewed are dispersed over the whole territory, supporting tools are necessary.

On the one hand, information technology and information systems (IT/IS) in urban planning and construction have advanced with the advance of IT in general. In South Korea, in particular, one of the leading countries for IT, there are moves to introduce IT/IS into management of megaprojects such as urban planning (Oh, 2001).

To encourage urban renewals, the South Korean government has funded research and development (R&D) projects. An organization named the Korea Urban Renaissance Center (KURC) was established to manage these R&D projects. The center aims to make an adjustment to urban renewals to follow the localized model in South Korea and create a long- or mid-term strategy, reinforcing the national capacity for urban renewal, analyzing the effect or performance of urban renewal projects, and implementing test beds and commercializing the output of R&Ds (KURC, 2007b).

Launched in 2007, the R&D project began with 41 sub-projects. One of these, "Development of Intelligent Program Management Information Systems" (i-PgMIS) aims to develop information systems to support decision making and information management for urban renewal mega projects (Kim et al., 2009). The authors have involved some of the research group, developing a framework of i-PgMIS according to their requirements. Integrating unit systems for functions such as cost/time management into i-PgMIS is also included. Unit systems are also being researched by other participants in the same R&D project.

This chapter introduces the framework of i-PgMIS managing the first version of a completed system and its development background.

2. Program management for urban renewal

2.1 Urban renewal mega project in Korea

In South Korea, urban renewal has been arisen as an issue in the area of urban planning. In accordance with the speed of industrialization, the urbanization of South Korea has been faster than in any other nation (Lee, 2000). Because new development was considered to be a faster way to accommodate people who moved to cities or existing people who wanted better urban infrastructure, earlier developed areas in cities remained vacant with no maintenance and became run-down.

Projects	A Cheongnyangni district renewal	B Sewoon fourth district renewal
Image of plan		
Location	Cheongnyangni fourth zone	Yeoji-dong, Jongro-gu
Site area	59 889m²	26 216m²
Gross floor area	594 889m²	336 026m²
Functions	Residence, business, office, sales, culture, hotel	Residence, business, office, sales, culture
Development method	Landowner-leading	Designated-developer-leading Residents-association-leading Landowner-leading
Stakeholders	KORAIL (client), private owners, Seoul city government, central government	Seoul urban development institute (client), resident associations, renting retailers, Seoul city government, central government

Resource: Cases of urban renewal project (KURC, 2007a)

Table 1. Examples of urban renewal projects in South Korea

As a regretful feeling formed about new development, the government of South Korea started to take action to encourage development in the form of urban renewal. Legislation was established to make old areas into areas for "The Urban Poor's Housing Environment Improvement" (Kim et al., 2009). So far in Seoul, 473 districts, including 297 districts downtown, are designated as areas for urban renewal (Seoul Urban Portal, 2010).

Development in the form of urban renewal can be executed using development methods classified corresponding to who leads the development: residents association, designated developer (usually public sector), or landowner (LH, 2006). For each method, the mandatory process is different. In the case of the landowner-leading method, approval by the resident association and the relevant processes are not required. Including the issues of social integration through renewal, public-private partnerships, and other stakeholder relationships further diversifies the stakeholders.

Projects in Seoul, the capital city of South Korea, are exemplified in Table 1. The two projects have complex facilities, and the portfolio of projects varies, so these projects show the trend of urban renewal across South Korea well. Not only the residence function, but also massive commercial and business functions are included.

The development type for project A is landowner-leading. Public and private owners, the central government, and the local government are stakeholders in these types of projects. Project B, which is led by a residents association, designated developer, and landowner simultaneously, has more complex stakeholder relationships.

2.2 Program management for urban renewal projects

The broad definition of a program is "a group of projects managed in a coordinated way to obtain benefits not available from managing them individually" (PMI, 2008). That is, program management is different from summing up the management of each project. There are risks that cannot be handled only by managing individual projects.

For example, if two projects are progressing at the same time, even if the budget of each plan is reviewed, the total cash flow can be more than the available cash. Two projects might have a related order so delays in one project might affect the later one. Sanghera (2008) exemplified managing projects in a company as a program. Though each project is well planned, failing to assign the total resources of the company can cause load for the company, which might bring bankruptcy.

Management comprises both non-soft and soft aspects (Kim et al., 2009). Non-soft aspects mean management that can be strictly controlled through quantitative data. Soft aspects include management of items that only can be described qualitatively but that affect the success of projects considerably.

Program management is differentiated from project management in the soft aspects of management: stakeholder management, benefit management, and decision-making governance. Urban renewal projects can be considered to be a program because they have these differentiated management factors.

A program includes various participants with conflicting interests. For instance, in the current urban renewal mega projects in South Korea, the government has started to engage on partnerships with the private sector, known as public-private partnership (PPP) (Seo, 2010). In urban renewal mega projects, the public sector pursues public advantages from the projects, while the private sector tries to maximize their return on investment in the projects. When it comes to resident associations, they make sure that their property is secured or that there is no additional expense for their share. Governance is considered important as well, as decision making in the projects in which various participants are involved, and deriving agreements, affect projects.

There is a case showing how the soft aspects affect a program. Management of these soft aspects can delay the whole program enormously. For example, the Sewoon fourth district renewal project in Korea has experienced design changes several times. Two organizations, the Seoul city office and a ward office, have had conflicted over the approval of high-rise buildings, related to the balance between profit from development and public benefit. Even though the project was initiated in 2003, even the approval for the plan has not yet been confirmed. Further, the migrating renting merchants who have resided in the area have not deduced agreements (Sung, 2010).

Stakeholders, the expected benefits for them, and decision-making governance should be supported for managing urban renewals. The existing project management systems are, however, focused on non-soft aspects of projects, such as cost or time management, which are presented in the form of numbers or indexes. Managing a program also requires numeric parts, but excluding other forms of management may lead to failure in the long term.

2.3 Vertical and horizontal program management

An urban renewal project can consist of several heterogeneous projects (Kim et al., 2009). Management of facilities with different functions can be implemented by separate practitioners. For example, while a hotel may be developed by one developer, a housing complex may be developed by another developer. In this case, there might be a manager employed by the city government who manages the whole project.

From the broader perspective, a group of programs can be a program, too. In this case, program management includes managing repetitive projects (Thomsen, 2008). Let's say a program manager is the local government. They might have several urban renewal projects in their cities, of which success will influence their public finance or resources. According to Thomsen (2008), program management in construction projects is an activity to gain advantages from managing repetitive similar projects. He exemplified managing a school-building project. Projects for building schools in a local area, which are usually led by an education administration, have a prototype. That is, the prototype has been set through the rotational experiences, and the next prototype would be improved with the experience.

In this paper, program management is introduced with both meanings. An urban renewal project is a horizontal program comprising several projects implemented almost simultaneously. Second, an urban renewal project is a vertical program that requires management of repeated projects with a time gap. When the concept is introduced to urban renewal mega projects in South Korea, a program can include projects with layered structure.

The program management information system for urban renewal projects, thus, should accommodate the structure of a program. First, for horizontal management, the system should be able to support the complicated process following from the combination of different projects. For vertical management, the system requires a structure for storing data to enable taking benefits from processing data from each project's perspective.

3. IT and IS in urban planning and construction

In order to investigate how an information system can include the technical concepts to support the described features of program management, exploring existing IT/IS and their trends is necessary. The information management of the urban planning and construction area has progressed (Kim et al., 2009) in two broad streams: spatial information processing (Stevens et al., 2007) and decision support (Chakrabarty, 2007).

Spatial information is based on geometric data. Software for spatial information is used not only to model spatial information, such as existing urban blocks in a city, but also to input a scenario for planning and outputting the effect of the plan. For instance, for urban planning and building planning, the Geographic Information System (GIS), a system designed to capture, store, manipulate, analyze, manage, and present geographically referenced data (Bolstad, 2005) has been introduced. Processed information from geometric data gives a clear understanding of prescriptive planning scenarios to review the effect of scenario based on geographical data reflecting the real physical environment (Yaakup et al., 2005).

In addition to spatial information management, management-based decision support has been introduced. Reviews of Project Management Information Systems (PMIS), a representative system supporting decision making in the construction project area, can provide a direction for implementing management information systems for urban renewal mega projects (Kim et al., 2009).

PMIS mainly focus on monitoring and controlling the implementation to keep up with the plan. Chitkara (1998) described PMIS well as a system that deals with data from the field. The raw and unsummarized data is processed into "information," which becomes the source for making decisions on an issue. For this reason, PMIS have functional menus. PMIS for construction projects usually consist of cost, duration, organization, and work scope management modules, and support integrated data processing among these. Communication on PMIS, thus, can be consistent with the integrated information.

PMIS are the trend in IT or IS for urban planning or construction projects in general. There have been three generations of the trend. First, a computer-aided information process was introduced. Computer-aided drawing (CAD) is the example. These technologies mainly support storing data in a digitalized way. The next trend focused on communication via the Internet. Recently, reuse of information utilizing integrated data arose as an issue in the IT/IS area. The concept was sharing information through a product and process model (Van Leeuwen & Van der Zee, 2005). With structured information and a database, the model can provide affluent information.

In construction IT, the building information model (BIM) is the example. BIM is a digitalized product model that has property in it. The "property" includes not only numeric data, such as the depth of an element, but also descriptive data, such as the phase to be constructed. The advantage of BIM is that the model can be used as a management model even in the operation and maintenance phase.

The trend of integrated information has involved an expanded scope of system use. Data from GIS is often limited to the urban planning phase. A protocol decided in this phase can affect management in the later phases or at a lower level, managing the planning for a building in this case (i.e., building limitations according to district planning). Some information needs to be successive among different management levels. For this requirement, the extent of BIM is stretching to the earlier phase and GIS to the later phase of project. Researchers have tried enhanced data usage between GIS and BIM through an integrated database server (Bentely, 2007), semantic data analysis (Isikdag et al., 2008), or standardized data format (Döllner & Hagedorn, 2007).

Another trend in IT/IS is moving its environment to the web. Cloud computing is a model providing ubiquitous network access to shared computing resources, including servers, storage, applications, and services. Stored information can be released with minimal management effort or interaction between systems (Mell & Grance, 2009). Introducing the concept of Web-based program management would create a consistent data storage environment. For stakeholders that are usually dispersed, it would provide easy access to information.

A management information system needs to take advantage of these technological trends to support program management. The authors introduce them into developing a program management information system. In the following section, specific requirements of the program management information system for urban renewal projects (i-PgMIS) are stated in relation to supporting program management.

The "i" indicates "intelligent," which is introduced to reduce complexity in this system. Notwithstanding the concept of intelligence, the system cannot make a decision instead of a human manager. The role is limited to helping participants make decisions through timely information. Functions that make humans perceive their status would take on that role.

4. Concept of i-PgMIS

4.1 Requirements according to program management for urban renewal

Specific requirements can be deduced from the characteristics of program management for urban renewal.

Compared to a project management information system, the scope of program management is expanded both horizontally and vertically. On the horizontal side, the program phase needs to include individual projects. On the vertical side, the management perspective is expanded from project to group of programs. In the management area, management of a group of programs is named portfolio management. In this paper, the word is used with a narrow meaning, which is limited to the monitoring of several programs.

From the soft aspects of program management, a program management information system is required, first, to create ease in allocating functions where required. As described, the program management system, dealing with entire lifecycle of projects, has a broad scope of management with various participants. Assigning access to the various functions and information on participation reflecting these diversified interests and work is challenging.

Second, planning activities for the urban renewal project need to be included. Compared to project management, the planning phase is weighed in program management considering that data from the planning phase becomes the source of management when the planning is implemented (Choi et al., 2010). While there have been information systems or software that support planning activities, usually the function is drawing up a planning scenario. Not only does the physical outline of planning need to be reviewed at this phase, but the scenario for benefits of stakeholders also needs to be evaluated. Thus, a comprehensive program plan should be enabled in the management system.

Third, briefing functions consistent with the stakeholders' perspective are required. The information required for a residents association to understand and for a government officer to document work would be different. Briefed information needs to, thus, have layers that each support appropriate manufacturing for adequate users. At the same time, information offered by the information system should be intuitive. For some stakeholders, who may be non-professionals on urban development or construction projects, visualized and general information will be more useful than Excel sheets indicating the cash flow.

In addition to the requirements to support soft aspects of program management for urban renewal projects, general management functions are essential for program management to earn practical data and control status. Management functions on cost, quality, communication, risks, and so on are included (PMI, 2010). Data in each management item need to be processed and briefed using the perspectives of various stakeholders, too.

In terms of the aspect of system structure, the database needs to be layered in accordance with the level of management, to store information and brief them as described.

4.2 Scope of i-PgMIS for managing urban renewal

In order to support the lifecycle of program management, the management information system needs to include information from the planning phase of a program through to the operation and maintenance phase of each project in the program.

In Fig. 1, the scope of program management and project management is described. Each program and project (or single facility) goes through the phases of conception, planning and feasibility, execution, and operation and maintenance. However, the meaning is different slightly, up to the level of management. While planning for a program is at a strategic level, planning for projects is at an executive level. For example, the preparation and execution phases at the program level correspond to the planning and construction phases at the project or facility level.

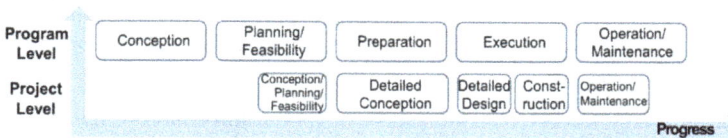

Fig. 1. Comparison between the scopes of program and project management

4.3 Technology for i-PgMIS: Plug and play

"Plug and play" is a technology for software engineering enabling a component-based development environment through layering, modularization, and information hiding (Bronsard et al., 1997). Modules of a specific function can be adjusted instead of newly developed just by being attached to the system.

In constructing an urban renewal management system, plug and play technology has powerful benefits. Where various kinds of functions are dependent on the type of diversified development methods, newly developing a system for every urban renewal project will take a lot of time. If a system manger can create a program management system with a plug and play method, it can bring enormous time saving. As well, data used in the same module can be filed into the same database, filling up historical data.

In i-PgMIS, most system-constructing activity can apply plug and play. The activities can include setting a system menu, breakdown structure, etc. Introducing plug and play technology requires standards for these items. The standards need to be established in i-PgMIS according to the type of urban development, decision-making governance, and any kind of unit comprising the main system.

Authority to use i-PgMIS also can be assigned the same way. By having default access or operating authority on i-PgMIS along with the stakeholders' area, i-PgMIS can assign the authority to organizations, thereby reducing the time required.

This principle can be applied to business process, too. Business process is usually determined up to the type of development methods. If a system manager calls the standardized business process from the database, tailors it, and then plugs it into the main system, the system will have its business process-based menu tailored.

4.4 Functions: Defining, controlling, and briefing program

Compared to PMIS, supports on planning and briefing are more weighted in i-PgMIS in relation to stakeholders, benefit management, and decision-making governance.

Planning accompanies a planning scenario involving cost and duration. The program definition system, a sub-system of i-PgMIS, takes part in creating a project scenario in the visualized way. The word 'Definition' was used to express the visualized defining activity.

The purpose of program definition is not only to create geometric models of scenarios, but also to create information structure with management hierarchy, which becomes the basic structure of the management of the projects. Urban renewal projects consist of several buildings of diversified use. The Cheongnyangni project, for example, consists of five housing buildings, one cultural complex, and one multi-complex building. The program manager would group facilities along with the building usage, the shape of the site, or as convenient.

After creating planning scenarios in the aspect of spatial information, the scenarios need to be reviewed in the aspects of cost and time. Cost- and time-management unit systems can contribute to this.

For general management requirements, unit systems are vital for i-PgMIS to perform as the control system. They support controlling the whole program with information gained from processing data input by the program manager.

Briefing is another function required for i-PgMIS to act as a monitoring system. Because every piece of information needs to be processed based on facts, or raw data, automatic processing is indispensable. With data processing transferred from practice in projects to the briefing system, the information will ensure the reliability of information and reduce the time required for gathering information.

4.5 The concept of i-PgMIS

i-PgMIS is a web-based system, adopting the advantages of cloud computing, such as easy access to the system and a unified data management environment.

The physical concept of i-PgMIS is shown in Fig. 2. Using the plug and play method, elements of systems can be plugged in a system.

A Web database has a hierarchy corresponding to the level of management perspectives. The raw data collected is processed and stored in a database (DB). The higher the level a database is used at, the more summarized and processed the information briefing. Extensible markup language (XML) format, which has strength in extensibility in data storage and interoperability between data interfaces (Ray, 2003), is used as a standardized data interface format when plugged systems interact.

i-PgMIS is based on the business process of the program. A process-based menu that suggests the work at the phase is introduced as intelligence. This is to simplify the use of system. Otherwise, users have to know all the required functions for program management at each process. The menu can also be plugged into a standardized business-process menu.

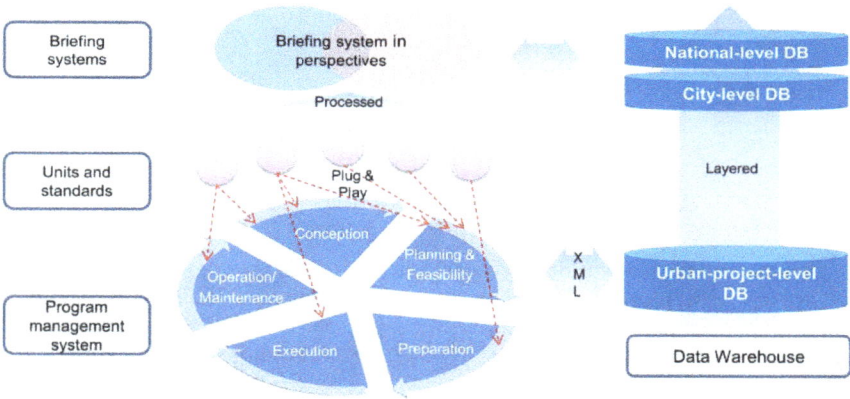

Fig. 2. The concept and system structure of i-PgMIS

5. Developing i-PgMIS

5.1 The structure of i-PgMIS

As shown in Fig. 3, i-PgMIS is composed of a program portal creation and control system, a program portal system, a participant's system, and a Web-based briefing system.

The program portal creation and control system is where "plug and play" technology is implemented. While the other three systems are where users do their management activity, the program portal creation and control system creates and controls the three systems.

The program portal system is a system for practical program management. The management activities in the lifecycle of a program are supported by a number of unit systems and functional modules. The program definition system is one of these. Unit systems consist of cost, duration, performance management, and process optimization. They are individual systems that function alone but share functional modules with program portal systems.

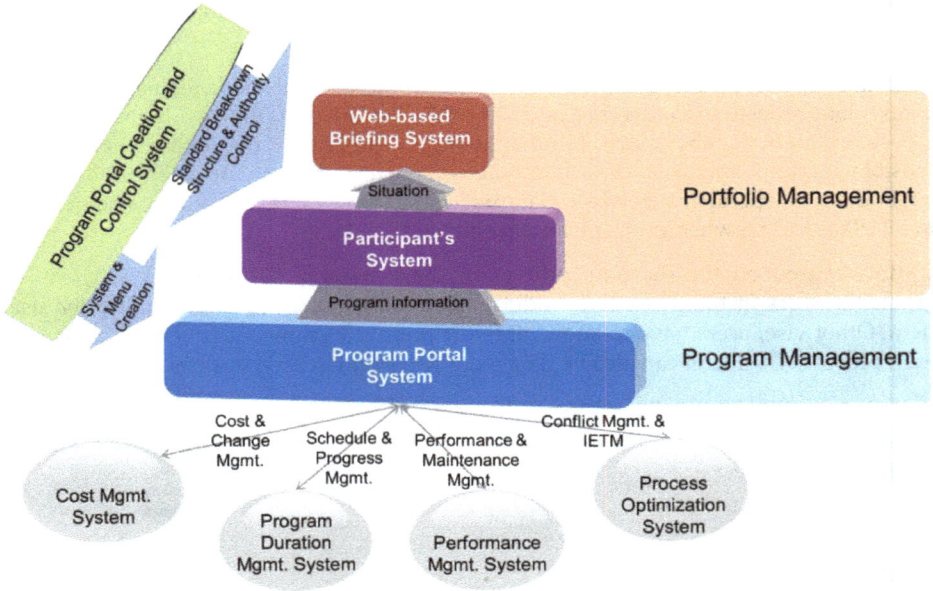

Fig. 3. The hierarchical structure of i-PgMIS

Briefing systems are separated into two parts, according to the managerial inspectors who are to be briefed with information. The participant's system and Web-based briefing system are for monitoring urban renewals at the local-area level and the national-area level, respectively. Data from the program portal system at the boundary of the management level are synthesized and briefed.

Information technologies are adopted to support the hierarchical and layered structure and Web-based policy. The layers of the system structure are shown in Fig. 4.

The strengths of Java, the development environment adopted by the i-PgMIS system developer, are portability across system platforms and scalability of data exchange. Considering that i-PgMIS requires the capacity to accommodate various users and the flexibility to integrate various functions, Java is better than other development languages.

The user interface (UI) of i-PgMIS uses HTML and CSS. If a user inputs data (e.g., numerical data for earned value at cost) through the UIs, the data is examined and processed using JavaScript.

The data accesses resources such as a database and internal or external systems. Interfaces between these resources and the i-PgMIS framework are of great importance. Internal systems include modules to support the functions of i-PgMIS. External systems indicates open systems that have already been developed and utilized in i–PgMIS for special purposes. For example, Google Earth 3D, which is used in the Web-based briefing system as a navigator showing the status of a project visually, is linked with the open API provided by Google. (The use of Google Earth will be described later in more detail.) Some of the external systems developed on local computers are loaded through Active X.

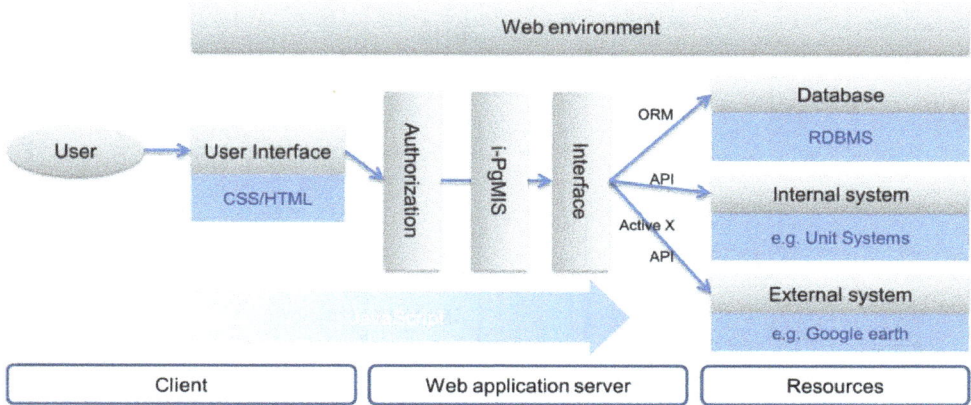

Fig. 4. Technological implementation for i-PgMIS

The i-PgMIS database is located on the Web. In order to manage the vast amount and kinds of data for urban renewal projects, a relational database management system (RDBMS) is introduced; Oracle 10g is used in this system. The Web application server connects separately created systems with the Web database. With object-related mapping (ORM) technology for the database, the queries needed to call and save XML-based data can be reduced.

5.2 Program portal creation and control system

The program portal creation and control system exists to control the other systems in i-PgMIS. The system administrator can operate this system and create other systems.

To create systems, the administrator has to click a checkbox for the required unit. Once a standardized property (i.e., business process) for each type is defined in advance, all the system operators have to do is tailor them slightly in accordance to a project. This process is described in Fig. 5. In this figure, a standard business-process-based menu used for a specific development method is tailored (Fig. 5 ①) in a system plugged (Fig. 5 ②) into the program portal system via program portal creation and control. Because tab menus for a specific process have a default setting, the tab menus used in the program portal system follow them (Fig. 5 ③). In the same way, the authority for accessing information or menus, functional modules, the breakdown structure of work, cost, or documents can be assigned.

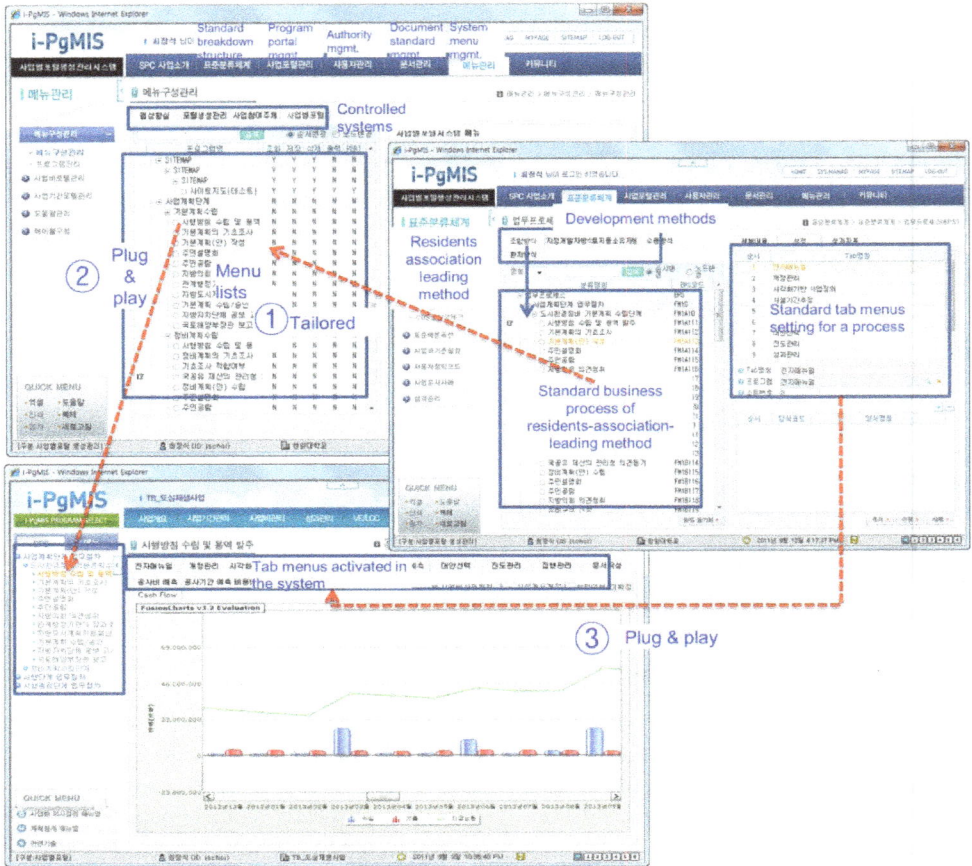

Fig. 5. Plug and play for creating menus in the program portal system

5.3 Program portal systems

A program portal system, in which program information is input and processed, has a user interface involving two types of menu: the top menu and the business-process-based menu. Fig. 6 shows the user interface of the program portal system. The top menu, so named because of its location, consists of a program definition system, links unit systems, and other supportive management modules. The business-process-based menu, located on the left side of the system, accompanies the tab menus that support the activities of each process. Thus, the tab menus can vary for each process. A business process acts as an indicator, showing the progress status. The last process worked on is remembered as the current process in the system.

The tab menus can vary up to the business phase because every phase has to be supported differently. For example, the tab menu for the business manual is always located at the front. This is based on the fact that a user must acknowledge how an urban renewal project is to be managed before proceeding.

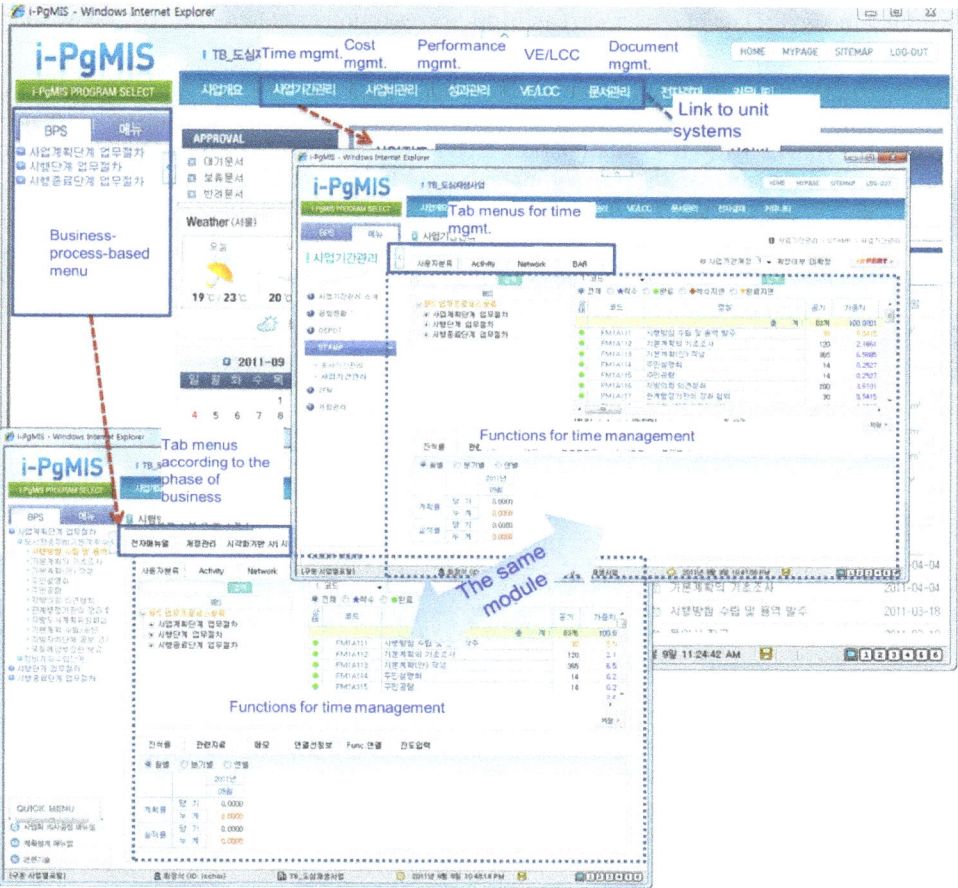

Fig. 6. User interface of program portal system

5.3.1 Program definition system

In the program definition system, as shown in the left side of Fig. 7, the user can define the program structure with an interface that enables the creation of a tree structure for managing an urban renewal project. The right side of the screen shot shows the space where the user can create a visualized model for districts and facilities. Though the principle is similar to 3D CAD, which provides 3D drawing, it is specialized for managing programs, enabling the user to relate the defined structure and visualized model. Thus, data such as area of a space, which is defined during visualization, can be categorized according to the tree structure.

Once a visualized model is created, it can represent a scenario. Several visualized scenarios can be stored as program alternatives in the Web database. Using data from the defined program model in unit systems, it is possible to compare scenarios in terms of cost and duration so decision makers can be supported. An integrated scenario for system use is illustrated in Fig. 8. Data of a scenario can include area, number of facilities, hierarchy, space use, etc.

With this data, the cost management unit system estimates the cost for implementing projects with the case-based reasoning method. Then, the time management system provides the functionality to create relationships among projects in terms of their duration. By doing that, a project manager can estimate the cost for the whole program.

Another function of a program definition system is checking the alternative for space planning through a visualized model. For example, in Fig. 7, the purple space indicates that the space is for cultural use. Whether a space is appropriately located within the building or in the whole building project can be reviewed. It is expected that features of an urban project that cannot be described using quantity data can be reviewed through the visualization function.

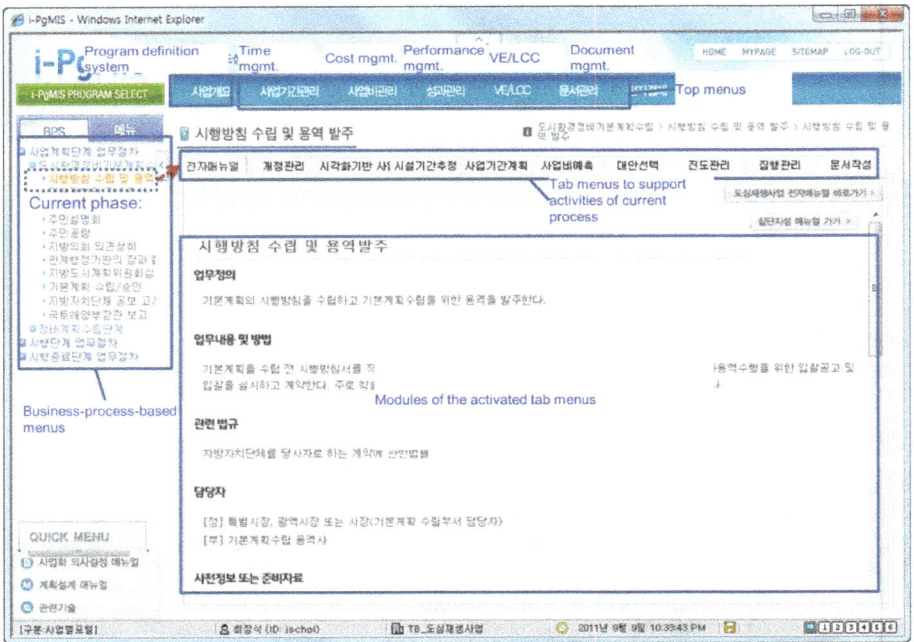

Fig. 7. Program definition in i-PgMIS

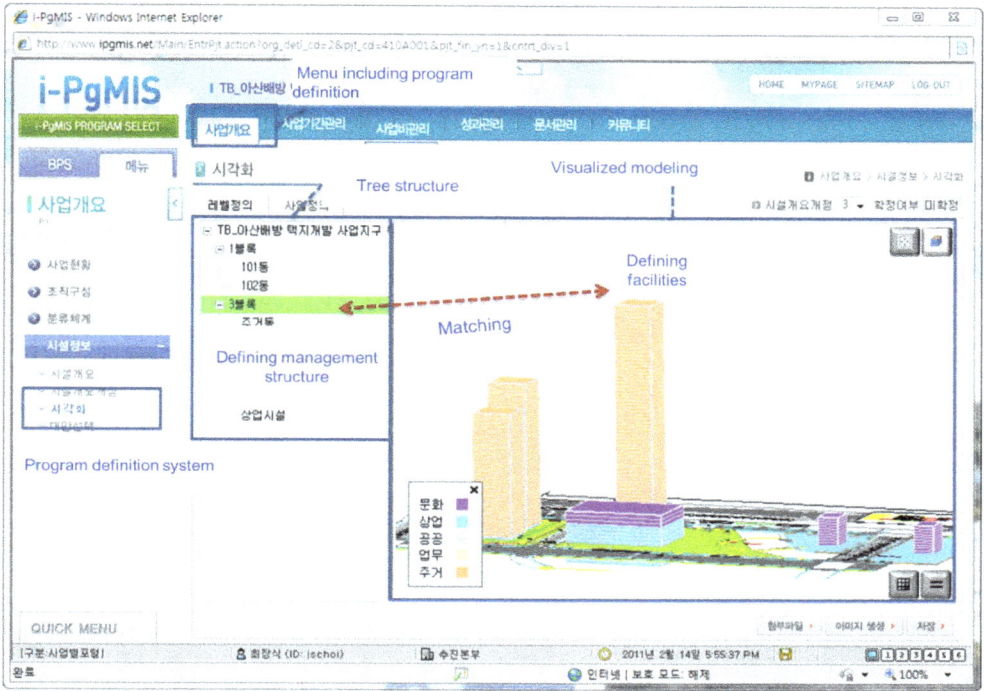

Fig. 8. Integrated planning with Program Definition System and Unit Systems

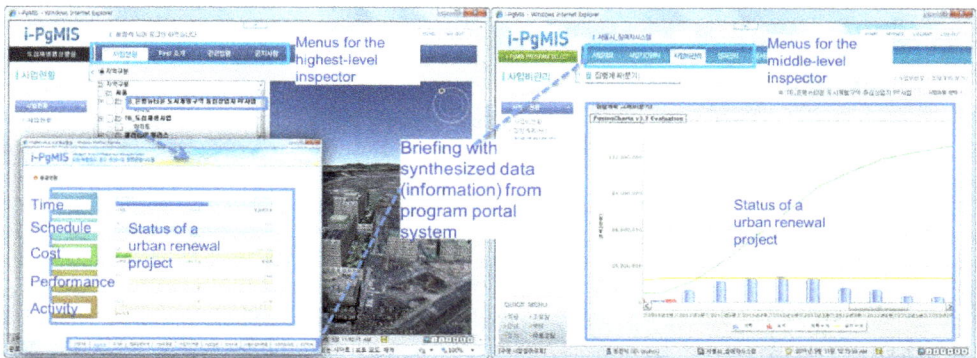

Fig. 9. Tab menus in business process menu

5.3.2 Unit systems and tab menus

Unit systems are individual systems that implement their own functions. There are time, cost, and performance management systems, and the interactive electronic technical manual (IETM). These systems were developed by co-researchers in the R&D branch.

The systems can exist on their own, but they share sub modules with the program portal system. The systems share functional tap menus with i-PgMIS. The "business manual" tab menu is shared with IETM. The "facility construction duration estimation" and "program duration planning" tabs are shared with the time management unit systems, and some tab menus are shared with a unit system. The shared module for time management is exemplified in Fig. 9. Also, some tabs function independently, such as document management.

In the tab menus at a specific phase of a project, the data and documents required for reporting and management are inputted. Texts and tables are inputted via Web pages. Documents created through Excel or a word processor can be uploaded. Some modules can activate a word processor with tables in it. If users input text or numeric data in the table, it is stored as an XML format document. Whenever the condition of a project changes according to the phase of the urban renewal project, the appropriate tab menus will be used to input data.

The specific mechanism and contents of the unit systems or tab menus are out of the scope of this paper, so we introduce them only briefly.

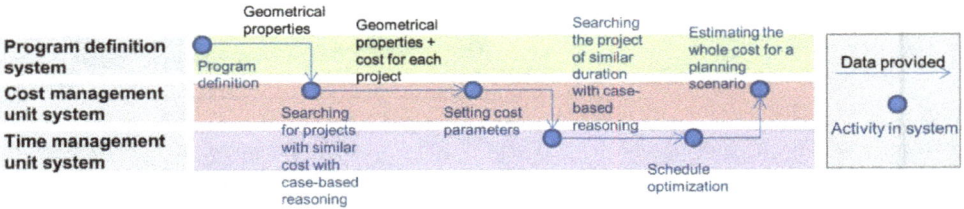

Fig. 10. Screenshot of Web-based briefing system (left) and Participant's system (right) in i-PgMIS

5.4 Briefing systems

The participant's system and Web-based briefing system both act as briefing systems. Each system reflects a different management level: local or national. The data of several projects input via each program portal system are processed and briefed. Briefed information is processed from data inputted in program portal systems by the project managers. If a local area has 10 urban renewal projects and the local government office manages them, the status of 10 projects, for example the information on resources that have been invested, is synthesized from 10 program portal systems. Data from projects including those in this area will be synthesized in Web-based Briefing System, because the projects in the area need to be monitored by central government too. Thus, the sources of information in the two systems are basically identical.

Fig. 10 shows screenshots of the two systems. As shown in the left image, the Web-based briefing system provides lists of projects using Google Earth with open API, so the location or appearance of the facilities is shown visually. Open API is expected to be used to develop further applications in i-PgMIS to extract geometric and facility model data in future. On other menus, the status of a project or projects under management is displayed with graphics in terms of time, cost, performance, and activity.

Information on the same project is illustrated in the participant's system, which supports more detailed monitoring. Menus in this system consist of those for middle-level inspectors. The status of cost, time, performance, and document management are displayed.

6. Conclusion

As urban renewal projects have become a major issue for urban planning in South Korea, R&D projects to support successful urban renewal have been funded. This paper described one of them, developing an information system for management urban renewal projects (i-PgMIS). The characteristics of program management and IT/IS trends were reviewed to give a direction for development. The program management information system has the policy of being Web-based, having a layered structure according to the level of management, and being created and controlled through the plug-and-play method.

At present, the preliminary test bed for i-PgMIS ver. 1 has been implemented. The purpose of the preliminary test bed is to examine whether systems attached using the plug-and-play method work flawlessly, which they do. i-PgMIS ver. 1 is supposed to be completed in April 2012, and the final version of i-PgMIS will be produced after two tests of the main test bed and a beta test, which is to be commercialized. Through this series of tests, i-PgMIS will become sophisticated in the aspects of user interface and data processing.

The limitation of this research is found in application to real projects, of which conditions are changeable. The business process for urban renewal projects can change as governing law is altered over time. i-PgMIS must be adjusted not only for the type of project but also for the time change. Further, the practicability of i-PgMIS and possible problems with data storage on the layered Web database need to be investigated with application to a real project.

Further studies to supplement these limitations will be accompanied with the test bed and beta test against practitioners in urban renewal projects.

7. Acknowledgement

This research is supported by a grant (07UrbanRenaissanceB03) from the High-Tech Urban Development Program funded by the Ministry of Land, Transport and Maritime Affairs of the Korean Government.

8. References

Bae, W., Lee, Y., Kim, Y., Jang, Y., Maeng, D., & Lee, S. (2008). The present conditions and tasks of urban renewal in Korea. *City Information*.

Bentely. (2007). Integrating CAD, GIS and BIM: An OGC Interoperability Project, from Bentely.com

Bolstad, P. (2005). *GIS fundamentals: a first text on geographic information systems*: Eider Press.

Bronsard, F., Bryan, D., Kozaczynski, W., Liongosari, E. S., Ning, J. Q., Ólafsson, Á. (1997). *Toward software plug-and-play*.

Chakrabarty, B. (2007). Computer-aided design in urban development and management--A software for integrated planning and design by optimization. *Building and environment, 42*(1), 473-494.

Chitkara, K. K. (1998). *Construction project management: planning, scheduling and controlling*: Tata McGraw-Hill Pub. Co.

Choi, H., Kwon, M., Kim, J., & Kim, J. (2010). *BIM-based Program Information Management System for Urban Renewal Mega Projects Planning*. Paper presented at the The First International Conference on Sustainable Urbanization, Hong Kong.

Döllner, J., & Hagedorn, B. (2007). Integrating urban GIS, CAD, and BIM data by service-based virtual 3D city models. *Urban and Regional Data Management UDMS 2007 Annual*, 157.

Isikdag, U., Underwood, J., & Aouad, G. (2008). An investigation into the applicability of building information models in geospatial environment in support of site selection and fire response management processes. *Advanced engineering informatics, 22*(4), 504-519.

Kim, J., Yoon, J., Kim, K., & Kim, J. (2009). Conceptual Model of Intelligent Program Management Information Systems (iPMIS) for Urban Renewal Mega Projects. *Journal of Asian Architecture and Building Engineering, 8*(1), 57-64.

KURC. (2007a). Cases of urban renewal project: Korea Urban Renaissance Center.

KURC. (2007b). The meaning of urban renewal projects Retrieved 1Sep, 2011, from http://www.kourc.or.kr

Lee, J. (2000). The Practice of Urban Renewal in Seoul, Korea: Mode, Governance, and Sustainability. Paper presented at the 2nd International Critical Geography Conference, Taegu, Korea.

LH. (2006). A handbook for redevelopment based on the law for Urban Poor's Housing Environment Improvement: Korea Land Hosing Corporation.

Mell, P., & Grance, T. (2009). The NIST definition of cloud computing. *National Institute of Standards and Technology, 53*(6).

Oh, K. (2001). LandScape Information System: A GIS approach to managing urban development. *Landscape and urban planning, 54*(1-4), 81-91.

PMI. (2008). *The Standard for Program Management*: Project Management Institute.

PMI. (2010). *A Guide to the Project Management Body of Knowledge(PMBOK Guide)*: Project Management Institute.

Ray, E. T. (2003). *Learning XML*: O'Reilly.

Sanghera, P. (2008). *Fundamentals of effective program management : a process approach based on the global standard*. Ft. Lauderdale, Fl.: J. Ross Pub.

Seo, J. R. (2010). A Study on the Analyzing Bottlenecks of PPP(Public-Private Partnership) in Mixed-use Development Projects. *Journal of Residential Environment Institute of Korea, 47*.

Seoul Urban Portal. (2010). The status of Urban Poor's Housing Environment Improvement Retrieved 1Sep, 2011, from http://urban.seoul.go.kr/

Stevens, D., Dragicevic, S., & Rothley, K. (2007). iCity: A GIS-CA modelling tool for urban planning and decision making. *Environmental Modelling & Software, 22*(6), 761-773.

Sung, S. (2010, 28Jan). Is Sewoon 4th district renewal project drifted again?, *Hankook encommomy*.

Thomsen, C. (2008). *Program Management: Concepts and Strategies for Managing Capital Building Programs*: Construction Management Association of America.

Van Leeuwen, J. P., & Van der Zee, A. (2005). Distributed object models for collaboration in the construction industry. *Automation in construction, 14*(4), 491-499.

Yaakup, A., Ludin, M., Nazri, A., Sulaiman, S., & Bajuri, H. (2005). GIS in urban planning and management: Malaysian experience.

6

Integrated Spatial Assessment (ISA): A Multi-Methodological Approach for Planning Choices

Maria Cerreta and Pasquale De Toro
University of Naples Federico II
Italy

1. Introduction

In decision-making processes for urban planning and design, evaluation can be considered a relevant *tool* to build choices, to recognize values, interests and needs, and to explore the different aspects that can influence decisions. Evaluation can be considered a *process* to integrate approaches, methods and models, able to support the different needs of the decision-making process itself. According to Trochim and Donnelly (2006), it is possible to define a *planning-evaluation cycle* with various phases requested by both planners and evaluators. The first phase of such a cycle, the so-called *planning phase*, is designed in order to elaborate a set of potential actions, programs, or technologies, and select the best ones for implementation. The main stages are related to (1) the formulation of the problem, issue, or concern; (2) the broad conceptualization of the main alternatives to be considered; (3) the detailing of these alternatives and their potential implications; (4) the evaluation of the alternatives and the selection of the preferable one; and (5) the implementation of the selected alternative. These stages are considered inherent to planning, but they need a relevant evaluation work, useful in conceptualization and detailing, and in assessing alternatives and making a choice of the preferable one. The evaluation phase also involves a sequence of stages that includes: (1) the formulation of the major goals and objectives; (2) the conceptualization and operationalization of the major components of the evaluation (program, participants, setting, criteria, measures, etc.); (3) the design of the evaluation, detailing how these components will be coordinated; the analysis of the information, both qualitative and quantitative; and (4) the utilization of the evaluation results. Indeed, evaluation is intrinsic to all types of decision-making and can take different meanings and roles within decision-making processes, especially if it is related to spatial planning (Alexander, 2006). "Evaluation in planning" or "evaluation within planning" seems to better interpret the concept of *planning-evaluation* proposed by Lichfield (1996) where the binomial name makes explicit the close interaction and reciprocal framing of evaluation and planning: evaluation is conceived as deeply embedded in planning, affecting planning, and evolving with it (Cerreta, 2010). Indeed, the evolution of evaluation methods reflects their evolving relationship with the planning process and the way in which they interact with the diversity and multiplicity of domains and values. To identify an analytic and evaluative structure able to integrate different purposes and multidimensional values within the decision-making processes means to develop evaluation frameworks not focusing only on the environmental, social and economic effects of different options, but also considering the nature of the stakes, selecting priorities and values in a multidimensional perspective. It is crucial to structure *complex decision-making processes* oriented to an integrated planning, that can support the selection, the monitoring and the management of different resources, and the interaction among decision-makers, decision-takers, stakeholders and local community.

In the above perspective, it is essential to adopt normative and instrumental approaches, but also "explorative" ones, open to plurality and dialogue among the different expertises involved (Fusco Girard et al., 2007). Facing the complexity of interacting perspectives, interests, and preferences (Wiek & Walter, 2009) means to identify a dynamic decision-making process, where *integration* represents the crucial point. An integrated approach to planning-evaluation involves many institutional and non-institutional stakeholders with divergent and conflicting values and mandates, with a high complexity of issues and interdependencies. According to Waddell (2011), the main challenges for and integrated approach are related to conflicting institutions, conflicting values, conflicting epistemologies, and conflicting policies. Different institutions have responsibility for different aspects of the domain, having narrow and often competing mandates; values differ among both institutional and non-institutional stakeholders, including the citizens, and they can be related to tangible and intangible dimensions; divergent epistemologies surely also are part of the assessment of the problems of integrating planning, using both quantitative and qualitative methods and models in order to overcome the gap between *implicit knowledge* and *explicit knowledge* (Te Brömmelstroet & Bertolini, 2010); conflicting policies at different levels and scales have to face legislation requirements and restrictions and need to find ways to open up the planning process on the project evaluation level to be consistent with broader normative guidance on integrating planning efforts.

In order to face the different levels of conflicts related to a spatial planning process, three main types of integration (Lee, 2006) can be considered:

-   *vertical integration of assessment*, which means to link together separate impacts, that are undertaken at different stages in the policy, planning and project cycles;
-   *horizontal integration of assessments*, which means to bring together different types of impacts (economic, environmental, social, etc.) into a single, overall assessment at one or more stages in the planning cycle. It means also an horizontal co-ordination between contemporaneous assessments for separate, and also interrelated, planning and project cycles;
-   *integration of assessments into decision-making*, that means to integrate assessment findings into different decision-making stages in the planning and project cycles.

The above types of integration can be helpful in facing the complexity of the planning environment, overcoming the limits of sectoral approaches and taking into account the multi-sectoral character and broadly defined content of many of the projects/plans to be assessed, the relative importance of complex impacts (indirect, induced and cumulative), the spatial and temporal complexity of their distribution, their multiple links, horizontal and vertical, and impacts from other projects or plans (Cerreta & De Toro, 2010; Lee, 2006) (fig. 1).

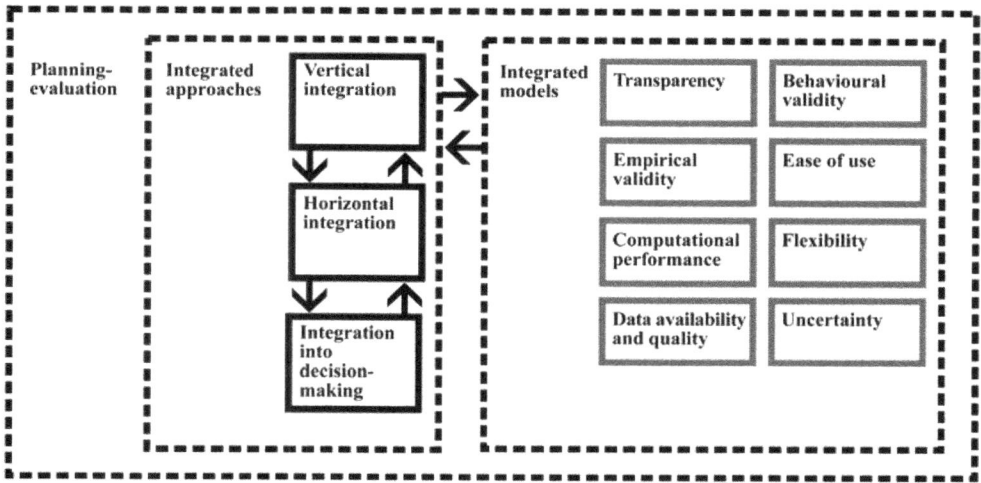

Fig. 1. Planning-evaluation: integrated approaches and integrated models

At the same time, some key challenges still facing integrated modelling and their application in practice include the following main characteristics (Waddell, 2011):

- *Transparency*: models will not be credible as tools for decision support in complex, conflict-laden domains such as land use, transportation and environmental planning, unless they can be explained with a sufficient degree of transparency;

- *Behavioural validity*: for a model to be credible in a contested domain, it must have sufficient behavioural validity to be believable as an independent artefact, within some clearly defined scope of applicability. Behavioural validity includes more common sense or intuitive understandings of how the world works;

- *Empirical validity*: models must be tested against observed data in order to assess their empirical validity. A model has to respond to input assumptions and make predictions that will reasonably correspond to observed reality. A model can be used to predict outcomes into the future, and it should be able to capture the essential trends in outcomes over some period of time;

- *Ease of use*: if a model is too complex to explain and implement, it also will ultimately not succeed in practice. A model system must strive to achieve a threshold of usability that makes it possible for staff within planning agencies to be able to use it, taking into account that complexity can lead to more mistakes;

- *Computational performance*: a model has to be characterised by a good computational performance able to define a valid simulation of reality modifications;

- *Flexibility*: a model has to be able to satisfy users in all cases and for all applications. Indeed, models and software platforms that are too rigid become a serious constraint, and limit applicability; models need to be adaptable to different users and different data and needs;

- *Data availability and quality*: in implementing a model a crucial point is developing the input data for it. In general, the science and tools to develop data usable in modelling are far from addressing the needs of users. Then data can be incomplete and error prone. Further, it is difficult to integrate them into a coherent database that is internally consistent. The difficulty of developing the data for a model system can be a very important obstacle to consider;

-   *Uncertainty*: only recently uncertainty has come into the lexicon of integrated modelling, but is becoming increasingly important in decision-making process related to spatial planning, especially in choosing among different alternatives.

The construction of suitable models is oriented to face complex problems that arise in socio-technical, socio-economic and socio-ecological contexts in order to transform an existing problem situation into a form that is more acceptable, understandable and manageable (Amin & Roberts, 2008). Often decision-makers and planners failure to fully understand such problems results in failures to formulate effective intervention strategies. In this research, Soft Operations Research (Soft OR) combined with System Dynamics (SD) modelling, Multi-Criteria Analysis (MCA), Multi-Group Analysis (MGA) and Geographical Information System (GIS) can help to improve stakeholders' understanding of a complex problem situation and to facilitate learning about it in a perspective of defining shared strategic actions.

2. Multi-methodological framework in decision support systems

According to Te Brömmelstroet & Bertolini (2010), the concept of *knowledge generation* is essential for building integrated strategies, where *socialization* (tacit with tacit: sharing experiences to create new tacit knowledge, observing other participants, brainstorming without criticism), *externalization* (tacit with explicit: articulating tacit knowledge explicitly, writing it down, creating metaphors, indicators and models), *combination* (explicit with explicit: manipulating explicit knowledge by sorting, adding, combining, looking to best practices) and *internalization* (explicit with tacit: learning by doing, developing shared mental models, goal based training) (Nonaka & Takeuchi, 1995; Nonaka et al., 2006) represent the main phases and the four key modes of knowledge conversion.

Through a process of knowledge generation iteratively acting in all four modes of knowledge conversion, interplaying between tacit knowledge end explicit knowledge, and by experiencing the four knowledge conversion modes, planners can develop a shared explicit language and use it to develop integrated strategies (Healey, 2007; Te Brömmelstroet & Bertolini, 2010). This approach to knowledge management to support strategy-making is also consistent with the epistemological structure of "post-normal science" developed by Funtowicz and Ravetz (1993), considering two crucial aspects: *uncertainty* and *value conflict*.

According to post-normal science, to recognize the importance of difference implies a different way to address complex systems and to face complexity means to take into account the self-organization chances, non-linear dynamics, non-continuous behaviours of complex systems and participated decision-making processes. This means to broaden the field of decision-makers and to involve new social actors in order to create an "extended community", able to elaborate new solutions (Funtowicz & Ravetz, 1994).

The approach of post-normal science forces decision-makers and planners to find solutions not only coming from the "expert knowledge", but also legitimated by "common knowledge", including uncertainty as part of the decision problem, and considering solutions based not only on exact scientific data (*hard data*), but also on public decisions, shared by the community (*soft data*). Indeed, facing and/or solving complex problems depends on the capability to consider them under different points of view, and to manage uncertainty, filling the gap between experts and community.

According to the above perspective, it stands out that "integrated evaluations" can be a key tool to support the decision-making process, especially when uncertainty, complexity and values of different social groups are many, different and conflicting (van der Sluijs, 2002). Integrated evaluations not only consider the inputs of data expressing the impacts of different solutions, but are also "open" to a wide public participation, so that they can offer more information for the evaluation itself and, in addition, can make the decision-making processes and the results more acceptable (Golub, 1997; Munda, 2008). Participation becomes essential not only to examine and evaluate choices on social, ethic, political, economic, environmental levels, but also to legitimate choices and make them acceptable for the community itself. Integrated evaluations constitute an ongoing process both, iterative and interactive, multi-disciplinary (respecting the issues addressed) and participative (respecting communities), able to recognize the relevance of *technical indeterminacy* and *value multiplicity*.

In this view, it is important to combine different approaches in the same framework, integrating different evaluation tools, such as environmental, social and ethical balance sheets, and also Economic Valuation, Input-Output Analysis, Life Cycle Assessment, Risk Assessment, Ecological Impacts, Ecological Footprint, Mass/Energy Valuation, Multi-Criteria Decision-Aid Methods, Future Studies (Finnveden et al., 2003).

Other relevant tools that could be useful to consider are those covering the possibility of combining Multi-Criteria Analysis and Multi-Groups Analysis with Geographical Information Systems (GIS), Internet Technology, Spatial Decision Support Systems, Cellular Automata Models. Integration of differing evaluation models with GIS (Malczewski, 1999) becomes decidedly important in the construction of a Spatial Decision Support System: a variety of territorial information (social, economic and environmental) may be easily combined and related to the characteristics of the different options of territorial use, facilitating the construction of appropriate indicators and improving impacts forecasting, leading up to a preference priority list of the various options. Integration among Multi-Criteria Analysis, Multi-Group Analysis and GIS may be exceptionally useful when there are strong conflicts, in which the role of local actors, their relations and objectives may be considered as a structuring element in the process of information construction in a spatial and dynamic evaluative model (Al-Shalabi et al., 2006; Joerin & Musy, 2000, Nekhay et al., 2009; Şener et al., 2010; Thirumalaivasan et al., 2003; Vizzari, 2011). In the recent years, theoretical research and new technologies have improved the identification and implementation of integrated approaches for building planning strategies and actions.

2.1 A selection of multi-methodological decision support systems

Some interesting examples of integration among different and complementary methods and techniques in spatial planning field have been proposed, where the application of GIS is combined with evaluation tools and Planning Support Systems (PSS). A multi-methodological decision support system can be considered as the integration of a *dynamic system* (able to consider the time evolution), a *deliberative system* (able to include all the stakeholders), a *comprehensive system* (able to take account of quantitative and qualitative aspects related to different components) and a *spatial system* (able to identify the territorial effects also through their visualization) (fig. 2). According to this approach, a multi-methodological decision support system should be characterized by the interaction of Knowledge Base (KB), Relational Database Management System (RDBMS), Graphical User Interface (GUI), Geographic Information System (GIS), Multi-Criteria Analysis (MCA), and Multi-Group Analysis (MGA).

Indeed, PSS include: visualization tools that make it possible to get a 3-D, visual sense of what one alternative future might look like; sketch-planning tools that allow users to enter rules and then to visualize the outcome of those assumptions; simulation systems trying to model the behaviour of urban agents and the potential effects of alternative policy actions. Some selected models were developed in the transport sector, considering the relevance of infrastructures and mobility in land use transformations.

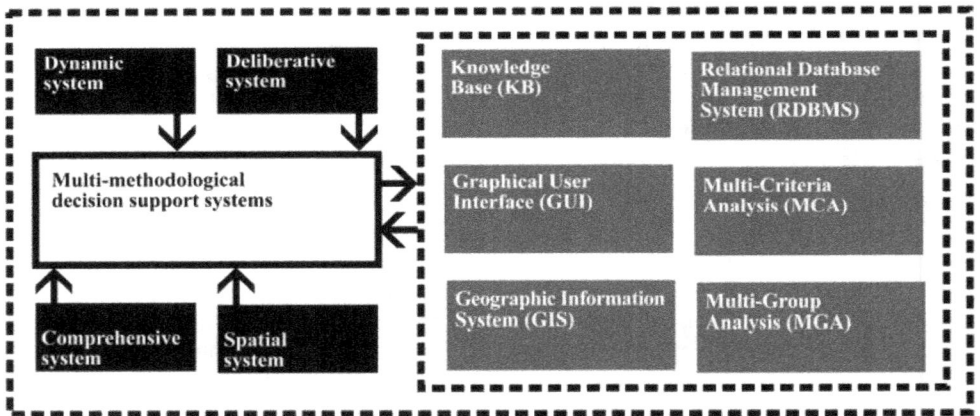

Fig. 2. Main characteristics of multi-methodological decision support systems

2.1.1 UrbanSim

UrbanSim, designed by Paul Waddell in the mid-1990s, falls in the third category of PSS, but also provide accessible visualization and stakeholder interaction (Waddell, 2002, 2011; Waddell et al., 2003) (fig. 3).

UrbanSim was developed as Land Use and Transport Interaction (LUTI) model in order to respond to a variety of needs to assessing the possible consequences of alternative transportation, land use, and environmental policies, trying to better inform deliberation on public choices with long-term, significant effects. The main reason was that the urban environment is so complex that it is not possible to anticipate the effects of alternative actions without some kind of analysis reflecting the cause and effect interactions that could have both intended and possibly unintended consequences. It is a software-based simulation system for supporting planning and analysis of urban development, incorporating the interactions between land use, transportation, the economy, and the environment. Since its initial release, UrbanSim has been increasingly adopted for operational planning use in the USA, Europe, Asia, and Africa, in planning agencies and in university research. The user community and research collaborators directly and indirectly support the application and refinement of UrbanSim. It is defined by an interactive web site that provides a virtual meeting ground for users and developers of the system, approximately half of them from the USA, and half from a rapidly growing list of countries. It can be used by cities, counties, non-governmental organizations, researchers and students interested in exploring the effects of infrastructure and policy choices on community outcomes.

Fig. 3. Example of UrbanSim application (source: http://www.uanalytics.com/urbansim)

2.1.2 Metropolitan Activity Relocation Simulator (MARS)

Metropolitan Activity Relocation Simulator (MARS) is another interesting model designed to improve decision-making process with specific attention to transport system (Emberger et al., 2006) (fig. 4). It is a dynamic Land Use and Transport Interaction (LUTI) model designed to support the decision-makers all through decision-making process (objective definitions, policy instrument identification, assessment of short and long-term impacts and appraisal), helping understanding of the concepts underlying the model and providing a transparent process. MARS is based on the principles of systems dynamics (Sterman, 2000) and synergetics (Haken, 1983), and is considered an ideal tool to model dynamic processes.

The MARS model environment allows to calculate a wide range of relevant indicators, and users can choose the set of indicators that fit the needs of their specific decision-making context. Then, MARS calculates the policy-dependent values for the key-indicators and allows the assessment and appraisal of the strategy, including also Cost-Benefit Analyses (CBA) and Multi-Criteria Analyses (MCA).

UrbanSim and MARS are only an example of the most advanced European LUTI models, that also include IRPUD (Wegener, 1998, 2004), DELTA (Simmonds, 1999, 2001), MEPLAN (Echenique et al., 1990), MUSSA (Martínez, 1996; Martínez and Donoso, 2001).

2.1.3 Land Allocation Decision Support System (LADSS)

Land Allocation Decision Support System (LADSS) (Matthews et al., 1999) is a tool developed at The Macaulay Land Use Research Institute (UK) for agricultural land use planning. More recently the term LADSS refers to the research of the team behind the original planning tool (fig. 5). Indeed, the focus of the research of the LADSS team has evolved over time from land use decision support towards policy support, climate change and the concepts of resilience and adaptive capacity. LADSS is the collective term for a farm-scale integrated modelling framework (IMF) that is being developed in order to simulate whole-farm systems. The acronym describes the projects original purpose as a land use planning tool back in the early 1990s. More recently, the project has expanded beyond its original remit to focus much more on deliberative processes involving decision-makers and other stakeholders. The LADSS framework core is biophysical simulation models overlaid by financial, social and environmental accounting modules. This framework provides a basis for the case-study assessment of how policy and environmental changes can impact upon land-use systems. Recently, these studies have centred around three main themes: Climate Change, CAP Reform and Agricultural Sustainability.

The focus of LADSS has changed in recent years from a tool designed to assist in the decision-making processes of land managers to a much wider framework that involves stakeholder groups as part of an integrated assessment approach, using a Decision Support System (DSS) as component of the process to explore options provides the decision-maker with a better understanding of the consequences of changes in land use and management. An integrated assessment approach is preferred, able to combine the DSS with deliberative processes involving stakeholders. The LADSS software runs on a Sun/Solaris platform and is made up of a Knowledge Base (KB), Graphical User Interface (GUI), Geographic Information System (GIS) and Relational Database Management System (RDBMS).

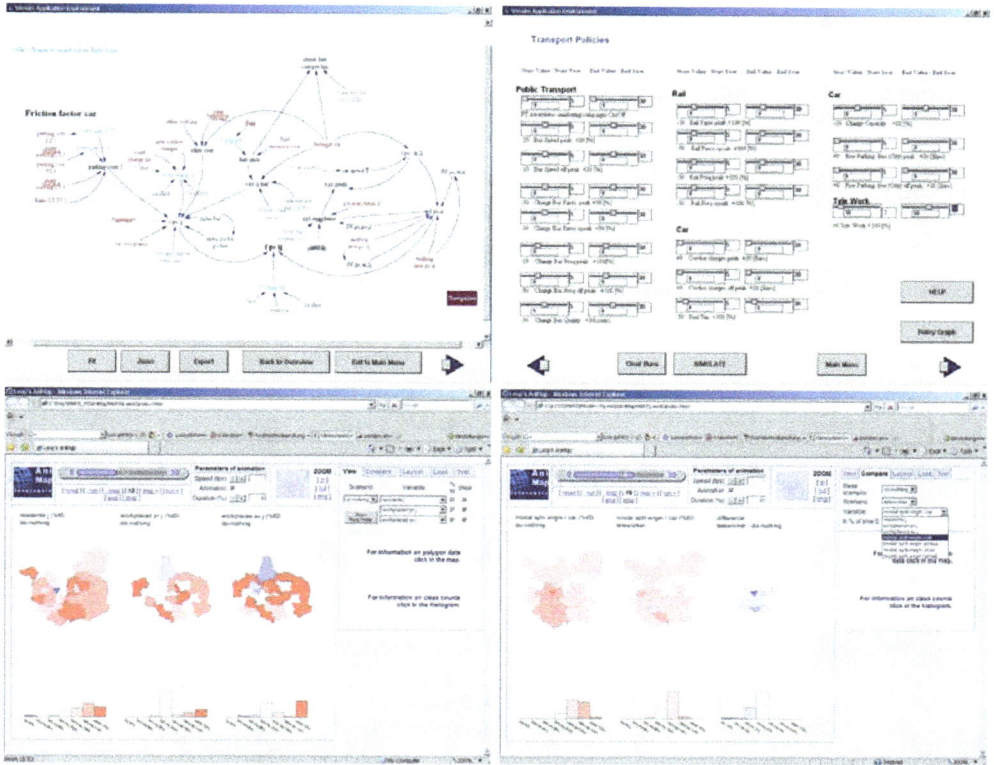

Fig. 4. Example of MARS application (source: http://www.ivv.tuwien.ac.at)

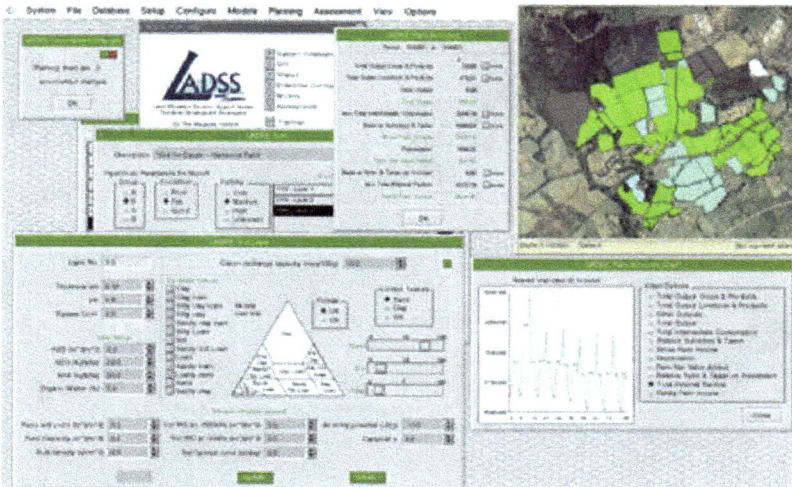

Fig. 5. Example of LADSS application (source: http://www.macaulay.ac.uk/LADSS)

2.1.4 LUCIS model

Another interesting approach is illustrated by the **LUCIS Model** (Carr & Zwick, 2007) that provides the information to understand and implement the Land-Use Conflict Identification Strategy (LUCIS) (fig. 6). LUCIS was developed over a period of ten years in a graduate design studio at the University of Florida for students from the Departments of Landscape Architecture and Urban and Regional Planning. Its conceptual basis was derived from Odum's Compartment Model (1969) that proposes four general land-use types for land classification. It evolved to use traditional land-use suitability analysis as a basis for projecting future land-use alternatives. Indeed, the LUCIS model uses the ArcGIS geo-processing framework to analyze suitability and preference for major land-use categories, determine potential future conflicts among the categories, and build future land-use

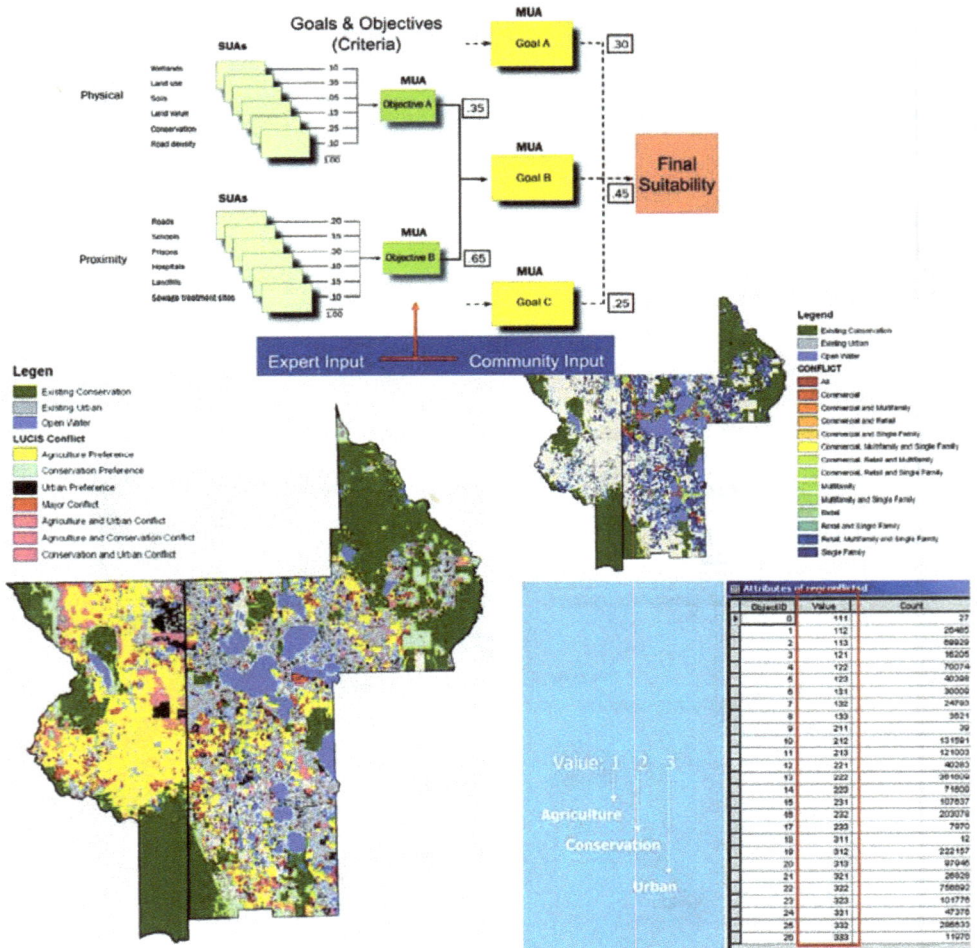

Fig. 6. Example of LUCIS Model application (source: GeoPlan Center, University of Florida)

scenarios. The basic concept is developing alternative future land use scenarios considered as a proactive approach to land management, resource management, and political and economic responsibility. With the help of technical tools such as Geographic Information Systems (GIS), regions across the United States are using scenario modelling to paint a picture of future development patterns. The selected methodology illustrates the impact of population increase and paves the way for developing more sustainable patterns of land use, producing a spatial representation of probable patterns of future land use for the following categories: existing conservation lands, existing urban lands, existing agricultural lands, areas for future conservation land use, areas for future urban land use, areas of probable future conflict between agricultural and conservation land uses, areas of probable future conflict between agricultural and urban land uses, areas of probably future conflict between conservation and urban land uses, areas of probable future conflict among agricultural, conservation and urban land.

2.1.5 What if?

In order to explore possible futures for a community **What if?** is an easy-to-use GIS-based Planning Support System (PSS) (Klosterman, 2001), that can be implemented to prepare long-term land use, population, housing and employment projections, political jurisdictions, and user-defined areas such as school districts, and traffic analysis zones (fig. 7). The

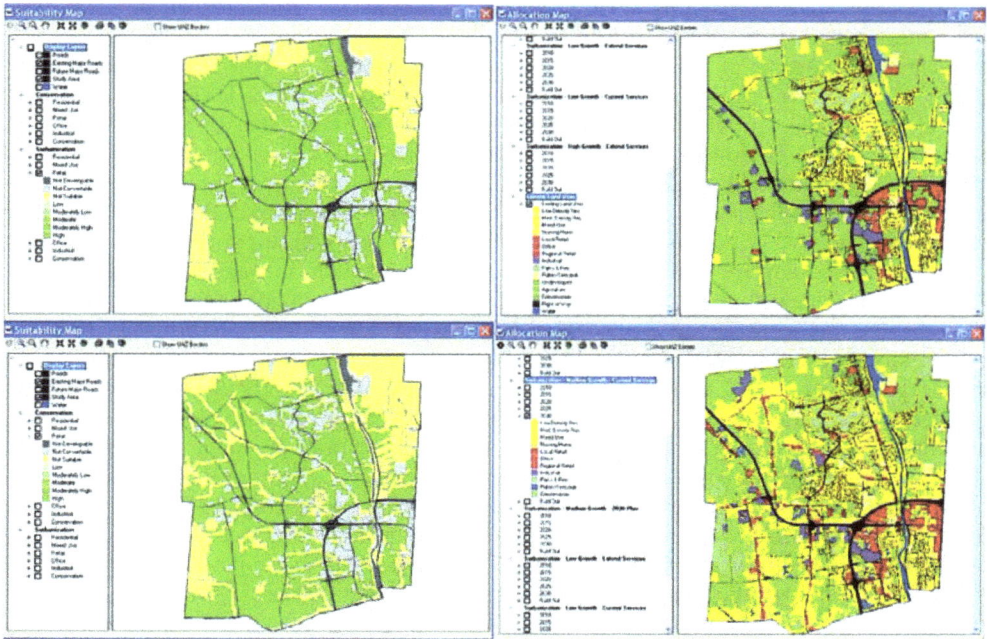

Fig. 7. Example of What if? application (source: Brail, 2008)

package is easy to use, customized to the user's GIS data and policy issues, and provides outputs in easy-to-understand maps and tables. Indeed, What if? can be used to prepare long-term land use, population, and employment projections for census tracts and user-defined areas such as political jurisdictions and traffic analysis zones. It allows users to determine quickly and easily the impacts of alternative policies to control urban growth, preserving agricultural land, or expanding public infrastructure in easy-to-understand maps and tables. What if? has been designed to be used in public settings by professionals, elected officials and private citizens. Local governments, regional planning organizations, and non-profit organizations across the United States and around the world have used it. As its name suggests, What if? allows planners, public officials, stakeholders, and private citizens to determine what would happen if public policy choices are made and assumptions about the future would prove to be true. Policy choices that can be considered in the model include the expansion of public infrastructure, the implementation of farmland or open space protection policies, and the adoption of land use plans, zoning ordinances, and other growth controls. What if? allows users to generate easily and quickly suitability maps and tables reporting the relative suitability of different locations for accommodating future land use demands.

2.1.6 Ecosystem Management Decision Support and Multi-scale Integrated Models of Ecosystem Services (MIMES)

An application framework for decision support of ecological assessments at any geographic scale is **Ecosystem Management Decision Support** (EMDS) (Reynolds et al., 1996) that integrates GIS, logic, and decision modelling to provide decision support for a substantial portion of the adaptive management process of ecosystem management (fig. 8). The NetWeaver logic engine evaluates data as respect to a knowledge base that provides a formal specification for the interpretation of data. The decision engine sets strategic priorities of landscape units, based on landscape condition derived from the logic model as well as any other management considerations pertinent to decision-makers. EMDS integrates state-of-the-art (GIS) as well as logic programming and decision modelling technologies in the Windows environment to provide decision support for a substantial portion of the adaptive management process of ecosystem management. EMDS uses **Criterium DecisionPlus (CDP)** from InfoHarvest, Inc. and **NetWeaver** from Rules of Thumb, Inc. as core components. The NetWeaver component performs logic-based evaluation of environmental data, and logically synthesizes evaluations to infer the state of landscape features. The Criterium DecisionPlus component prioritizes landscape features as respect to user-defined management objectives, using summarized outputs from NetWeaver as well as additional logistical information considered important to the decision-makers (InfoHarvest, 2001). In particular, Criterium DecisionPlus (CDP) decision management system helps structuring and communicating complex decisions among alternatives. It is a graphical Windows Desktop application that includes multi-criteria decision analysis (AHP and SMART) and uncertainty management. CDP manages both qualitative and numerical inputs, and helps eliciting preferences from decision-makers, and then provides contributions, sensitivity and tradeoffs analysis in order to validate those preferences.

According to the necessity to implement an integrated approach in planning, the **Multi-scale Integrated Models of Ecosystem Services (MIMES)** (Gund Institute for Ecological

Economics, 2007) is a suite of models for land use change and marine spatial planning decision-making (fig. 9). The models quantify the effects of land and sea use change on ecosystem services and can be run at global, regional, and local levels. The MIMES use input data from GIS sources, time series, etc., to simulate ecosystem components for different scenarios defined by stakeholder input. These simulations can help stakeholders evaluating how development, management and land/sea use decisions will affect natural, human and built capital. Building interactive databases for regional, integrated decision-making is an important aspect of implementing MIMES.

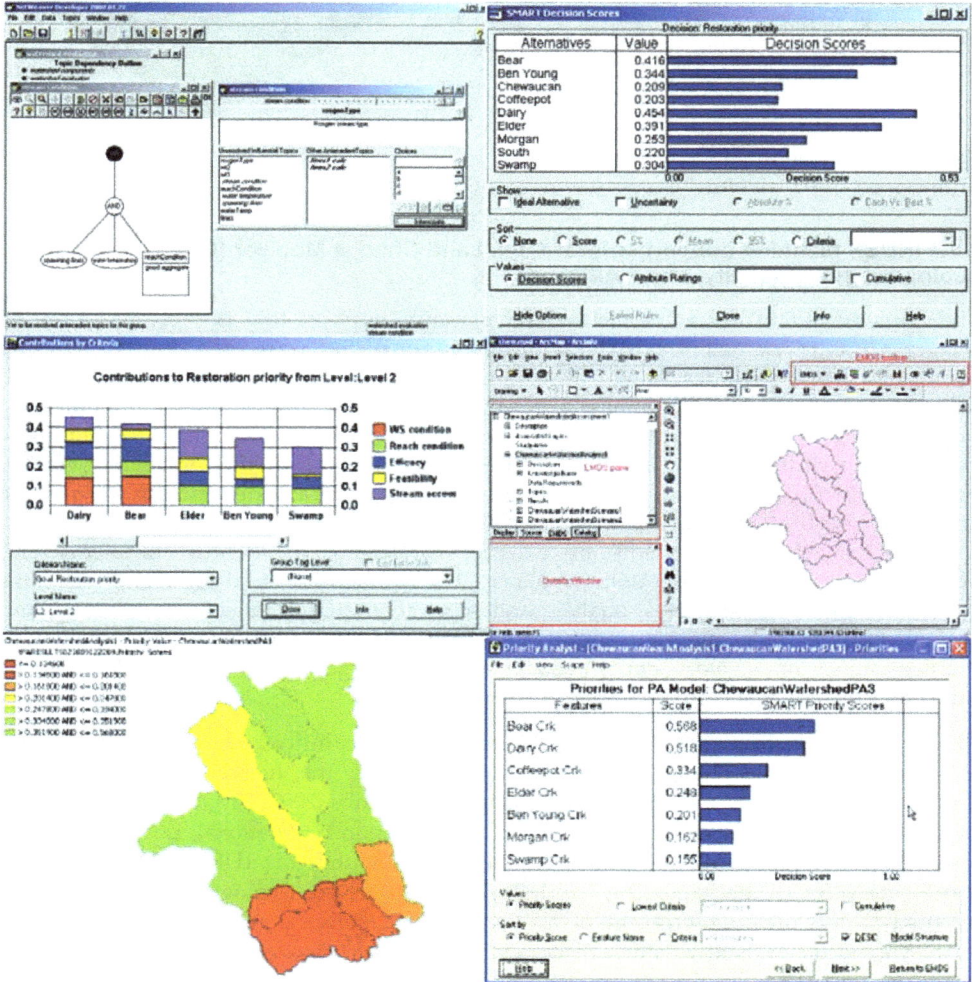

Fig. 8. Example of EMDS application (source: Reynolds et al., 1996)

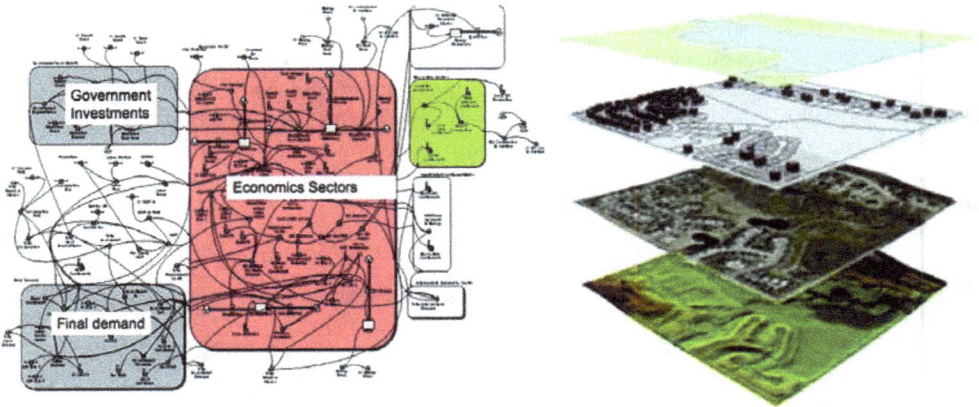

Fig. 9. Example of MIMES application (source:
http://www.afordablefutures.com/services/mimes)

2.1.7 INDEX planning support software, the Land Change Modeler (LCM) for ecological sustainability, and MAPTALK

The implementation of an interactive GIS planning support tool is represented by an integrated suite named **INDEX Planning Support Software** useful for assessing community conditions, designing future scenarios in real time, measuring scenarios with performance indicators, ranking scenarios by goal achievement, monitoring implementation of adopted plans (Allen, 2001) (fig. 10). Introduced in 1994, it is now supporting a wide variety of planning processes across the United States, with over 150 organizations in 35 states equipped with the software. INDEX is designed to support the entire process of community planning and development, and applications often begin with benchmark measurements of existing conditions to identify problems and opportunities reserving attention in plans. INDEX is used to design and visualize alternative planning scenarios, analyze and score their performance, and compare and rank alternatives. Once plans are adopted, INDEX supports implementation by evaluating the consistency of development proposals against plan goals. Over time, achievements can be periodically measured with progress reports.

The Land Change Modeller (LCM) for Ecological Sustainability is an integrated software to analyze land use change, projecting its trend into the future, and assessing its implications for habitat and biodiversity change (Clark Labs, 2007) (fig. 11). Commissioned by the Andes Conservation Biology Center of Conservation International, LCM is a vertical application developed by Clark Labs and integrated within the IDRISI GIS and Image Processing software package. The Land Change Modeler for Ecological Sustainability is oriented to the pressing problem of accelerated land conversion and the very specific analytical needs of biodiversity conservation. LCM is organized into five main areas: analyzing past land use change, modelling the process of change, predicting the changes into the future, assessing implications for biodiversity, and their evaluating planning interventions for maintaining ecological sustainability.

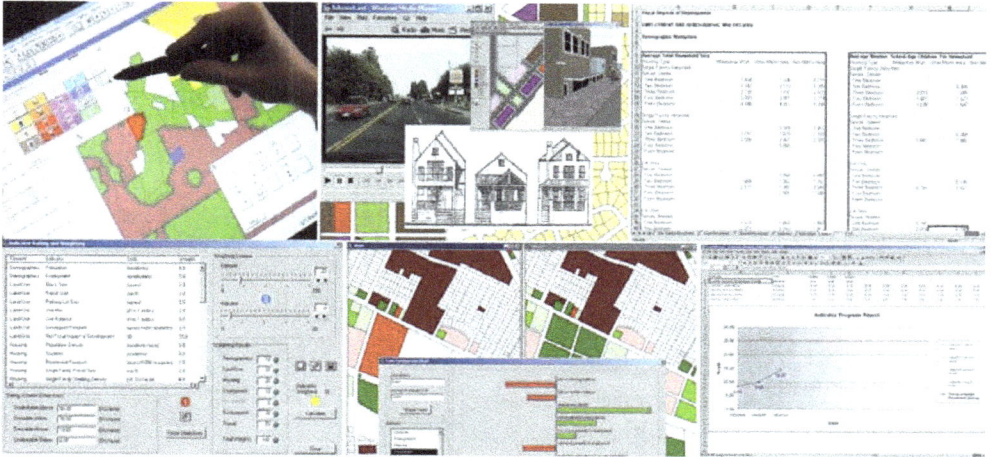

Fig. 10. Example of INDEX Planning Support Software application (source: www.crit.com)

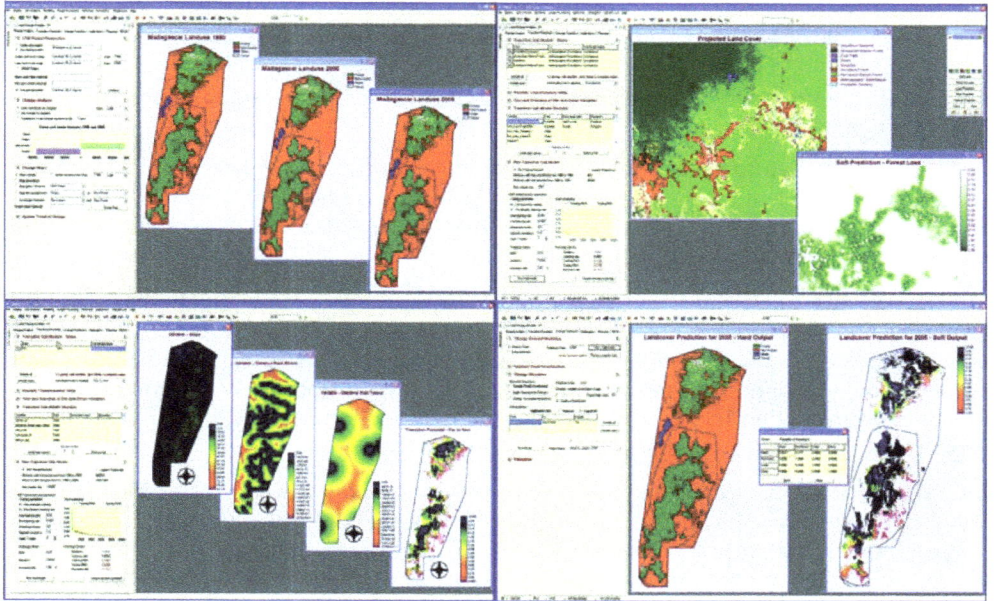

Fig. 11. Example of LCM application (source: Clark Labs, Clark University, www.clarklabs.org)

Another approach related to dynamic and spatial decision-making is developed by **MAPTALK** (W!SL, 2003), that offers a mutual GIS able to make an efficient use of this geographic information in spatial decision-making processes were stakeholders feel no qualms using it (fig. 12). Stakeholder participation and group decision-making is effectively supported by digital support of spatial brainstorms, discussions and (geographic) information sharing.

MAPTALK thereby facilitates dialogue, decision-making and constructive engagement, and can be considered as an accelerator for spatial planning processes. The use of MAPTALK results in a directly available cohered plan, with a well-documented process. Together with the Landscape center of Alterra and Wageningen Interactive Network Group (WING) a service is provided to facilitate interactive spatial planning processes.

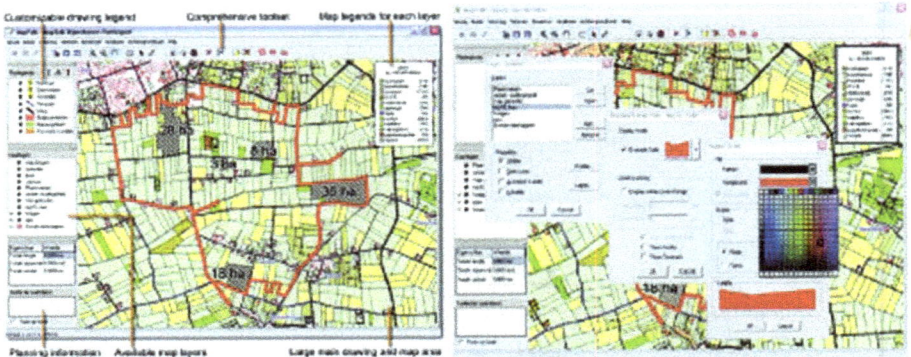

Fig. 12. Example of MAPTALK application (source: http://www.maptalk.nl)

2.1.8 Integrated planning for resilient communities: A technical guide to integrating hazard, ecosystem and land use planning

A toolkit is designed to support integrated planning that addresses hazard mitigation and ecosystem-based planning within a land use planning context: **Integrated Planning for Resilient Communities: A Technical Guide to Integrating Hazard, Ecosystem and Land Use Planning** (Hittle, 2011) (fig. 13).

The toolkit crosses disciplines and jurisdictional boundaries and can highlight the benefits of ecosystem conservation. It is characterized by the integration of three decision-support tools and methods, and the implementation of an integrated analysis of hazards, ecosystem conservation goals, and land use measures. A skilled expert team for community resiliency toolkit implementation is necessary, even if the required expertise will vary according to project size, complexity, and timeline.

Indeed, successful integration and implementation of ecosystem-based management, hazard analysis, and land use planning requires public and stakeholder involvement. It is relevant that each tool has a particular role in the toolkit, even if some roles are shared or overlap, giving the user flexibility in how the tools are applied:

- **CommunityViz** is the primary tool used to depict land use scenarios and summarize indicators across all tools. It is used to model future growth, to change suitability to create different future use patterns, and to associate hazard and ecosystem data with specific polygons or parcels. It also produces outcomes in terms of many socio-economic indicators.
- **NatureServe Vista** takes the land use scenarios from CommunityViz and depicts additional scenario details important for ecological analyses, such as hazards or land management activities. Vista then assesses different land use scenarios to determine

how well they meet conservation goals for a set of conservation elements. The results are a set of performance measures as respect to goals, and maps of areas where the scenarios are compatible with or conflict with conservation goals. Vista also supports generation of alternative scenarios for assessment in CommunityViz.

- The **Roadmap** calculates the exposure of various populations and facilities to hazards. The outputs of the calculations are percentages of various populations or facilities impacted by various hazards. It is also possible to create a spatial representation of where hazard risk and vulnerability overlap. These spatial representations can guide the creation of alternative future scenarios, which can then be assessed as respect to vulnerability, if some assumptions are made about where future populations or facilities are likely to be.

- **Iterative Assessment and Planning**: the tools are linked using a series of scenarios, such as (1) current land use and other conditions; (2) expected "business-as-usual" land use at a future time, and (3) preferred or alternative future land use(s). The alternatives identified are entered into CommunityViz, either by specifically changing land use characteristics for polygons, or by changing the rules for build-out, suitability scores for parcels, or the way growth is allocated. Several iterations may be required to develop a preferred scenario that meets as many objectives as possible. This process is also educational for stakeholders as the tools can demonstrate tradeoffs among objectives, or actions that satisfy multiple objectives at once.

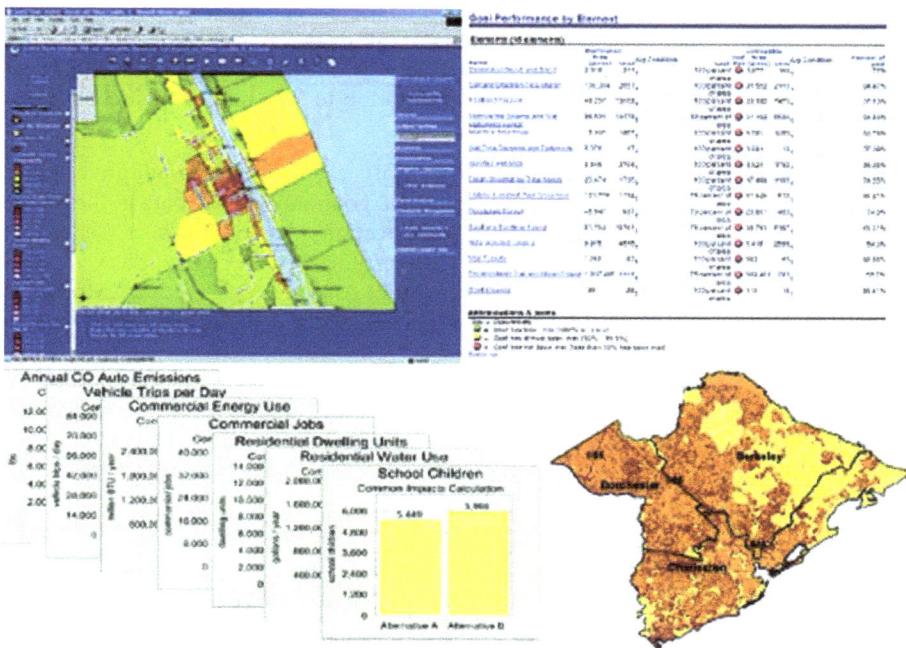

Fig. 13. Example of Integrated Planning for Resilient Communities: application (source: http://resilient-communities.org)

The result of these steps will be the "proposed scenario" that can be presented to decision-makers and stakeholders for review. Undoubtedly, further requests for changes will be made and the toolkit can be used to assess the ramifications of any of these proposed changes. The result of that process will be the "accepted scenario" or plan that will become the basis of a revised "future land use scenario" in the toolkit.

The relevance of the selected models is related to their potential to implement in a synergic way different planning and evaluation tools, in order to support decision-making process oriented to the elaboration of strategic planning choice and situated actions.

3. The Integrated Spatial Assessment (ISA) approach

Taking into account of the above potentials and critical aspects of multi-methodological decision support systems, it is relevant to identify an integrated approach for planning-evaluation where the process and its phases are able to understand the local needs and guide situated decision-making process.

The proposal of a multi-methodological evaluative framework, that includes the cognitive skills and habits of the stakeholders and experts involved in mutual, joint and dynamic learning processes, can help generating more efficient and effective results than sectoral approaches, where interdisciplinarity and transdisciplinarity are essential. In the above perspective the Integrated Spatial Assessment (ISA) approach has been proposed, in which the recognition of tangible and intangible values is the basis for a collective decision-making that includes: the development of goals, the sharing of knowledge, negotiation and compromise, problem-posing and problem-solving, the evaluation of needs, and the definition of goals, but also the attention to questions of justice and equity (Sinclair et al., 2009).

The proposed approach may help communities clarify values, be more adaptive and pro-active, respond to change, set personal and communal goals, and participate in the planning decision-making process. At the same time, the application of spatial tools, as Geographical Information Systems, is a useful support to identify territorial references and link values and planning choices. The integration of Multi-Criteria Analysis (MCA) and Multi-Group Analysis (MGA) and Geographical Information Systems (GIS) is remarkably fruitful in land management where the role of local agents, their relations and objectives may be considered as a structuring element for the process of information construction in a spatial and dynamic evaluative model (Joerin et al., 2001). Compared to traditional forms of GIS utilization, it should be possible to evaluate data covering not only the current situation but also:

1. the spatial characteristics of options proposed;
2. the temporal modification of data following the options implementation;
3. the expressed preferences of local agents;
4. the conflict analysis among the various stakeholders;
5. the evaluation of various options in order to obtain a preference priority list.

Taking into account the previous steps, we defined a methodological process that combines the contribution of different methods and tools. In particular, the first methodological step that we propose is the application of Problem Structuring Methods (PSMs) combined with Public Participation Geographic Information Systems (PPGIS) for the construction of a shared knowledge framework. The PSMs are methods that provide a useful support to information structuring within Decision Support Systems, and are able to deal with a variety of non-structured problems and situations, prevailing over traditional approaches and following communicative conceptions of planning (Rosenhead & Mingers, 2001). In particular, non-structured problems are characterised by a multiplicity of agents; a multiplicity of points of view; incommensurable interests, important intangible values, and uncertainty. In these situations, through PSMs it is possible to visualize a problem so that participants can clear up their positions and converge in one or more potential issues aimed at building consensus. Through PSMs it is possible to represent graphically the complexity of the issues examined, explore the space-solutions, compare discrete alternatives, face uncertainty in terms of "possibilities" and "scenarios" rather than in terms of probability and prediction. PSMs are based on an explicit modelling of cause-effect relations, and their technical simplicity allows them to be used in "facilitated groups" and workshops.

At the same time, the PPGIS is defined by Sieber (2006) as the use of GIS to broaden public involvement in policy-making as well as the value of GIS to promote the goals of nongovernmental organizations, grassroots groups and community based organizations. PPGIS is meant to bring the academic practices of GIS and mapping to a local level in order to promote knowledge production. The idea behind PPGIS is empowerment and inclusion of marginalized populations, who have little space in the public arena, through geographical technology education and participation. PPGIS uses and produces digital maps, satellite imagery, sketch maps, and many other spatial and visual tools, to change geographical involvement and awareness at a local level. The local participatory management of urban neighbourhoods usually comes from "claiming the territory", and has to be made compatible with national or local authority regulations in managing and planning urban territory (McCall, 2003).

The second methodological step combines Multi-Criteria and Multi-Group Decision Support Systems with GIS in order to overcome the limitations of specific techniques through the application of different methods, coming from different disciplines and define a more complete and integrated framework of analysis and evaluation. Many experiences of integration of Multi-Criteria Analysis, Multi-Group Analysis and GIS have been developed referring to different sectors and using different evaluation methods. This type of integration creates a "spatial multi-criteria and multi-group analysis". Spatial multi-criteria decision-making problems typically involve a set of geographically defined alternatives from which a choice of one or more alternatives is made as respect to a given set of evaluation criteria (Jankowski, 1995; Malczewski, 1999). Spatial multi-criteria analysis is very different from the conventional multi-criteria techniques due to the inclusion of an explicit geographic component. It requires information on criterion values and the geographical locations of alternatives in addition to the decision-makers' preferences for a set of evaluation criteria. This means that analysis results depend not only on the geographical distribution of attributes, but also on the value judgments involved in the decision-making process. Therefore, two considerations are fundamental for spatial multi-criteria analysis: the GIS component (i.e., data acquisition, storage, etc.); and the multi-criteria analysis component (i.e., aggregation of spatial data and decision-makers' preferences into discrete decision alternatives) (Al-Shalabi et al., 2006). Spatial analysis combined with multi-criteria methods has been used in recent years to support evaluation, especially in the field of land-use planning. For example, GIS technology was used to assess the criteria requested to determine the suitability of land for housing. Because the required criteria were heterogeneous and measured on various scales, GIS was integrated with an outranking multi-criteria method called ELECTRE-TRI (Joerin et al., 2001). Integration between GIS and multi-criteria analysis using Analytical Hierarchy Process (AHP) was applied in selecting the location for housing sites in a complex process, involving not only technical requirements, but also physical, economical, social, environmental and political requirements (Al-Shalabi et al., 2006). GIS and Multi-Criteria Analysis provided also a better insight into the consequences of alternative water regimes on the performance of wetland functions, supporting stakeholders participation. In particular, Multi-Criteria Analysis was performed using the software package DEFINITE (Janssen et al., 2005).

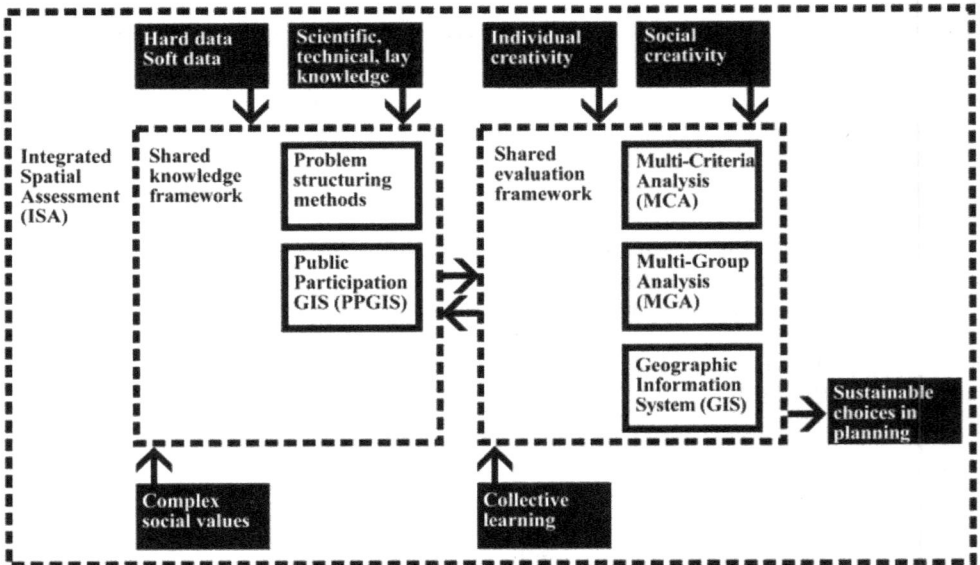

Fig. 14. Integrated Spatial Assessment approach

In general, in the last decade, a wide range of applications was experimented for decision-making, linking multi-criteria assessment and GIS, considering both different methods and different fields: urban and territorial planning, nature conservation, risk management, etc. (Chen et al., 2001; Geneletti, 2004; Malczewski, 2004).

We propose to extend this integration in the perspective of "Integrated Assessments" in order to consider not only the technical aspect of the decision-making problem but also the involvement and participation of the local community in planning choices. Indeed, integration between Multi-Criteria Analyses, Multi-Group Analyses and Geographical Information Systems can be useful when facing conflicts, keeping in mind the local agents' role, the existing relationships and the pre-selected objectives as a structural part of the information building process within a spatial and dynamic evaluation model. As respect to the traditional use of GIS we are able to take into account not only the status-quo data, but also the spatial characteristics of the proposed options, the changing data over time, the elicitation of agents' preferences, the conflict analysis, the impact assessment of the different options (Fusco Girard et al., 2008).

Therefore, it is possible to structure a decision support system that includes "social creativity" (Fischer et al., 2005) as the key component for the decision-making process, and considers the "reflexive community" as a necessary interlocutor to interact with. In this way, individual and social creativity can be integrated to face complex problems through innovative approaches. In this perspective, "Integrated Spatial Assessment" (fig. 14) – which is a participative approach – is a useful tool for decision-making, including technical and political evaluations. Furthermore, it refers to articulated and complex value systems, inserted in conflicting and changing realities, where it is necessary to operate consistently with sustainability principles.

The integration of Problem Structuring Methods, Public Participation GIS, Multi-Criteria and Multi-Group Decision Support Systems and Geographic Information Systems identifies a decision-making process that allows the analysis of the complexity of human decisions for a flexible environment in which collective knowledge and learning has a significant role in decisional processes, and the possibility to explore the transformation strategy definition in spatial planning field according to sustainable and complex values.

4. Cava de' Tirreni Masterplan: An example of Integrated Spatial Assessment (ISA) application

For the Masterplan of Cava de' Tirreni[1], in the Province of Salerno (Italy), the Strategic Environmental Assessment (SEA) process was elaborated to give significant support to planning activity and to help the local government building the suitable choices for the territory. The SEA was seen as an interactive and dynamic approach throughout the whole planning process (Fischer, 2007), allowing to:

- define the status and the evolution trends of human and natural systems, including *hard* (objective, related to real world stuff where things are measured, are fixed in dimensions and location in space) and *soft* (subjective, related to the world of ideas, where the characteristics of a thing can change and specifications are malleable) data, thus creating a complete frame of their interactions to support decision-making process;
- assume the environmental, territorial and social goals, the landscape restoration and environmental protection as stated in the current law and territorial plans, and to find goals and main strategic choices according to a bottom-up approach in planning;
- evaluate the effects of protection policies and significant transformations of territory designed in the plan, and consider the possible alternatives;

---

[1] The working group was thus organized: Urban planning and scientific coordination, Carlo Gasparrini with Cinzia Panneri, Paolo D'Onofrio, Mirella Fiore, Vincenzo Rizzi, Luigi Innamorato, Alessia Sannolo, Anna Terracciano, Pasquale Inglese, Daniele Cannatella; Geomorfology, Silvana Di Giuseppe; Agronomy, Maurizio Murolo; Landscape, Vito Cappiello with Anna Aragosa; Economic-financial feasibility, Ettore Cinque with Andrea Mazzella; Infrastructures and Mobility, Giulio Valfrè with Vincenzo Cerreta (D'Appolonia SpA); Strategic Environmental Assessment, Maria Cerreta, Pasquale De Toro, Saverio Parrella. We thank for support and collaboration the technical staff of Cava de' Tirreni Municipality.

-   find the measures to avoid the possible negative effects and to mitigate, reduce and/or compensate the impacts of the preferable planning choices;
-   define the pressure factors and the necessary indicators to evaluate and control the plan effects referring to the goals and the expected results.

In the Cava de' Tirreni case SEA process was carried out to support the development of the Masterplan and was a practical opportunity to test the Integrated Spatial Assessment (ISA) approach (Fusco Girard et al., 2008). This approach was developed to integrate multi-dimensional aspects within a complex development of strategies and choices in planning, acknowledging the importance of the environmental, social, and economic effects of a decision-making process focused on the creation of alternative transformative options (fig. 15).

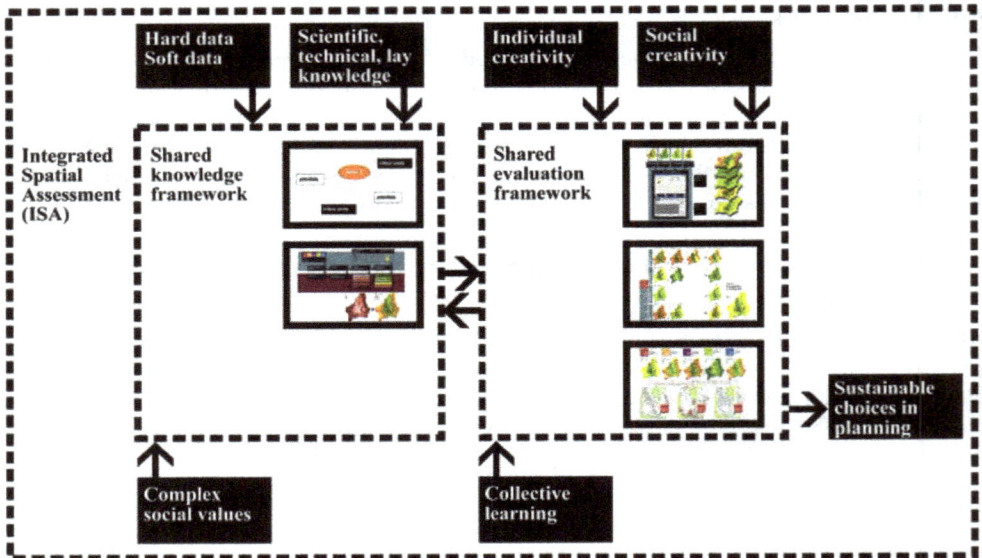

Fig. 15. Integrated Spatial Assessment approach in Cava de' Tirreni Masterplan

In ISA, the recognition of complex social values (Fusco Girard & Nijkamp, 1997) is the basis for a collective decision-making process, that includes the steps of problem-setting, problem-posing and problem-solving, and the sharing of different forms of knowledge, and that takes into account issues of justice and equity. Different analyses are combined to manage conflicts and include various levels of uncertainty.

For the Cava de' Tirreni Masterplan there was a continuous and dynamic interaction between assessment context and assessment process during the whole decision-making process, allowing to select each time the most appropriate methods and techniques based on the goals and considering the results of each step.

Public meetings, in depth interviews, and data and information collection were implemented, mainly aimed at defining a permanent interaction "platform" supporting dialogue and mutual learning between citizens, experts and administrators, in coherence also with the national and European guidelines on Strategic Environmental Assessment. The interaction platform is based on a relational frame supported by a Geographic Information System (GIS); it evolves together with the planning process and allows the creation and development of all plan-related decisions. The participation and consultation steps were fundamental for the application of a sustainable territorial development principle, since finding and recognizing values and resources is vital to enhance local potentials and select approaches and tools for a good governance process. Public meetings created a direct dialogue with citizens and stakeholders and a constant common ground of discussion among citizens, professionals and Municipality.

The main goal was to broaden the knowledge of Cava de' Tirreni, with a special care for the most relevant issues in future urban, social, economic and cultural transformations of the territory, and to single out the collective needs. Thus, there was a continuous interaction process between "common knowledge" (citizens, associations, civil society, etc.) and "expert knowledge" (technicians and administrations), considering SEA as a "joint factor" among the actors. Three main topics were considered during the meetings about the territorial development of Cava de' Tirreni: What is the vision of future? Which strategies to use? Which actions to undertake? In the long run, it is very important to decide how to direct future development, considering not only the scenarios coming from collective expectations, but also significant strategies and actions, to find the best ways of intervention on the territory. For the public consultation a questionnaire was formulated through which associations and citizens could express their point of view regarding present and future of the city. Starting from ten visions designed in earlier meetings, the discussion focused on five main topics: "Cava as a beautiful and identity-bearing city"', "Cava as a regenerated and friendly city", "Cava as a modern and productive city", "Cava as a territorial hub city", "Cava as a ecological city". The visions reflect the community perception of complex social values of territory and express the relevant resources at different levels. They were examined with the Strategic Options Development and Analysis (SODA) approach (Rosenhead & Mingers, 2001), a decision-support system that allows to face complex problems with non-structured qualitative data starting from the elaboration of "cognitive maps". Using the software Decision Explorer 3.1.0, cognitive maps were elaborated starting from verbal protocols, structuring the contents under a formal and methodological point of view (fig. 16). The elaboration of the cognitive maps explained the structure of argumentations carried out during the meetings, keeping the rich amount of data and managing the complexity of information. Through different links identifying the connections of concepts, the main issues were related to one another distinguishing among "visions", "potentials", and "critical points". The kind and number of each link express the importance given to the topics by the different groups. The chain of argumentations allowed to express the expectations, the preferences and the critical points singled out during the meetings, through a "strategic cognitive map", whose topics were classified according to a chromatic scale:

- orange: visions of the future;
- green (three different shades): environmental, infrastructural and settlement potentials;
- purple (three different shades): environmental, infrastructural and settlement critical points.

Starting from the argumentations and the identification of the links in the strategic cognitive map, the whole cognitive model was analyzed to find the preferable vision. Through the

Domain Analysis and the Central Analysis it was possible to evaluate recurrent topics that are relevant to decide the guidelines of future scenarios.

The final rank was obtained comparing the results of Domain Analysis and of Central Analysis. The favourite vision is "Cava as a ecological city", followed by "Cava as a modern and productive city", "Cava as a territorial hub city", "Cava as a regenerated and friendly city" and "Cava as a beautiful and identity-bearing city". Indeed, the different visions are related to one another and can be seen as complementary and synergic in a Plan that cares for the complex objectives of sustainability. Potential and critical points were analyzed in the same way, highlighting the most significant ones to solve or enhance. The results identify some essential issues useful to define the transformations to be included in the Masterplan.

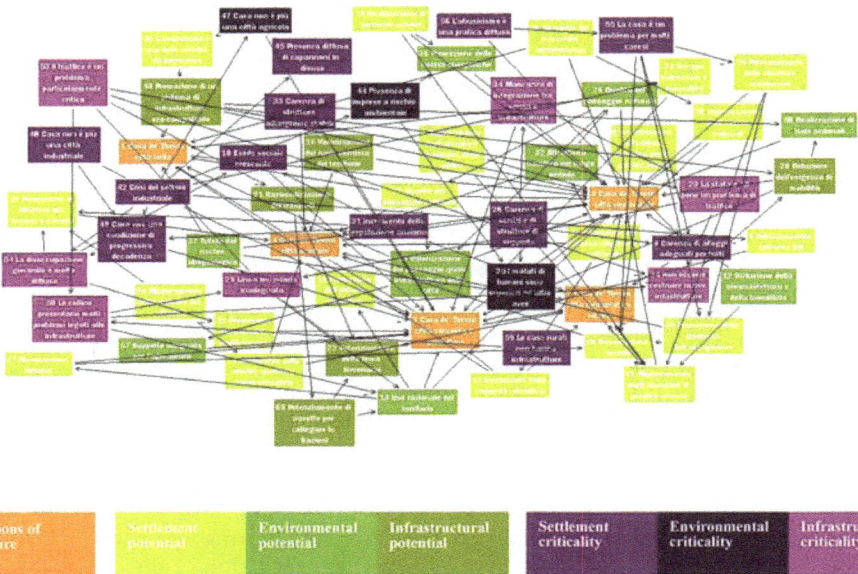

Fig. 16. An example of cognitive map: visions, potentials and criticalities

Consistently with the hierarchical structure of the decision-making process, the visions were articulated into general goals, strategic lines, and strategic actions. In details, strategic actions were linked to three guide-projects that are the main reference to direct planning in the operative phase. The guide-projects are the synthesis of issues coming from the participative and consultative process and they identify the most relevant transformation and conservation interventions within an infrastructural, spatial, functional and symbolic relations system.

To decide the possible placement of different planning choices, the multicriteria method called Analytic Hierarchy Process (AHP) (Saaty, 1980, 1992) was used in combination with Geographic Information System (GIS) elaborations (Marinoni, 2004). The application of AHP into GIS allowed to go beyond the simple overlay of different themes, making a pairwise comparison of the criterion of each hierarchical level. For each of the five visions a "susceptibility map to localization" was generated, expressing the attitude of the territory to accept a given strategic action, considering potential impacts. The lower are the territorial and environmental impacts caused by an action, the higher the susceptibility of the territory to receive that action.

To find alternative locations of strategic actions and of related guide-projects, a three level hierarchical structure was made for each vision ("environmental themes", "criteria" and "values/characteristics") expressing the last level through a five points scale associated to a chromatic one. The criteria were given the same weight for all the visions, while the environmental issues were compared in pair creating five matrixes, one for each vision. The AHP method allows to combine the weights of the criteria coming from the comparisons with the scores associated to different classes of susceptibility to localization obtaining, within GIS, the susceptibility maps of each planning action (fig. 17).

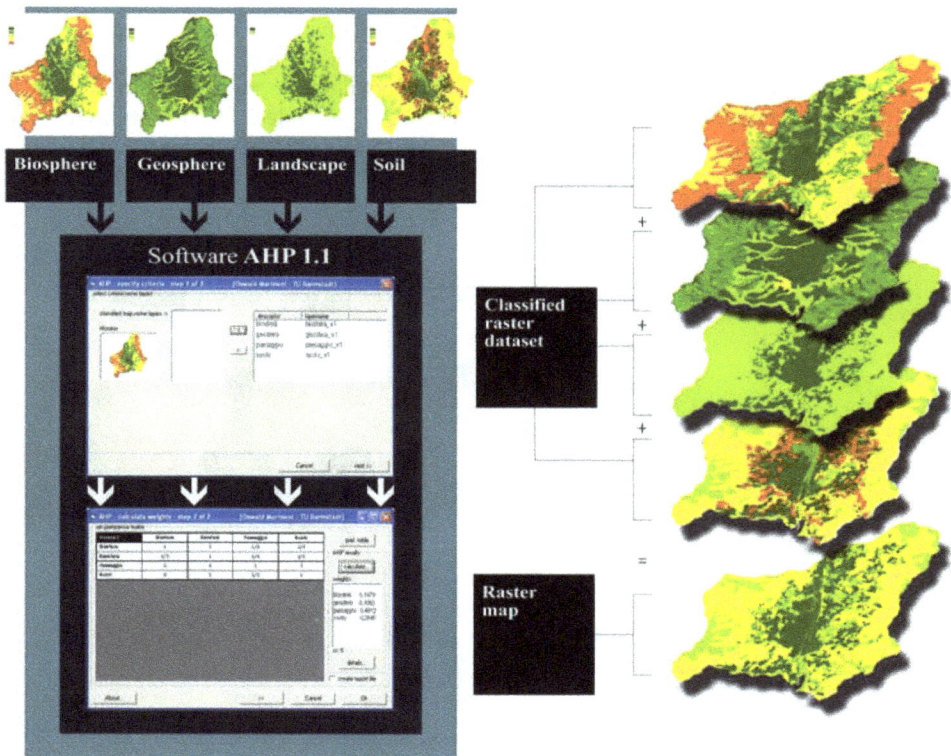

Fig. 17. The elaboration of maps of susceptibility to localization

For each Vision we have obtained the susceptibility map to localization related to *biosphere* (territorial biopotential index, biodiversity degree, infrastructural fragmentation index); *geosphere* (slopes stability, seismic zoning); *landscape* (landscape units); *soil* (land use, cultivations productivity); and overall susceptibility map to localization. The indicators selected for each environmental theme were elaborated starting from the studies and analysis made by the different experts of the working group and the database structured by the technical staff of Cava de' Tirreni Municipality.

Therefore, for each vision, it was possible to have the relative map of susceptibility to localization and it was possible to pass from the Visions to three technical "Guide-projects" oriented to the city transformation (fig. 18). It is clear that the evaluation supported the planning phases enhancing the characteristics of each area and, most of all, placing activities where it is pre-emptively possible to minimize territorial and environmental impacts, creating the whole strategic planning frame. Through the interaction among visions identification and maps of susceptibility to localization it has been possible to develop shared and complementary guide-projects, where the use of a combination of techniques penetrates and includes informal, 'soft spaces' of decision, able to complement the more formal process, combining flexible and functional approaches with formal development plan strategies (Allmendinger & Haughton, 2009; Cerreta, 2010). In the Cava de' Tirreni masterplan, the opportunities that emerged from the interactions focused mainly on the preservation of the identity of a context wishing to regenerate itself. The integrated use of SODA, MCDA and GIS shaped the different phases, acting as a powerful combination for providing decision support in strategic decisions. SODA helps decision-makers in devising visions and exploring possible effects, while MCDA and GIS support an in-depth performance assessment of each strategic vision and related actions, as well as the design of more robust options.

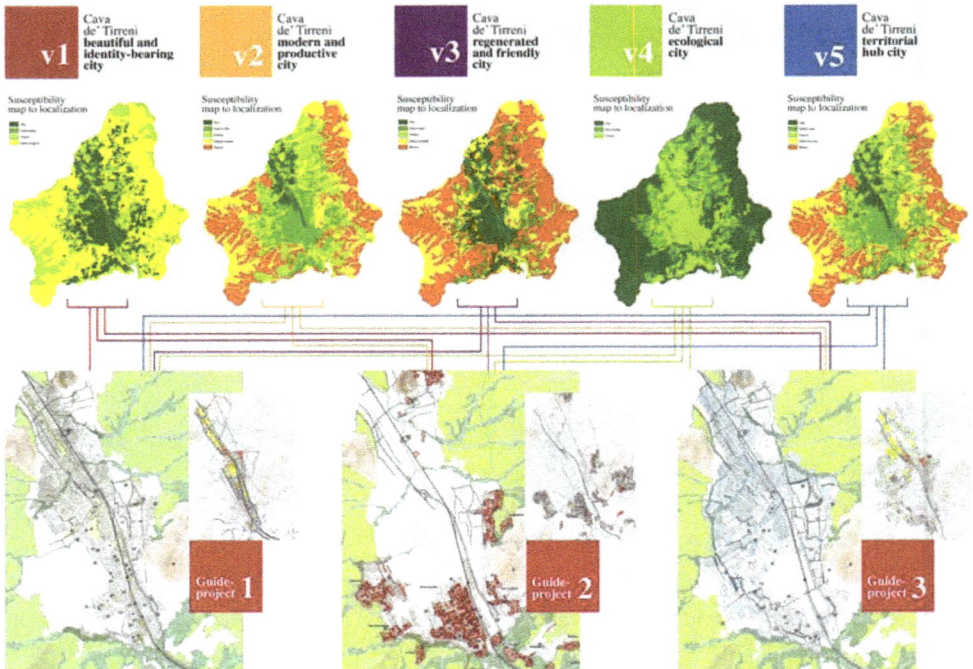

Fig. 18. The maps of susceptibility to localization and the three guide-projects

The implementation of ISA approach helps to overcome the limits of each single method, to accommodate a multi-dimensional and plural perspective and improve the quality of the decision-making process. Indeed, by using the ISA approach, we aimed to integrate social, territorial and environmental aspects in the development of strategies and planning choices, while recognizing the important role of stakeholder perceptions and environmental effects within the collective decision-making process for the creation of alternative opportunities. ISA approach may enable the interpretation of material and immaterial relations characterizing a context, the acknowledgement of existing tangible and intangible values, and the creation of strategies aimed at the production of new values and at the sustainable development of many local resources in a multi-dimensional perspective.

5. Conclusion

The selection of models illustrated highlights that many computer based tools and instruments have been developed to try and provide a common language for integrated visioning or strategy development in planning, even if these instruments face serious implementation problems in overcoming the gap between instrument development (by consultants and/or universities) and daily planning practice.

In most cases, the present technology focus produces instruments based above all on scientific rigor rather than also on practical relevance; not adapted to the complex and dynamic planning context; not transparent; not user friendly and not flexible (Te Brömmelstroet & Bertolini, 2010). Therefore, such instruments cannot link-up with the context specifics and do not contribute to implement the Planning Support Systems (PSS) (Geertman, 2006; Uran & Janssen 2003; Vonk et al., 2005) and to improve communicative planning practice (Timms, 2008; Willson, 2001).

In order to understand how to structure and improve integrated planning-evaluation processes, it is relevant to analyze how it is possible to implement the interaction among the assessment context, the assessment process and the assessment methods, how to select different approaches and techniques, and how to choose them considering the decision context specificity and the type of plan or project.

The ISA approach (Cerreta & De Toro, 2010) proposed let us explore the tools of the integrated evaluations helping to recognize their technical effectiveness and, at the same time, improving the transparency of evaluation process, to build the decision able to reflect the different needs and expectations. Through such planning-evaluation, it is possible to help communities become more aware not only of their own opinions and preferences, but also of those of other subjects, helping to find participated and shared solutions.

In this perspective, ISA can be a useful tool for decision-making, including technical and political evaluations and referring to articulated and complex value systems, in a conflicting and changing reality. The integration of Problem Structuring Methods, Public Participation GIS, Multi-Criteria and Multi-Group Decision Support Systems and Geographic Information Systems identifies a decision-making process that allows the analysis of the complexity of human decisions for a flexible environment in which collective knowledge and learning assume a significant role in decisional processes, and the possibility to explore the transformation strategy definition in spatial planning field according to sustainable and complex values.

6. References

Alexander, E.R. (Ed.). (2006). *Evaluation in Planning. Evolution and Prospects*. Ashgate, ISBN 9780754645863, Aldershot, United Kingdom

Allen, E. (2001). INDEX: Software for community indicators, In: *Planning Support Systems: Integrating Geographic Information Systems, Models and Visualization Tools*, R.K. Brail, R.E. Klosterman, (Eds.), pp. 229-262, ESRI, ISBN 1589480112, Redlands, California

Allmendinger, P., Haughton, G. (2009). Soft Spaces, Fuzzy Boundaries, and Meta-governance: the New Spatial Planning in the Thames Gateway. *Environment and Planning A*, Vol.41, pp. 617-633

Al-Shalabi, M.A., Bin Mansor, S., Bin Ahmed, N., & Shiriff, R. (2006). GIS based multicriteria approaches to housing site suitability assessment, *Proceedings of XXIII FIG Congress, Shaping the Change*, Munich, Germany, October 8-13, 2006

Amin, A., & Roberts J. (Eds.). (2008). *Community, Economic Creativity and Organization*. Oxford University Press, ISBN 9780199545490, Oxford, UK

Brail, R.K. (2008). Planning Support Systems for Cities and Regions. Lincoln Institute of Land Policy, ISBN 978-1-55844-182-8, Cambridge, Massachusetts

Carr, M.H. & Zwick, P.D. (2007). *Smart Land-use Analysis: the LUCIS Model Land Use Identification Strategy*, ESRI Press, ISBN 9781589481749, Redlands, California

Cerreta, M. (2010). Thinking through complex values, In: *Making Strategies in Spatial Planning. Knowledge and Values*, M. Cerreta, G. Concilio, V. Monno, (Eds.), pp. 381-404, Springer, ISBN 9789048131068, Dordrecht, The Netherlands

Cerreta, M., & De Toro, P. (2010). Integrated Spatial Assessment for a Creative Decision-making Process: a Combined Methodological Approach to Strategic Environmental Assessment. *International Journal of Sustainable Development* (IJSD), Vol.13, No.1/2, pp. 17–30, ISSN 0960-1406

Chen, K., Blong, R., & Jacobson, C. (2001). MCE-RISK: integrating multicriteria evaluation and GIS for risk decision-making in natural hazards. *Environmental Modelling and Software*, Vol.16, pp. 387–397. ISSN 1364-8152

Clark Labs, (2007). The Land Change Modeler for Ecological Sustainability, IDRISI *Focus paper*, Clark University, Worcester, MA, www.clarklabs.org

Echenique, M.H., Flowerdew, A.D.J., Hunt, J.D., Mayo, T.R., Skidmore, I.J., & Simmonds D.C. (1990). The MEPLAN Models of Bilbao, Leeds, and Dortmund. *Transport Reviews*, Vol.10, No.4, pp. 309–322, ISSN 0144-1647

Emberger, G., Ibesich, N., & Pfaffenbichler, P. (2006). Can decision-making processes benefit from a user friendly land use and transport interaction model?, *Proceedings of 8th International Conference on Design & Decision Support Systems in Architecture and Urban Planning*, Eindhoven, The Netherlands

Finnveden, G., Nilsonn, M., Johansonn, J., Personn, Å., Moberg, Å., & Carlsonn T. (2003). Strategic Environmental Assessment Methodologies – Application within the Energy Sector. *Environmental Impact Assessment Review*, Vol.23, pp. 91-123, ISSN 0195-9255

Fischer, G., Giaccardia, E., Edena, H., Sugimoto, M. & Ye, Y. (2005). Beyond Binary Choices: Integrating Individual and Social Creativity. *International Journal of Human-Computer Studies*, Vol.63, pp. 482-512, ISSN 1044-7318

Fischer, T.B. (2007). *Theory and Practice of Strategic Environmental Assessment: towards a More Systematic Approach*, Earthscan, ISBN 9781844074525, London, UK

Funtowicz, S., & Ravetz, J.R. (1993). Science for the Post-Normal Age. *Futures*, Vol.25, pp. 568-582, ISSN 0016-3287

Funtowicz, S., & Ravetz, J.R. (1994). Emergent Complex Systems. *Futures*, Vol.26, pp. 739-755, ISSN 0016-3287

Fusco Girard, L., & Nijkamp, P. (1997), *Le valutazioni per lo sviluppo sostenibile della città e del territorio*, Angeli, ISBN 9788846401823, Milano, Italy

Fusco Girard, L., Cerreta, M., De Toro, P. & Forte, F. (2007). The Human Sustainable City: Values, Approaches and Evaluative Tools, In: *Sustainable Urban Development. The Environmental Assessment Methods*, Deakin M., Mitchell G., Nijkamp P., Vreeker R. (Eds.) (Vol. 2), pp. 65-93, Routledge, ISBN 9780415322171, London, UK

Fusco Girard, L., Cerreta, M., & De Toro, P. (2008). Valutazione Spaziale Integrata. Il Puc di San Marco dei Cavoti. In: *Urbanistica Digitale*, Moccia F.D. (Ed.), pp. 469-487, Edizioni Scientifiche Italiane, ISBN 9788849515978, Napoli, Italy

Geertman, S. (2006). Potentials for planning support: a planning-conceptual approach. *Environment and Planning B: Planning and Design*, Vol.33, No.6, pp. 863–880, ISSN 0265-8135

Geneletti, D. (2004). A GIS-Based Decision Support System to Identify Nature Conservation Priorities in a Alpine Valley. *Land Use Policy*, Vol.21, pp. 149-60, ISSN 0264-8377

Golub, A.L. (1997). *Decision Analysis: An Integrated Approach*, John Wiley, ISBN 9780471173342, New York, USA

Gund Institute for Ecological Economics (2007). *Multiscale Integrated Models of Ecosystem Services (MIMES)*, http://www.uvm.edu/giee/mimes/

Haken, H. (1983). Advanced Synergetics—Instability Hierarchies of Self-organizing Systems and Devices. *Springer Series in Synergetics*, Vol.20, Springer, ISBN 3540121625 9783540121626, Berlin, Germany

Healey, P. (2007). *Urban Complexity and Spatial Strategies: towards a Relational Planning for our Times*. Routledge, ISBN 9780415380355, London, UK

Hittle, J. (2011). *Integrated Planning for Resilient Communities: A Technical Guide to Integrating Hazard, Ecosystem and Land Use Planning*. EBM Tools Network. www.ebmtools.org

InfoHarvest (2001). *Criterium Decision Plus User Guide*, Seattle WA, InfoHarvest, Inc.

Jankowski, P. (1995). Integrating Geographical Information Systems and Multiple Criteria Decision Making Methods. *International Journal of Geographical Information Systems*, Vol.9, pp. 251-273, ISSN: 1365-8816

Janssen, R., Goosen, H., Verhoeven, M.L., Verhoeven, J.T.A., Omtzigt, A.Q.A. & Maltby, E. (2005). Decision Support for Integrated Wetland Management. *Environmental Modelling and Software*, Vol.20, pp. 215-229, ISSN 1364-8152

Joerin, F., & Musy, A. (2000). Land Management with GIS and Multicriteria Analysis. *International Transactions in Operational Research*, Vol.7, pp.6 7-78, ISSN 1475-3995

Joerin, F., Theriault, F. & Musy, A. (2001). Using GIS and Outranking Multicriteria Analysis for Land-Use Suitability Assessment. *International Journal of Geographical Information Science*, Vol.15, pp. 153-174, ISSN 1365-8816

Klosterman, R.E. (2001). TheWhat if? planning support system, In: *Planning Support Systems: Integrating Geographic Information Systems, Models and Visualization Tools*, R.K. Brail, R.E. Klosterman (Eds.), pp. 263-284, ESRI Press, ISBN 1-5894-011-2, Redlands, CA

Lee, N. (2006). Bridging the Gap between Theory and Practice in Integrated Assessment. *Environmental Impact Assessment Review*, Vol.26, pp. 57-78, ISSN 0195-9255

Lichfield, N. (1996). *Community Impact Evaluation*, UCL Press, ISBN 9781857282375, London, UK

Malczewski, J. (1999). *GIS and Multicriteria Decision Analysis*, John Wiley, ISBN 9780471329442, New York, USA

Malczewski, J. (2004). GIS-based land-use suitability analysis: a critical overview. *Progress in Planning*, Vol.62, pp. 3–65, ISSN 0305-9006

Marinoni, O. (2004), Implementation of analytic hierarchy process with VBA in ArcGIS. *Computational Geosciences*, Vol.30, pp. 637–646, ISSN 1420-0597

Marinoni, O., & Hoppe, A. (2006). Using the Analytic Hierarchy Process to Support the Sustainable Use of Geo-Resources in Metropolitan Areas. *Journal of Systems Science and Systems Engineering*, Vol.15, No.2, pp. 154-164, ISSN 1004-3756

Martínez, F.J. (1996). MUSSA: A Land Use Model for Santiago City. Transportation Research, *Transportation Planning and Land Use at State, Regional and Local Levels*, pp. 126–134

Martínez, F.J., & Donoso, P. (2001). MUSSA: Un Modelo de Equilibrio del Uso del Suelo con Externalidades de Localización, Planos Reguladores y Políticas de Precios Optimos, *Proceedings of X. Congreso Chileno de Ingeniería de Transporte*, Concepción, 2001

Matthews, K.B., R. Sibbald, A.R., & Craw, S. (1999). Implementation of a spatial decision support system for rural land use planning: integrating GIS and environmental models with search and optimisation algorithms. *Computers and Electronics in Agriculture*, Vol.23, pp. 9-26. ISSN 0168-1699

McCall, M.K. (2003). Seeking Good Governance in Participatory-GIS: A Review of Processes and Governance Dimensions in Applying GIS to Participatory Spatial Planning. *Habitat International*, Vol.27, pp. 549-573. ISSN 0197-3975

Munda, G. (2008). *Social Multi-Criteria Evaluation for a Sustainable Economy*, Springer, ISBN 9783540737025, New York, USA

Nekhay, O., Arriaza, M., & Guzmán-Álvarez, J.R. (2009). Spatial analysis of the suitability of olive plantations for wildlife habitat restoration, *Computers and Electronics in Agriculture*, Vol.65, pp. 49–64, ISSN 0168-1699

Nonaka, I. & Takeuchi, H. (1995). *The Knowledge-creating Company: how Japanese Companies Create the Dynamics of Innovation*. Oxford University Press, ISBN 9780195092691, New York, USA

Nonaka, I., von Krogh, G., & Voelpel, S. (2006). Organizational knowledge creation theory: evolutionary paths and future advances. *Organization Studies*, Vol.27, No.8, pp. 1179–1208, ISSN 0170-8406

Odum, E.P. (1969). The Strategy of the Ecosystem Concept, *Science*, Vol. 164 (18 April 1969), p.266. ISSN 0036-8075

Reynolds, K., Cunningham, C., Bednar, L., Saunders, M., Foster, M., Olson, R., Schmoldt, D. Latham, D., Miller, B., & Steffenson, J. (1996). A knowledge-based information management system for watershed analysis in the Pacific Northwest US. *Artificial Intelligence Applications*, Vol.10, pp. 9-22., US Forest Service

Rosenhead, J., & Mingers, J. (2001). *Rational Analysis for a Problematic World Revisited*, John Wiley & Sons, ISBN 9780471495239, Chichester

Saaty, T.L. (1980). *The Analytical Hierarchy Process*, McGraw Hill, ISBN 9780070543713, New York, USA

Saaty, T.L. (1992). *Multi-Criteria Decision-Making. The Analytic Hierarchy Process*, RWS Publications, ISBN 9780962031724, Pittsburgh, PA, USA

Şener, Ş., Şener, E., Nas, B., Karagüzel, R. (2010). Combining AHP with GIS for landfill site selection: A case study in the Lake Beys_ehircatchment area (Konya, Turkey), *Waste Management*, Vol.30, pp. 2037–2046, ISSN 0956-053X

Sieber, R. (2006). Public Participation and Geographic Information Systems: A Literature Review and Framework. *Annals of the American Association of Geographers*, Vol.96, pp. 491-507, ISSN 0004-5608

Simmonds, D.C. (1999). The Design of the DELTA Land-use Modelling Package. *Environmental Planning B Planning Design*, Vol.26, pp. 665–684 ISSN 0265-8135

Simmonds, D.C. (2001). The Objectives and Design of a New Land Use Modelling Package: DELTA. In: *Regional Science in Business*, Clark, G.P., Madden, M. (Eds.), Springer, ISBN 9783540417804, Berlin, Germany

Sinclair, A.J., Sims, L. & Spaling, H. (2009). Community-based approaches to strategic environmental assessment: lessons from Costa Rica. *Environmental Impact Assessment Review*, Vol.29, pp. 147–156, ISSN 0195-9255

Sterman, J.D. (2000). *Business Dynamics-Systems Thinking and Modeling for a Complex World.* McGraw-Hill, ISBN 9780072389159, New York, USA

Te Brömmelstroet, M. & Bertolini, L. (2010). Integrating land use and transport knowledge in strategy-making. *Transportation*, Vol.37, No.1, pp. 85-104, ISSN 0049-4488

Thirumalaivasan, D., Karmegam, M., & Venugopal, K. (2003). AHP-DRASTIC: Software for specific aquifer vulnerability assessment using DRASTIC model and GIS. *Environmental Modelling and Software*, Vol.18, pp. 645–656, ISSN 1364-8152

Timms, P. (2008). Transport models, philosophy and language. *Transportation*,Vol.35, No.3, pp. 395–410 ISSN 0049-4488

Trochim, W.M.K., & Donnelly, J.P. (2006). *The Research Methods Knowledge Base*, 3e. Atomic Dog Publishing, ISBN 9781592602919, Mason, Ohio, USA

Uran, O., Janssen, R. (2003). Why are spatial decision support systems not used? Some experiences from the Netherlands. *Computers, Environment and Urban Systems*, Vol.27, pp. 511-526, ISSN 0198-9715

van der Sluijs, J.P. (2002). A Way Out of the Credibility Crisis of Models Used in Integrated Environmental Assessment. *Futures*, Vol.34, pp. 133-146, ISSN 0016-3287

Vizzari, M. (2011), Spatial modelling of potential landscape quality. *Applied Geography*, Vol.31, pp. 108–118, ISSN 0143-6228

Vonk, G., Geertman, S., Schot, P. (2005). Bottlenecks blocking widespread usage of planning support systems, *Environment and Planning A*, Vol.37, pp.909–924, ISSN 0308-518X

W!SL (2003). http://www.maptalk.nl/

Waddell, P. (2002). UrbanSim: Modeling Urban Development for Land Use, Transportation and Environmental Planning. *Journal of the American Planning Association*, Vol.68(3), pp.297-314. ISSN 0194-4363

Waddell, P. (2011). Integrated Land Use and Transportation Planning and Modelling: Addressing Challenges in Research and Practice. *Transport Reviews*, Vol.31, No.2, pp. 209-229, ISSN 0144-1647

Waddell, P., Borning, A., Noth, M., Freier, N., Becke, M., & Ulfarsson, G. (2003). UrbanSim: A Simulation System for Land Use and Transportation. *Networks and Spatial Economics*, Vol.3, pp. 43-67, ISSN 1566-113X

Wegener, M. (1998). *Das IRPUD-Modell: Überblick*. http://www.raumplanung.uni-dortmund.de/irpud/pro/mod/mod.htm Zugriff: 27.09.2005

Wegener, M. (2004). Overview of Land Use Transport Models. In: *Handbook of Transport Geography and Spatial Systems*, Hensher, D.A. (Ed.), p. 672, Elsevier, ISBN 9780080441085, Amsterdam, The Netherlands

Wiek, A. & Walter, A.I. (2009). A Transdisciplinary Approach for Formalized Integrated Planning and Decision-making in Complex Systems. *European Journal of Operational Research*, Vol.197, pp. 360-370, ISSN 0377-2217

Willson, R. (2001). Assessing communicative rationality as a transportation planning paradigm. *Transportation*, Vol.28, pp. 1–31, ISSN: 0049-4488

7

Quantitative Methods in Environmental and Visual Quality Mapping and Assessment: A Muskegon, Michigan Watershed Case Study with Urban Planning Implications

Di Lu[1], Jon Burley[1], Pat Crawford[1], Robert Schutzki[2] and Luis Loures[3]
[1]School of Planning, Design, and Construction
[2]Department of Horticulture, Michigan State University
[3]CIEO - Research Centre for Spatial and Organizational Dynamics
University of Algarve
[1,2]USA
[3]Portugal

1. Introduction

For generations, people have been in pursuit of at least two conflicting goals. On one hand, people seek to make life physically easier. Especially during the past few centuries, the results include the development of new products, ease of transportation, and ease of communications. On the other hand, people have growing concerns about the condition of the natural environment, including water, air, soil, plants, and animals. These concerns even include perceptual values about the quality of the environment such as aesthetics/visual quality. As the human population grows, these competing concerns require thoughtful planning, design, and management to efficiently and effectively facilitate the use and protection of the environment. However, it has been at times, difficult for professionals, governmental agencies, and citizens to develop thoughtful measures and means concerning people's perception and reaction associated with the environment.

Assessing landscape aesthetics and perceived environmental quality are often quite intangible tasks and can be difficult to be described quantitatively. However, investigators around the world have explored approaches to evaluate the value of landscape aesthetics by numerous mathematical methods. In the United States, during the 1970s and 1980s legislative Acts stimulated the evolution of manuals and experiments for assessing and managing landscape resources and scenic quality (Zube et al., 1982). Many governmental institutes and organizations produced manuals and guidelines to assess and manage landscape resources, for example, Visual Management System (VMS) from US Forest Service (USFS, 1973), Visual Resource Management (VRM) from Bureau of Land Management (BLM, 1980), and Landscape Resource Management (LRM) from US Soil Conservation Service (Schauman & Adams, 1979). Environmental planners and designers are also greatly interested in developing methods and procedures to evaluate and predict the visual and ecological quality on wild and scenic rivers, scenic highways, scenic and recreational parks, trials, and wetlands (Burley, 1997).

Visual quality assessment approaches in planning and design facilitate professionals to analyze existing conditions and proposed spatial treatments. The approach often requires the use of photographic images or digital drawings to assess the quality of the environment. Photographs have been tested in many studies, and investigators have demonstrated that photographs could be used as substitutes for site visits, as there was strong concurrence between photos and real landscape by respondents (Boster, 1974; Zube, 1974). In addition to landscape planners and professional resource managers, a significant number of individuals, including ecologists, geographers, environmental experts and psychologists are engaging in landscape perception and assessment research, and all of them have introduced and explored different sets of methods and models from their disciplines. The literature in this area of research has become extensive and vast, beyond the scope of this chapter; however Smardon et al., (1986) and Taylor et al., (1987) provide insightful overviews of the fundamental knowledge base. In general, landscape perception and assessment is often considered as a function of the interaction of humans and the landscape including the urban landscape.

Some scholars have proposed and debated around the existence of four general approaches in understanding and describing landscape perception and quality: the expert paradigm, the psychophysical paradigm, the cognitive paradigm and the experimental paradigm (Taylor et al., 1987). The expert paradigm is founded upon the idea that trained professionals are highly suited to assess landscape quality either heuristically or through contrived indexes that actually have no statistical basis. This approach is highly aligned with the arts and formal art theory of past centuries. This approach is often based upon many normative theories that have little scientific backing. The psychophysical paradigm is a respondent based perception approach where preferences are statistically analyzed with various visual treatments. This approach is much more aligned with the sciences. The psychophysical paradigm contains numerous predictors to quantify environmental preferences, but is often weak in theory (explanations) concerning why certain preferences exist. The cognitive paradigm is in many ways similar to the psychophysical statistical approach but address another set of variables that may be more difficult to construct and establish, such as defining mystery in the landscape. The experience paradigm includes measures of attitudes, feelings, and impressions as one experiences the environment. During the past two decades, variables from the psychophysical, cognitive, and experiential paradigms have been studied together to form a more unified approach. Besides those four fundamental ways to examine landscape quality, there is an increasing trend to engage in mapping these landscape perceptions in both rural and urban contexts in the same manner that one can map other physical attributes. This chapter focuses upon the connections between measuring environmental/visual quality, mapping these qualities, and potential uses.

2. Mapping visual/environmental quality

It is suggested that the modern era of visual quality assessment began with a recreation scientist Elwood Shafer and colleagues, concerning their publication of predictive visual quality equations (Shafer, 1969; Shafer et al., 1969; Shafer & Tooby, 1973) that employed the psychophysical approach. Before this time, the expert approach dominated professional practice around the world and the expert approach is still widely used today. Shafer employed contemporary social science research methods to numerically obtain a perception based evaluation of black and while rural landscape photographs. He then measured various properties of the photographs such as the area of water in the image and the perimeter of foreground vegetation, statistically relating several of the properties with the preferences of the respondents. Shafer's equation was then demonstrated in a forest management situation to illustrate the practical application of the research (Brush & Shafer, 1975). While there were precedents for the spatial representation of visual quality before Brush and Shafer's publication, their map is an important example concerning the quantification of environmental/visual quality (Figure 1). These types of maps portray the predicted   respondent   impression   of   similar   land-cover   and   similar   surface geomorphological features.

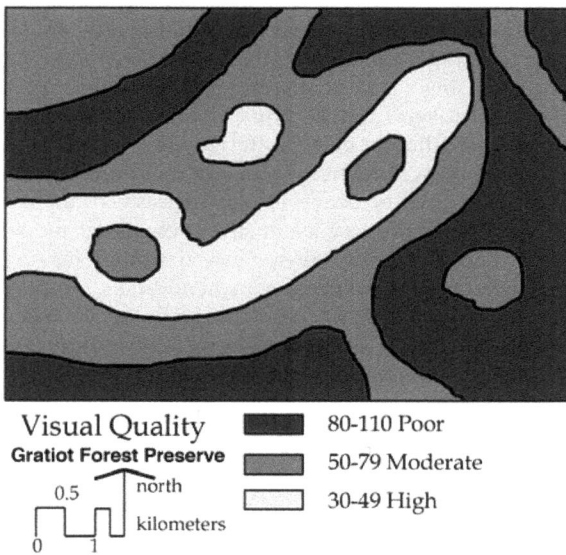

Fig. 1. A map illustrating the general character of maps produced by Brush & Shafer (1975). Larger numerical scores indicate poor visual quality and lower scores indicated preferred environments.

Any type of work like Shafer's (meaning highly statistical, not based in art theory, or difficult to apply) was highly open to criticism. Shafer's equation was seriously criticized by Bourassa (1991), Carlson (1977), and Wienstein (1976), especially because the statistical model seemed unlinked to any formal, predictive theory to explain the relationships between the physical variables measured in the photographs and the preference ratings of the respondents. Burley (1997:54) notes, "Essentially their criticisms are valid; yet in some respects Shafer and colleagues were somewhat unfairly denounced because engineers, ecologists, economists, and agronomists often gain wide acceptance for developing statistical relationships between variables without having a single theory to explain the relationships, developing a theory is not a prerequisite for developing an equation.

Nevertheless, Shafer's equation was difficult for some social scientists to accept and investigators examined other directions in environmental landscape preference research." In many respects it may take years of model testing, theory development, replication, and validation before research is usually accepted; yet this seminal effort seemed not to be widely regarded or fully appreciated. After all, before Shafer's efforts, visual quality experts had little statistical evidence to support their claims, equipped with primarily speculative theories and educated guesses.

Since the creation of the Brush and Shafer map (1975), numerous spatial tools and approaches have been employed to study visual quality mapping. Lu (2011) reviews many of these approaches by investigators applying geographical software systems (GIS), remote sensing tools, digital visualization tools across numerous landscape types from urban settings to rural settings such as Crawford (1994). In addition to Lu's cited sources concerning the mapping of visual quality, Fuente de Val et al., (2006) demonstrated that the heterogeneity of the environment might be an important spatial character in determining visual aesthetic quality for Mediterranean landscapes. Panagopoulos and Vargues (2006) addressed the relevance of using visual maps in golf course development in the Algarve, enabling the assessment of adverse visual impacts of golf courses and the creation of mitigation measures and rehabilitation design alternatives. Arriaza et al., (2004) noted that in Spain, the perceived factors that influence visual quality in rural settings are the degree of wilderness of the landscape, the presence of well-preserved man-made elements, the percentage of plant cover, the amount of water, the presence of mountains and the color contrast. Van den Berg and Koole (2006) suggest that motives in recreational intent influenced respondent preference for environments. In addition, Loures et al., (2008) studied how the use of visual quality maps could influence and improve post-industrial redevelopment. The research activity of these investigators indicate that studies in visual quality landscape assessment continues to increase the knowledge base.

Based upon the literature above, many researchers have explored varied approaches to measure visual quality and make landscape visual quality maps of particular areas. However, there are few experiments to actually validate visual quality maps. The new research presented in this chapter attempts to make a landscape visual quality map of "Lower Muskegon Watershed", which can be also statistically validated. If successful, we believe that we may be entering a new era of visual quality and environmental measurement and monitoring. We believe that the link between spatial visual properties and depictions on maps may be within practical reach.

3. A recent visual/environmental quality mapping investigation

Recently we conducted and completed an investigation that is previously unreported in any other publication to see if we could construct and validate a visual quality map in our study area in Michigan, the lower portion of the Muskegon River Watershed (Figure 2). The Muskegon River Watershed is one of the largest watersheds in the State of Michigan and spans across nine counties: Wexford, Missaukee, Roscommon, Osceola, Clare, Mecosta, Montcalm, Newaygo and Muskegon. There are several reasons that the Lower Muskegon Watershed was chosen as the study area: (1) it is composed of several land cover types to study from urban to agricultural and woodland; (2) updated landscape land-use and cover type information was readily available (from 1998); and (3) it is a study area that has

external construct validity (meaning that the equation can be applied to Michigan) to
employ a relatively robust and somewhat predictive equation (Equation 1), that is several
generations and iterations from the seminal efforts of Shafer.

USA

Michigan

Lower Muskegon Watershed

Fig. 2. A series of imbedded maps illustrating the location of the study area, the Lower
Muskegon Watershed.

Photographs were collected in this study on the Lower Muskegon Watershed in May and
August of 2010. A total of 131 photographs from Lower Muskegon Watershed were
recorded, and each of them was positioned and tagged on the map of Lower Muskegon
Watershed by Global Positioning System (GPS). The objectives and principles to collect
photographs was guided by trying to obtain different types of environments across the
study area. The images contain physical attributes such as people, wildlife, water, roads,
flowers, vegetation, buildings, industrial facilities, urban savanna, farmland, forest, lakes
and rivers. Two sets of 30 photos were chosen from the original 131 photos to analyze. One
set to create a predictive map and the other to validate or refute the predictive map by
applying a statistical procedure to compare the second set of images with the predicted
scores based upon the generated map of visual quality scores.

The basic GIS raster data in this study were downloaded from an Arc GIS database from
Grand Valley State University, containing 1998 land-use data. The project of "Sustainable
Futures for the Muskegon River Watershed" from Grand Valley State University provides
all the "Updated 1998 Land Use Data" in Lower Muskegon Watershed of six counties:
Muskegon, Lake, Mecosta, Montcalm, Newaygo, and Osceola counties. By merging all the
land use data of these counties, a land-use map of Lower Muskegon Watershed was
generated.

This study utilized Burley's experimental method (Burley, 1997; Lee & Burley, 2008) to measure the variables of the photos (which uses a grid of 30 rows and 38 columns). Each of the 30 photos from set one and set two were measured according to Equation 1 by Burley (1997), which is based on physical variables and environmental quality index. The equation explains 67 percent of respondent preference, has all predictors with a $p \leq 0.05$ and is not over-specific despite having a fair number of predictor variables. The equation contains an overall p-value of ≤ 0.0001. This equation is representative of the current state-of-the art in constructing visual/environmental quality predictive models, where a set of regressors explain between 50 and 70% of respondent's preference. In the future it is expected investigators may improve the predictive power of visual quality/environmental quality equations by exploring new variables, by expanding the number of the images studied, and increasing the number of respondents studied.

$$Y= 68.30 - (1.878*HEALTH) - (0.131*X1) - (0.064*X6) + (0.020*X9) + (0.036*X10) + (0.129*X15) - (0.129*X19) - (0.006*X32) + (0.00003*X34) + (0.032*X52) + (0.0008*X1*X1) + (0.00006*X6*X6) - (0.0003*X15*X15) + (0.0002*X19*X19) -- (0.0009*X2*X14) - (0.00003*X52*X52) - (0.0000001*X52*X34)$$

(1)

Where:

HEALTH= environmental quality index (Table 1)

X1= perimeter of immediate vegetation

X2= perimeter of intermediate non-vegetation

X3= perimeter of distant vegetation

X4= area of intermediate vegetation

X6= area of distant non-vegetation

X7= area of pavement

X8= area of building

X9= area of vehicle

X10= area of humans

X13= area of herbaceous foreground material

X14= area of wildflowers in foreground

X15= area of utilities

X16= area of boats

X17= area of dead foreground vegetation

X19= area of wildlife

X30= open landscapes = X2+X4+(2*(X3+X6))

X31= closed landscapes = X2+X4+(2*(X1+X17))

X32= openness = X30-X31

X34= mystery = X30*X1*X7/1140

X52= noosphericness = X7+X8+X9+X15+ X16

Environmental Quality Index	
Variable	Score
A. Purifies Air	+1 0 -1
B. Purifies Water	+1 0 -1
C. Builds Soil Resources	+1 0 -1
D. Promotes Human Cultural Diversity	+1 0 -1
E. Preserves Natural Resources	+1 0 -1
F. Limits Use of Fossil Fuels	+1 0 -1
G. Minimizes Radioactive Contamination	+1 0 -1
H. Promotes Biological Diversity	+1 0 -1
I. Provides Food	+1 0 -1
J. Ameliorates Wind	+1 0 -1
K. Prevents Soil Erosion	+1 0 -1
L. Provides Shade	+1 0 -1
M. Presents Pleasant Smells	+1 0 -1
N. Presents Pleasant Sounds	+1 0 -1
O. Does not Contribute to Global Warming	+1 0 -1
P. Contributes to the World Economy	+1 0 -1
Q. Accommodates Recycling	+1 0 -1
R. Accommodates Multiple Use	+1 0 -1
S. Accommodates Low Maintenance	+1 0 -1
T. Visually Pleasing	+1 0 -1
Total Score	_____

Table 1. An index adapted from Smyser (1982) that has been examined as a predictor in visual/environmental quality studies.

In Burley's Equation 1 there is a set of regressors with negative coefficients. Like in Shafer's studies, this set of regressors positively relate to visual quality (Burley, 1997). They include the presence of immediate vegetation and distant non-vegetation, the presence of wildlife, the presence of flowers and the presence of openness (X1, X6, X14, X19 and X32). Images containing these features may score in the 40s or 30s. These regressors are perceived as positive enhancements by respondents. There is also a set of regressors with positive coefficients (Burley, 1997). This set of regressors are strongly associated with poor visual quality (Burley, 1997). These regressors include the presence of vehicles, humans, utility structures and overall noospheric (man dominated) features (X9, X10, X15 and X52). This means that the more humans, vehicle, building and artificial structures in a photograph or real landscape, the worse the visual quality becomes. Images with abundant noospheric features can score as high as about 110. There is a third set of variables to be considered: neutral variables (Burley, 1997). Typical neutral variables are sky, clouds, sun, moon, water, ice, snow and so on. They affect the presence of ·both positive and negative variables. The more area these neutral variables occupy in a photograph, the more likely the score is close to a neutral value, which is 70. Figures 3 and 4 present examples of images where equation 1 has been applied to the image with resulting scores.

Fig. 3. A farmland image with a predictive score of 56.356 — Copyright © 2010, Di Lu, all rights reserved, used by permission.

Fig. 4. A typical downtown image with the score of 80.4456 — Copyright © 2010, Di Lu, all rights reserved, used by permission.

Another feature of the equation is the inclusion of cultural, economic, and ecological variables in an index derived from the work of Carol Smyser (1982) (Table 1). This index has not been employed as a potential regressor by others as most investigators emphasized to test somewhat aesthetic physical variables, not culture, economics, nor ecology. When the index is included in statistical analysis, it usually is one of the more important variables to predict visual quality for North Americans, French, and Portuguese respondents (Mo et al., 2010). Since the index is more than just a visual measure and includes other environmental considerations, investigators have begun to reconsider the values and perception of respondents. When respondents examine and experience the environment, they appear not to separate a mixture of aesthetic, cultural, economic, and ecological criteria and instead appear to consider the complete environmental Gestalt of the image or space. Because the respondents seem to be considering more than just aesthetic criteria, metrics such as the one by Burley (1997) seem to be a combination of aesthetic and other environmental assessments. Nevertheless, the Smyser derived index still is a somewhat qualitative index; yet when applied and tested with groups of over 100 respondents in the years 1985, 1992, 1993, 1994, 2009, and 2010 to examine the variability of the index, respondents usually are within plus or minus 2 points when employing the index. In addition, the collective mean scores for each year that the index was examined are within decimal points of each other. Thus the index seems to have some surprising consistency.

With the equation, it is possible to compare various photographs by constructing a plot of the predicted mean score for a statistical equation and then calculating the 95% confidence tables for the mean scores (Burley, 1997). A graph was constructed to illustrate the confidence plots (Figure 5). Comparisons between scores from various images were made horizontally by determining whether there is an overlap between the two tails (Burley, 1997). The confidence tails for Figures 3 and 4 do not overlap in Figure 5, and thus it is possible to conclude the two images are significantly different.

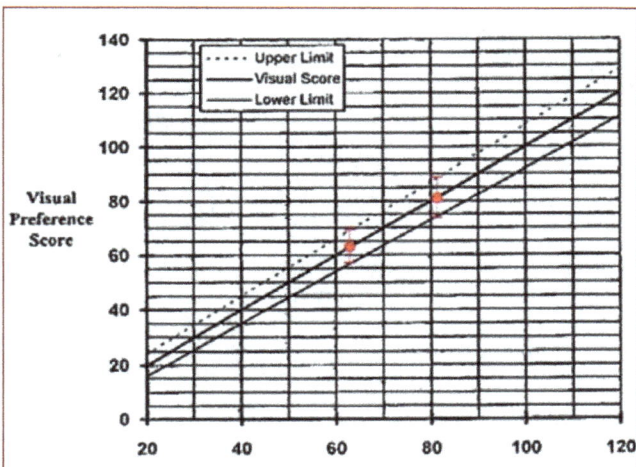

Fig. 5. A comparison of the images in Figures 3 and 4.

With the equation presented by Burley, the intent of the study was to see if this equation could be used to create a map and then test the validity of the map. To create the map, the visual quality scores of the first set of photographs were measured and matched to land-uses. The average visual quality score for each land-use type generated a map of average predicted visual quality scores for the study area (Figure 6). The GPS locations of the second set of images allowed one to determine the predicted score on the map for these locations. Then the predicted and actual score could be compared. The statistical analysis between the two sets of scores can be conducted using Kendall's Coefficient of Concordance (W) (Daniel, 1978), one of the few tests for similar agreement (significant similarity). The test is a nonparametric test and quite flexible in use and application. The test uses simple algebra and can be computed on a spreadsheet.

Legend- Visual Quality

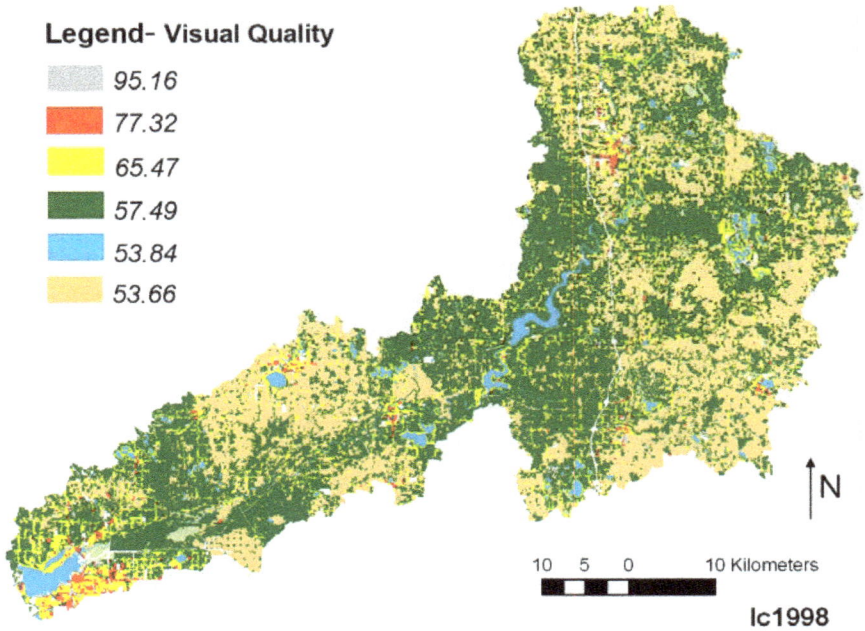

Predictive Visual Quality Map

Fig. 6. A map of predicted visual/environmental quality scores for the lower Muskegon watershed. Lower values indicate better quality, higher scores indicate poor quality.

Suppose that a score *i* is given the rank $r_{i,j}$ by treatment number *j*, where there are in total *n* scores per treatment and *m* treatments. In our case there are 30 images (n=30) in each treatment (n=2) where predicted map score and the score when measuring the photograph with an equation represent the treatments. Then the total rank (R_j) given to a treatment is the sum of the ranks for a treatment. Then computationally the test statistic W is computed in equation 2.

$$W = 12 \sum_{j=1}^{n} \left(R_j \right)^2 - \left[3m^2 n (n+1)^2 \right] / \left[m^2 n \left(n^2 - 1 \right) \right] \tag{2}$$

The test statistic *W* is between 0 and 1. If *W* is 0, there is no overall trend of agreement among the respondents. If *W* is 1, the responses might be regarded as essentially unanimous. Intermediate values of *W* suggest a degree of concordance among different responses. W approximates a Chi-square statistic with n-1 degrees of freedom, as illustrated in equation 3.

$$X^2 = m (n-1) W \tag{3}$$

Set Two Image NO.	Measured Score	Measured Score's Ranking	Predictive Score From A Map	Predictive Score's Ranking
21	107.805	1	95.160185	1.5
4	93.2968	2	95.160185	1.5
28	83.52013	3	77.32490333	4
22	81.84413	4	77.32490333	4
11	80.84248	5	77.32490333	4
26	73.11433	6	65.46785333	7.5
16	72.98182	7	65.46785333	7.5
1	72.91732	8	65.46785333	7.5
30	68.89192	9	58.92745	10
24	67.92477	10	65.46785333	7.5
20	61.07647	11	53.83573333	17.5
2	60.12997	12	53.660695	26
17	59.10933	13	53.660695	26
13	56.192	14	57.478774	12
14	55.29468	15	53.660695	26
18	54.71145	16	57.478774	12
3	54.04903	17	53.83573333	17.5
7	52.3262	18	57.478774	12
10	51.93535	19	53.660695	26
8	51.06412	20	53.660695	26
19	50.506	21	53.83573333	17.5
12	48.79608	22	53.660695	26
27	48.7268	23	53.83573333	17.5
15	48.2258	24	53.660695	26
29	48.1872	25	53.83573333	17.5
25	46.87449	26	53.83573333	17.5
6	44.4542	27	53.83573333	17.5
5	43.3638	28	53.660695	26
9	42.1124	29	53.83573333	17.5
23	41.91216	30	53.660695	26

Table 2. The scores and the rankings of the measured images and the predicted scores and rankings derived from a map.

In Table 2 the measured scores are ranked from high to low, with 107.805 assigned as the highest score and 41.91216 assigned as the lowest score for the measured images; while the predictive map scores are ranked according to corresponding expected value based upon land-use/land-cover. In Kendall's Coefficient of Concordance, a W value of 0.851112347 was generated. A corresponding Chi-Square table was consulted to determine if the derived value for Chi-Square was significant ($p \leq 0.05$) at twenty-nine degrees of freedom (Daniel, 1978). Since the derived value of 49.36 is greater than the table value of 42.56, the null hypothesis was rejected and the hypothesis that the two sets of numbers are in concordance ($p \leq 0.05$) was accepted. It was determined, through statistical analysis, that the relationship of predictions (land-use map based scores) and the real photographs are in concordance and significant to a (95%) confidence level. In other words, for the study area, it was possible to generate a map that was relatively reliable and one could predict the visual/environmental quality of a place without actually having been there. Rather, based upon a sample of the study area, the visual/environmental quality can be predicted with some level of assurance.

In this research investigation, many more photographic samples could be taken to create a landscape visual quality map; but for this research we were more interested in determining if only a few (30) number of images would generate significant results. The reason was that by using fewer photos, this research could examine the methodology to test for significant concordance (95%) under less than ideal data collection conditions, but if this sample indicated significant results, it was then possible to save money and time to create maps with only small sample sizes. Thus, only 30 pairs of images were chosen but with high variation (from rural to urban landscape) to test the ability of the methodology in assessing landscape visual quality. The results suggest that this methodology has practical possibilities to quantitatively and reliably measure the environment.

4. Urban planning applications

Many previous investigations demonstrated the validity of using surveys, such as respondent groups in landscape evaluation experiments. Nevertheless, in this experiment, no respondent group was employed to evaluate landscape images, although the equation utilized (Burley's Equation 1) was generated from a respondent group study (Burley, 1997). For planning and design studies and assessment, it might be faster and just as reliable not to gather more information with people-based surveys, but rather rely upon previous research results and predictive models.

In the context of landscape planning and design, landscape visual quality assessments are sometimes considered not important because they lack substantial evidence or due to their subjectivity. The GIS based land use-map might be used in this context to facilitate a reinforcement of visual quality assessment concerning aesthetic changes in the environment. The results of this study suggests that a GIS based land-use map could serve in visual quality assessment as well as in the professional practices of landscape planning. Land-use maps could be used to measure landscape quality instead of real images. With the help of GIS based land-use maps, initial site surveys might be reduced. Designers and planners are able to use predictive equations and GIS data during the early design and assessment phases. However, predicting site-assessed visual quality does not mean that replacing public and expert assessment is always advocated. Instead these procedures are a tool to gage the impact of planning, design, and management impacts.

In an environmental impact assessment, visual quality assessment is an important factor as well. Many of the environment impact properties could not be measured easily because these indicators are very subjective. With the help of GIS based land-use maps to predict potential visual impact, it might supply a quantitative method for environmental impact assessment. For example, as it is known that the predictive visual quality score of farmland is significantly better than the predictive score of industry in Lower Muskegon Watershed. If the developers initiate to build a factory at a farmland area of Lower Muskegon Watershed, it is obvious that the report of visual impact assessment would be negative according to the predictive model presented in this study.

The results of the experiment could also help manage viewsheds. A viewshed is an area of land, water, or other environmental element that is visible to the human eye from a fixed vantage point. In this paper, the area covered by each recorded photograph is defined as a viewshed. As well as illustrated in this example, all the photo data collected and located in this paper could be used for later viewsheds management research.

While the equation explains a fair amount of variance and can be applied quantitatively, the regressors are not necessarily fixed. New equations may arise with more predictive power and with a different set of regressors. In combination with the equations, general principles and guidelines as the ones reported by Burley (2006a), Kaplan et al., (1998), and Kaplan and Kaplan (1989) may be more helpful to planners and designers in creating and managing the environment than numerical equations. The equations are simply numerical affirmations of these general ideas and recommendations.

For urban planning applications, the most basic application would be to compare images between before and proposed design treatments from key or important viewpoints. The quantitative measure could indicate which treatments perceptually increase the quality of the environment. This technique is explained by Mazure and Burley (2009) and also illustrated by Burley (2006b), but in a more rural setting. For historic preservation purposes, improving the quality may not always be appropriate but to rather maintain a specific quality of an historic setting. Then the quantitative measure could be employed to test treatments for maintaining a specific perceived condition.

In urban settings, sectors of the city could be mapped to establish a base-line and monitor urban environmental quality in the same manner other environmental characteristics are mapped and evaluated. This approach could also be employed at the site/parcel level. Numerical goals and objectives could be established in development plans to maintain a specific level of visual/environmental quality. In addition, the quantitative method could be employed to evaluate corridor/experiential plans and proposals. Kendall's Coefficient of Concordance can be used to test for significant similarity and Friedman's Analysis of Variance (Daniel, 1978) can test for significant differences between plans and treatments. There may be numerous practical approaches to aid the planner and design in crafting the design and management of urban places. However, time will tell whether this integration of social science methods, landscape metrics, and statistical tests will yield productive results in the planning and design area. Many researchers have hope and aspirations for their work, yet often it remains unrealized.

5. Conclusion

The development of measuring aesthetic/environmental quality has made progress over the past 50 years. Investigators now understand and can predict about 65% to 70% of the variance concerning respondent perception to the spatial quality of the environment. In addition, it is possible to create statistically reliable maps to predict visual quality across the urban and rural environment. The process is relatively efficient and effective. Planners, designers, and citizens can measure the perceived effects of spatial treatments and can assess the perceived impact of various proposals and plans. This approach is one more tool in a toolbox of expert and statistical measures to understand the impacts proposals and plans may have upon the environment.

6. Acknowledgements

The authors wish to thank Wang Sihui a graduate of Nanjing Forestry University and the Masters of Arts in Environmental Design at Michigan State University for her GIS assistance, Grand Valley State University for creating a current land-use map that made this study possible, to Dr. Trisha Machemer from the School of Planning, Design, and Construction at Michigan State University for allowing access to her GIS database, and to Shawn Partin for his advice during his review of the manuscript.

7. References

Arriaza, M., Cañas-Ortega, J., Cañas-Madueño, J., & Ruiz-Aviles, P. (2004). Assessing the visual quality of rural landscapes. *Landscape and Urban Planning*, Vol.69, No.1, pp. 115-125, ISSN 0169-2046

Bourassa, S.C. (1991). *The Aesthetic of Landscape*, Belhaven Press, ISBN 1852930713, London, UK

Boster, R.S., & T. C. Daniel. (1972). Measuring public responses to vegetation management. *16th Annual Arizona Watershed Symposium Proceedings*, Arizona Water Commission, Phoenix, Arizona, September 20, 1972, Phoenix, AZ Report Number 2, pp. 38-43.

Brush, R.O., & Shafer E.L. (1975). Application of a landscape-preference model to land management. In: *Landscape Assessment: Values, Perceptions, and Resources*, Zube, E.H., R.O. Brush, R.O., & and Fabos J.G., pp. 168-182, Dowden, Hutchinson and Ross, Inc., ISBN 0-470-98423-6, Stroudsburg, Pennsylvania, USA

Bureau of Land Management. (1980). *Visual Resource Management Program*, Department of Interior, Bureau of Land Management, Division of Recreation and Cultural Resources, Washington, D.C.

Burley, J.B. (1997). Visual and ecological environmental quality model for transportation planning and design. *Transportation Research Record*, Vol.1549, pp. 54–60, ISSN 0361198

Burley, J.B. (2006a). The science of design: green vegetation and flowering plants do make a difference: quantifying visual quality. *The Michigan Landscape* Vol.49, No.8, pp. 27-30.

Burley, J.B. (2006b). A quantitative method to assess aesthetic/environmental quality for spatial surface mine planning and design. *WSEAS Transactions on Environment and Development*, Vol.5, No.2, pp. 524-529, ISSN 17905079

Carlson, A. (1977). On the possibility of quantifying scenic beauty. *Landscape Planning*, Vol.4, No.2, pp. 131-172, ISSN 0169-2046

Crawford, D. 1994. Using remotely sensed data in landscape visual quality assessment. Landscape and Urban Planning, Vol.30, No.1-2, pp 71-81, ISSN 0169-2046

Daniel, W.W. (1978). *Applied Nonparametric Statistics*, Houghton Mifflin Company, ISBN 0-395-25795-6, Boston, Massachusetts, USA

Fuente de Val, G., Atauri, J. & de Lucio, J. (2006). Relationship between landscape visual attributes and spatial pattern indices: A test study in Mediterranean-climate landscapes. *Landscape and Urban Planning*, Vol. 77, No. 4, pp. 393-407, ISSN 0169-2046

Kaplan, R., & Kaplan S. (1989). *The Experience of Nature: A Psychological Perspective*, Cambridge University Press, ISBN 0521341396, Cambridge, UK

Kaplan, R., Kaplan, S., & Ryan, R. (1998). *With People in Mind: Design and Management of Everyday Nature*, Island Press, ISBN 1559635940, Washington, D.C., USA

Lee, E.J., & Burley, J.B. (2008). Assessing visual quality change 25 years after post-mining housing development in Plymouth, Minnesota, *New Opportunities to Apply Our Science*, 2008 National Meeting of the American Society of Mining and Reclamation, Richmond, VA,June 14-19, 2008, R.I. Barnhisel (Ed.) American Society for Mining and Reclamation, Lexington, Kentucky, USA

Loures, L., Vargues, P., Horta, D. (2008). Landscape aesthetical and visual analysis facing the challenge of the development of sustainable landscapes – the case study of the post-industrial area to the left margin of the Arade River. *International Journal of Design & Nature and Ecodynamics*, Vol.3, No.1, pp. 65-74, ISSN: 1755-7445

Lu, Di. (2011). *Visual Quality Assessment at the Lower Muskegon Watershed*, Masters of Arts in Environmental Design Thesis, Michigan State University, E. Lansing, Michigan, USA

Mazure, A., & Burley, J.B. (Chinese translation by Wang H., & S. Wang, S.) (2009). Balancing art and science of landscape architecture: hand-drawing in China and computer-aided visual quality assessment in Michigan. *Journal of Nanjing Forestry University* (Humanities and Social Sciences Edition), Vol.9, No.2, pp. 84-93, ISSN 1671-1165

Mo, F., Le Cléach, G., Sales, M., Deyoung, G. and Burley, J.B. (2011). Visual and environmental quality perception and preference in the People's Republic of China, France, and Portugal. *International Journal of Energy and Environment*, Vol.5, No.4, pp 549-555, ISSN 2076-2895

Panagopoulos, T., & Vargues, P.M. (2006). Visual impact assessment of a golf course in a Mediterranean forest landscape. In: *Patterns and Processes in Forest Landscapes, Consequences of Human Management*, Lafortortezza, L., Sanesi, G., pp. 279–285, Accademia Italiana di Scienze Florestali, Firenze, Italy

Shafer, E. Jr. (1969). Perception of natural environments. *Environment and Behavior*, Vol.1, No.1, pp. 71-82, ISSN 0013-9165

Shafer Jr., E., Hamilton Jr., J.F., & Schmidt, E.A. (1969). Natural landscape preference: a predictive model. *Journal of Leisure Research*, Vol.1, pp. 1–19, ISSN 0022-2216

Shafer Jr., E., & Tooby, M. (1973). Landscape preference: an international replication. *Journal of Leisure Research*, Vol.5, pp. 60–65, ISSN 0022-2216.

Schauman, S., & Adams, C. (1979). Soil conservation service landscape resource management. *Proceedings of our national landscape: a conference on applied techniques for analysis and management of the visual resource [Incline Village, Nev., April 23-25, 1979]*, Elsner, G. H., & Smardon, R. C. (technical coordinators), pp 671-673, General Technical Report PSW-GTR-35. Berkeley, CA, Pacific Southwest Forest and Range Experiment Station, Forest Service, U.S. Department of Agriculture,

Smardon, R., Palmer, J., & Felleman, J. (1986). *Foundations for Visual Project Analysis*, A Wiley-Interscience Publication, ISBN 0471881848, New York, New York

Smyser, C. (1982). *Nature's Design: A Practical Guide To Natural Landscaping*, Rodale Press, ISBN 9780878573431, Emmaus, Pennsylvania

Taylor, J. G., Zube, E. H., & Sell., J. L. (1987). Landscape assessment and perception research methods, In: *Methods in Environmental and Behavioral Research*, Bechtel, R.B., Marans, R.W., & William Michelson, W., p. 361-393, Van Nostrand Reinhold Company, ISBN 0442222570, New York, New York.

United State Forest Service. (1973). *National Forest Landscape Management*, U.S. Department of Agriculture, Agriculture Handbook no. 434, U.S. Government Printing Office, Washington, D.C.

Van den Berg, A.E. & Koole, S. (2006). New wilderness in the Netherlands: An investigation of visual preferences for nature development landscapes. *Landscape and Urban Planning*, Vol.78, No.4, pp. 362-372, ISSN 0169-2046

Weinstein, N.D. (1976). The statistical prediction of environmental preferences: problems of validity and application. *Environment and Behavior*, Vol.8, No.4, pp. 611-626, ISSN 0013-9165

Zube, E.H. (1974). Cross-disciplinary and intermode agreement on the description and evaluation of landscape resource. *Environment and Behavior*, Vol.6, pp. 69-89, ISSN 0013-9165

Zube, E.H., Sell, J.L., & Taylor, J.G. (1982). Landscape perception: research, application and theory. *Landscape Planning*, Vol.9, pp. 1–33, ISSN 0169-2046

8

Identification and Analysis of Urbanization and Suburbanization in Olomouc Region – Possibilities of GIS Analytical Tools

Jaroslav Burian and Vít Voženílek
Palacký University, Olomouc
Czech Republic

1. Introduction

In the last few years, we could see a significant increase in the importance of spatial planning as a part of the social process and development of affected regions in developed countries. It is suitable to use geospatial technologies (GIT) like geographical information system (GIS) of global positioning system (GPS) for high-quality spatial planning on the level of municipalities and regions.

Unfortunately, spatial planning sometimes strives for the best distribution of human activities at the expense of environmental conditions. However, the environmental conditions should play an essential role in the spatial planning. In practice, it often happens that new buildings are built at places unsuitable for human activities (floodplains, landslide areas), thus leading to a conflict between physic-geographic conditions and human activities both existing and proposed.

In the course of the last two centuries urbanization and suburbanization has brought about substantial changes in layouts of municipalities, concerning not only municipalities themselves but also their surroundings. Some municipalities are not prepared for the process of suburbanization and that is why many problems can appear. Without well done urban (municipality) plan, based on good quality results of spatial analysis, the city development can be chaotic and can leads to urban sprawl.

Therefore, it is extremely useful to recognize the phases of the urbanization process in due time and respond with appropriate means of municipal planning. With the help of geographical information technologies it is possible not only to recognize but also to determine, quantify, analyze or model particular phases of the urbanization process. At first this article deals with the ways and possibilities of studying the urbanization processes on a general level; further on it focuses on these processes in the Olomouc region.

2. The process of suburbanization

In the course of the last decades efforts have been made to come up with a theory which could describe development of municipalities within the scope of the population system. One of the most significant theories is, for instance, the concept of the differential urbanization which is described by Geyer and Kontuly (1993). Currently the van den Berg's theory of the stages of urban development is accepted as a general developmental model of municipalities (Ouředníček, 2000). Van den Berg (Van den Berg et al., 1982) recognizes four stages of urbanization: urbanization, suburbanization, deurbanization and reurbanization. These stages are characterized by changes in proportion of inhabitants living in municipalities and their suburbs, which is caused by different directions of population movement.

During all the stages of urban development – concurrently with the changing allocation of inhabitants – there are changes in layout and utilization of space inside urban units. This is why the stages of development of contemporary modern municipalities are characterized by changes of their inner structure.

Van den Berg (van den Berg et al., 1982) describes suburbanization as the second development stage of a city during which inhabitants move towards its outskirts. This process has greater demands for suburban space; a municipality is spatially spreading to its surroundings and also to more distant areas occupying new adjacent pieces of land. There is a population movement from inner cities to suburbias. This stage is influenced mainly by housing standards and traffic development.

Two forms of suburbanization are usually differentiated – residential and commercial (Ouředníček et al., 2008). The first one describes a settling of city outskirts by means of family housing and low-storey development (construction of satellite towns). Commercial suburbanization comprises growth of new commercial, manufacturing, storage and logistic activities. It results in moving industrial activities to less objectionable areas and creating more attractive housing localities. A new traffic infrastructure is being built and accessibility of a city centre from its suburbs is improved. Simultaneously there is an increase in mobility of labour and individual commuting to the centre of municipalities (Ouředníček, 2000).

2.1 Demarcation of the space of the suburbanization process

Studies of the urbanization processes indispensably require a demarcation of a centre and its suburbs – that is the area in which the particular processes take place. However, this demarcation is not unambiguous and it can be reached by number of approaches. Nowadays there is no accurate and unambiguous demarcation of urbanized and suburbanized areas.

Mostly, administrative boundaries are used during the process of demarcation of urban areas. It is first and foremost because of the easy accessibility of information for these units (Frey, 2001). This demarcation is often neither accurate nor adequate; however, regarding the accessibility and accuracy of statistical data it is frequently the only possible one.

While studying suburbanization, the administrative demarcation of a municipality against its suburbs is not as important as a morphological demarcation given by the boundaries of contiguous development (Sýkora, 2003). On the other hand, virtually all the statistical data are provided only for administrative boundaries.

For a demarcation of a municipality and its suburbs Mulíček (Mulíček, 2002) uses its administrative boundaries which more or less copy its factual boundaries demarcated by the contiguous development. In numerous municipalities the administrative boundary differs more or less substantially from the boundary demarcated by the actual contiguous development which is, moreover, often different from the built-up area demarcated by the municipal plan. Often there are also vast areas inside the cities which divide it into more parts and disrupt its compactness. The area between the Povel and Slavonín quarters in the southern part of the city of Olomouc can serve as a perfect example.

Suburbs of a city generally comprise of municipalities which have the most intensive relations (industrial, traffic and economical) with the city. Most often a centre and its suburbs which are connected to it by strong bonds are referred to as a functional urban area (FUA). Maier (Maier et al., 2007) state that there is no official demarcation of FUA in the Czech Republic. In their study the authors describe the demarcation of this area on the basis of the actual commuting to the centre of the settlement from the 2001 population and housing census. The decisive criterion was the limit of 25% of the economically active inhabitants commuting to the centre of the region. The municipalities which did not reach this limit but were inside the area that fulfilled the criterion were also included in the FUA.

2.2 Localization and identification of suburbanization

There is a substantial irregularity in disposition of both new developments and particular suburban areas (Ouředníček & Temelová, 2008). Therefore, there is no overall, continuous growth of big cities' suburban areas; new developments are more likely to be found in suitable locations along roads, in the vicinity of bigger urban settlements with a well developed infrastructure or in more attractive locations which meet the crucial requirements for "healthy" housing within the reach of a city.

AGE STRUCTURE OF EMIGRANTS FROM OLOMOUC CITY TO FUA OLOMOUC in 1991-2009

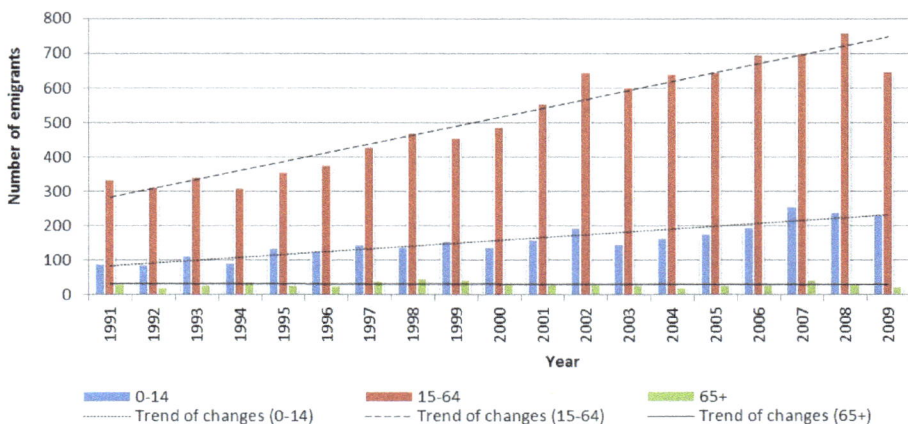

Fig. 1. Changes of the age structure caused by the suburbanization

Suburbanization is mostly caused by moving younger, more educated people that have enough money to build house. In suburbs average age and average rate of unemployment is decreasing, number of young and more educated people is increasing. One of this phenomenon is clearly seen on the figure 1.

Besides the suburbanization itself there is another suburban development taking place in the suburbs (suburbias – sometimes referred to as satellite towns), e.g. building activities or revitalization of economic functions (Ouředníček, 2008).

Within the boundaries of a municipality it is extremely difficult to determine whether it is suburbanization or urbanization which is taking place (Sýkora, 2003). The boundary between these two phenomena is frequently fuzzy. In case a municipality is spatially expanding as a homogenous body gradually growing along its edges, it is rather an ongoing urbanization. If, however, the development takes place in locations that are spatially separated from the so-far-urbanized areas, even though it has strong functional connections with the city, we refer to it as suburbanization. Yet in both cases the city undergoes an overall growth and spatial expansion.

3. Studies of urbanization processes

Sýkora (Sýkora, 2003) defines two basic approaches to the monitoring of suburbanization. The first one notices changes in spatial distribution of inhabitants based on a comparison of a core area and its periphery. The second one employs an evaluation of the selected features of the morphologic structure of metropolitan districts. From the perspective of geographical information technologies the first approach consists mainly in processing statistical data in GIS, whereas the second one relies on processing aerial or satellite photographs.

3.1 Research by methods of GIS

Geographic information system (GIS) is currently used for identification, description and visualization as well as for an analysis and modeling of the urbanization processes (especially urbanization and suburbanization). GIS is most frequently used to study the urbanization processes indirectly. In can be used to monitor changes of a built-up area, changes in migration and movement of inhabitants (Dobešová & Křivka, 2011) or in the spatial distribution of cities/towns.

One of the most significant advantages of GIS is a possibility to quickly and schematically combine information about spatial properties of specific objects with their descriptive attributes (Voženílek, 2005). A subsequent visualization can, however, be dealt with in a highly inappropriate manner (more in Burian & Šťávová, 2009).

GIS tools in the form of analytical overlaying are used e.g. for multi-criteria decision making about an ideal utilization of an area or for a detection of changes caused by a temporal evolution of various spatial systems (Hlásný, 2007). Maantay and Ziegler (2007) introduce numerous examples of an application of GIS analytical tools for an urban environment. Other authors (Yeh & Li, 2002) utilize cellular automata to simulate development of population density for municipal planning. Vorel (Vorel, 2006) describes cellular automata and multi-agent systems (sometimes also referred to as free-agent models) as one of the most frequently used models for urban/regional planning. In the Czech Republic it is Grill

(Grill et al., 2008) who devoted themselves to this topic. In their article the authors describe an effort to create a model which would be capable of predicting the effects of municipal planning on the future environmental quality of areas designated for housing and commercial applications.

Significant aspects on connection land use changes, cellular automata and neural networks are described by Pechanec (Pechanec et al., 2011b). Clarke and Gaydos (Clarke & Gaydos, 1998) describe the possibilities of connecting GIS with cellular automata to create a model which could simulate the process of growth of cities. The authors state that currently it is extremely difficult to incorporate a cellular automaton directly into the GIS environment; therefore it often stays outside this environment.

One of the possible ways to determine the urbanization processes is to employ indeterminacy (fuzzy logic). This approach is described e.g. by Krabegovic (Krabegovic et al., 2006) as exemplified by employing fuzzy logic in GIS to perform multi-criteria decision making. The most detailed account of the problems of the implementation of information systems (GIS in narrower conception) in regional/urban planning is given by Laurini (Laurini, 2001). He considers GIS to be not only a tool for professional visualization, but first and foremost a tool for processing spatial analyses, modeling, making prognoses and scenarios of regional development or multi-criteria decision making.

Laurini also stresses the importance of DSS (Decision Support Systems), i.e. tools for supporting decisions, which can be employed in regional planning in bigger territories. The issue of DSS is dealt with on a general level, e.g. in the work of Pechanec (Pechanec, 2005). More detailed description for regional planning is provided by Batty and Desham (Batty & Desham, 1996). Hlásný (Hlásný, 2007) describes the issue of „Time GIS" which adds the temporal dimension to the spatial one. In future, TimeGIS could be an important tool in the area of urbanism, e.g. for modeling and predicting regional development.

3.2 Models and software for monitoring the urbanization processes

While analyzing various natural and socio-economic systems it is highly desirable to create their simplified models – so called conceptual models which, considering the given level of abstraction, enable modeling and predicting the behavior of these systems (Hlásný, 2007).

Every model is, to a certain extent, a generalized reality. Therefore, the results provided by a model can never be on a level of actual real results. This fact has to be taken into account while using the results for various purposes (e.g. regional/urban planning). A model can never reach the complexity of a real system and this fact has to be counted on while assessing the results. If one chooses a block of buildings as the lowest resolution level, he/she can not use the results of the analysis for a smaller object (e.g. a building).

Correctly and logically made data model guarantees an effective implementation of a digital regional/municipal plan in geographic information projects (Voženílek, 2005).

Current GIS products (ArcGIS, MapInfo, GRASS) provide a wide range of analytical tools suitable for regional/municipal planning. An illustrative example of this feature is e.g. the ArcGIS extension described by Schaller (Schaller, 2007), when a regional planning of the area around Munich was administered with the help of an extensive toolbox created in Model Builder. Another ArcGIS extension is mentioned by Schaal (Schaal, 2004) who describes the PLANanalyst add-on. The LUCIS model (Zwick & Carr, 2007) is a good example of a combination of GIS analytical tools into the ArcGIS extension which can be used by regional/municipal planners. Kumar and Sinha (Kumar & Sinha, 2006) demonstrate the possibilities of the analytical tools of freeware programs. Their research employed the GRASS (Geographic Resource Analysis Support System) software as a tool for municipal planning. In their book Brail and Klosterman (Brail & Klosterman, 2001) describe several programs which are regularly used for the needs of regional planning (chiefly in the USA but also in other parts of the world). It is for instance the METROPILUS software built as an application for the ESRI ArcView GIS software. A very similar principle is used e.g. by the INDEX software which has a vast number of functions for evaluating and planning changes in landscape. On the other hand, there are stand-alone products like TRANUS, CUF I, CUF II or CUBRA. The editors of this publication provide a series of articles focused not only on PSS (Planning Support Systems) but also on making development scenarios and simulations and, last but not least, on visualization of the results.

Furthermore, in his numerous publications (Klosterman, 1999) Klosterman describes his own software solution for making development scenarios – the What if? software tool which ranks among the PPS group (Planning Support Systems). It is also an extension of the ESRI products. A detailed description of the above mentioned models is provided by Burian (Burian, 2008). In his article he closely examines the LUCIS, LADSS, Gogracom 5W, Urban SIM, MUSE and SUDSS models and suggests a concept of an ideal scheme of the software used for urban planning. Pechanec (Pechanec et al., 2009 and Pechanec et al., 2011a) describe the LOREP model created to analyze and predict danger of floods for the needs of urban planning.

All the above mentioned models deal with the issues of urbanization on a simplified level (built-up/open areas) and do not distinguish between the particular stages of urban development.

3.3 Statistical data for urbanization processes modeling

One of the main manifestations of the residential suburbanization is multi-family or larger housing development in country municipalities. Therefore, a documentation of a suburbanization most frequently uses data containing numbers of constructed housing units (number of constructed housing units per 1000 inhabitants). This is often joined by another indicator – monitoring the population growth – represented by migration balance (Fig. 2). The analysis of these simple maps can bring the basic overview of urban processes in the region. The municipalities like (Hlušovice, Bystrovany, Bukovany or Dolany) with high values can be easily identify.

Description of suburbanization with the help of socio - economic data is provided in details by Stuchlíková (Stuchlíková, 2009). The author depicts the process of suburbanization in the Hradecko-Pardubická agglomeration since the 1990's till the present day in connection with the decrease of migration rates and massive housing development. This issue is also described by Ouředníček (Ouředníček, 2006) or Mulíček and Olšová (Mulíček & Olšová, 2002). One can easily deduce that the suburbanization processes that are taking place in a number of Czech towns and cities are inconclusive and ever-changing. Stuchlíková (Stuchlíková, 2009) illustrates the manifestations of suburbanization not only with migration of population from the centre to its suburbs but she also describes the change in the age and education structure in connection with the suburban newcomers. That is to say that migration caused by suburbanization tends to populate suburban areas with well educated and upper income inhabitants. Virtually the same conclusion is reached also by Ouředníček (Ouředníček, 2008).

Sýkora (Sýkora, 2003) warns of certain risks connected with the use of aggregated statistical data. Such data do not cover migration of inhabitants who, for various reasons, have permanent residence in a central municipality with better and more accessible services. Moreover, not every transfer of residence (e.g. from an older house to a newer one) is a demonstration of suburbanization.

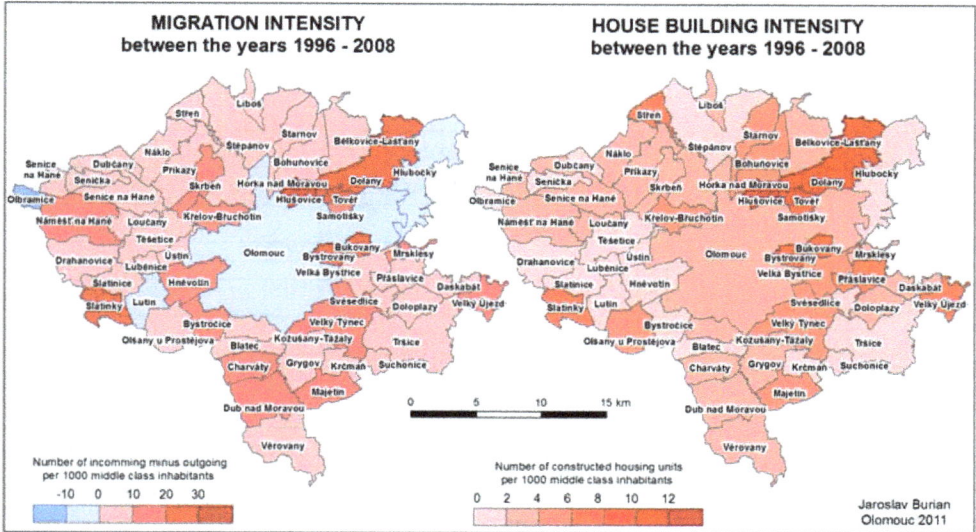

Fig. 2. Visualization of the most frequently used statistical data

In some cases only the migration of inhabitants from a central city to its suburban areas is being monitored; in others it is the general migration, not only from the centre of a municipality but also from other municipalities and parts of a state.

An absolute indicator of immigrants appears to be more appropriate for demarcating suburban areas because it takes into account the actual housing development and the sum of new suburban migrants better than a comparative indicator of net migration or its relative crude net migration rate. These indicators should rather be used for an evaluation of the intensity and spatial structure of suburbanization.

Nowadays a considerable amount of satellite towns grow on the outskirts of central municipalities and their territory lies within their administrative boundaries; therefore they are statistically almost imperceptible. In the Olomouc region we can find such an example for instance in Chomoutov, Nedvězí, Topolany, Slavonín, Droždín, Svatý Kopeček, Lošov and Radíkov. The family housing development in these areas is a typical demonstration of suburbanization. Yet these localities are former independent municipalities which are currently only quarters located within the city of Olomouc.

Table 1 provides a brief summary of statistical indicators which the author considers to be the most important for the study of the urbanization processes. MOS data (data from the municipal statistics) are accessible annually but only for municipalities and not for basic settlement units, that would be better for detailed analysis. Census data (population and housing census) are accessible for municipalities and basic settlement units only once in a decade at the date of consensus. All data in Table 1 are available for whole Czech Republic in the same quality, same data model and are distributed online for free.

Group	Indicator	Data source
Demography	Population	MOS
	Population (citizens under 14 years of age)	MOS
	Natural increase	MOS
Population distribution	Educational structure	MOS
	Employment according to economic activity	Census
Migration of population	Immigrants	MOS
	Emigrants	MOS
	Net migration	MOS
	Direction of migration	MOS
	Cause of migration	MOS
	Commuting to work and to school	Census
	Commuting according to means of transport	Census
	Commuting according to frequency	Census
Household facilities	Ownership of a passenger vehicle	Census
Municipal facilities	Built-up area in Ha	MOS
	Water-supply system	MOS
	Sewer system	MOS
	Gas supply	MOS
House building	Constructed housing units in total	MOS
	Constructed housing units in family houses	MOS
	Permanently inhabited housing units according to time construction	Census

Table 1. The most important statistical indicators for the study of the urbanization processes

4. Analysis of the suburbanization of the Olomouc region using GIS

A method (a conceptual model of the suburbanization intensity assessment – Fig. 3) for the modeling of the suburbanization process in the Olomouc region has been created on the basis of studying the specialized literature described in the previous chapter. From the third step this method has been implemented as a set of automatic tools for the ArcGIS software.

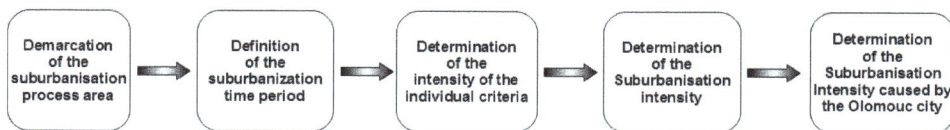

Fig. 3. Conceptual model of the suburbanization intensity assessment

The model is based on processing tabular data from the Czech Statistical Office and it employs the principle of combining vector data layers (shapefiles or other native formats of ArcGIS – Personal Geodatabase or File Geodatabase). It uses chiefly the Join (joining the tables), the Add Field (adding new attributes) and the Calculate Field operations (calculation of the intensities). It has been developed by the authors to use the ArcGIS 9.x interface in which it has been implemented as the Suburban Analyst toolbox (Fig. 4). This toolbox contains two sets of tools – Calculate Intensity and Calculate Suburbanization. The Calculate Intensity set of tools contains four independent models for partial intensities calculations (Built-up Areas Intensity, Commuting Intensity, Housing Intensity and Migration Intensity).

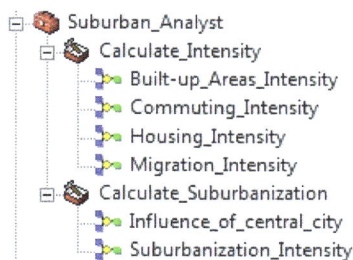

Fig. 4. The Suburban Analyst toolbox

The second set of tools (Calculate Suburbanization) contains two models. The first one (Influence Of Central City) performs the calculation of a central city's influence on suburbanization. The second model (Suburbanization Intensity) employs weighted combination of vector layers; it is an analogy of the raster operation Weighted Overlay. To make it work properly it is necessary to launch all the previous models for calculating the specific values first. The final result of the whole model is a vector layer of all the municipalities in the given area which contains the values of the suburbanization intensity during the years 1996–2008 influenced by the sum of all immigrants; furthermore, it contains the values of the suburbanization intensity in the same period influenced only by the immigrants from Olomouc.

The main advantage of this model is its intuitive graphical form (Fig. 5) which enables all the component models to be launched independently and repeatedly with the possibility of changing all the input parameters (input and output layers, attribute values, weight of particular parameters).

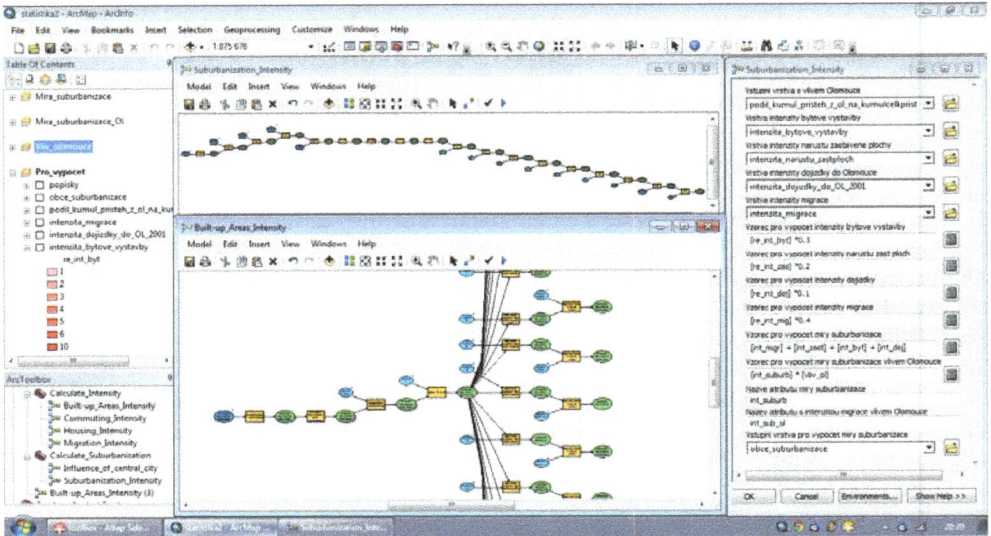

Fig. 5. Graphical interface of the ArcGIS software with Suburban Analyst Toolbox

4.1 Demarcation of the suburbanization process area

In the first phase the area of the process of suburbanization has been demarcated. This demarcation is most frequently performed with the help of commuting to a central city or net migration. The area is demarcated on the basis of the so-called FUA – Functional Urban Area (Maier et al., 2007). The decisive criterion in this particular research was the limit of 25 % of the economically active inhabitants of a municipality commuting to the centre of the region. The municipalities which did not reach this limit but were inside the area that fulfilled the criterion were also included in the FUA. Moreover, the Hlubočky village – the only municipality in the Olomouc administrative region that was not included in the FUA – was added to the area demarcated in this manner. Regarding the fact that this village is adjacent to the city of Olomouc in the west and to the Libavá military zone in the south, one can assume that it is the Olomouc city which is the most important one concerning commuting and suburbanization. The area demarcated in the previously mentioned way (FUA Olomouc+) includes the total of 53 municipalities and the city of Olomouc.

The main advantage of this model is its intuitive graphical form (Fig. 5) which enables all the component models to be launched independently and repeatedly with the possibility of changing all the input parameters (input and output layers, attribute values, weight of particular parameters).

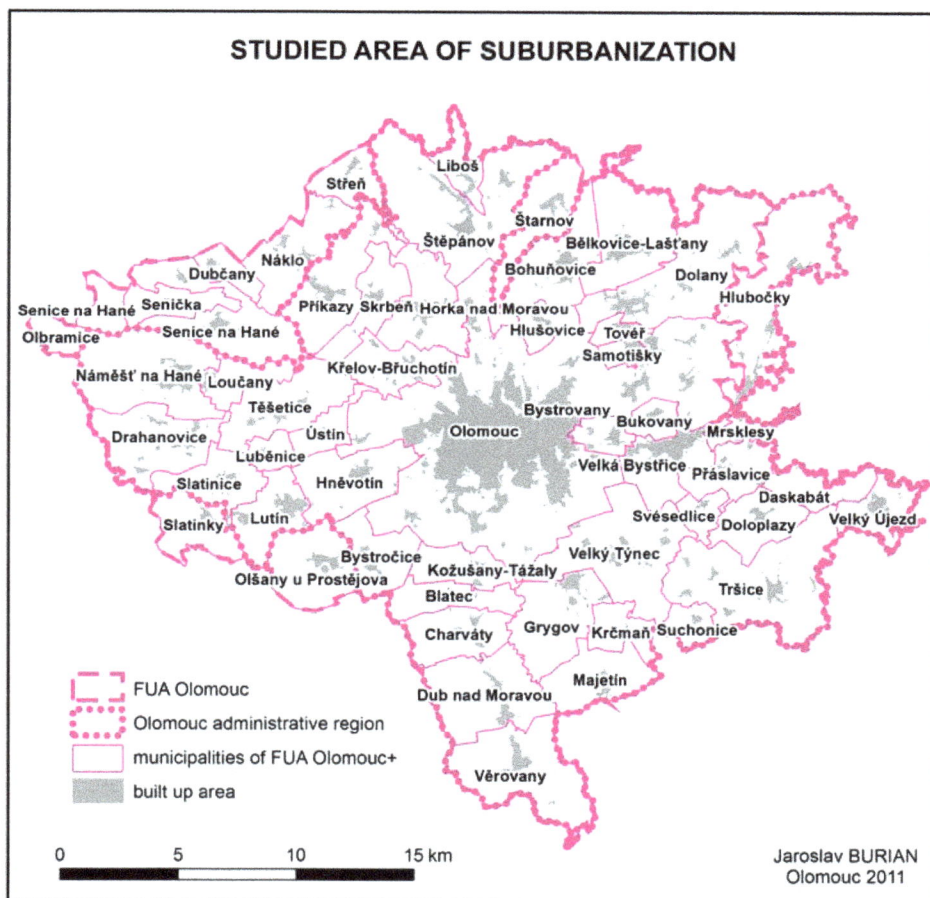

Fig. 6. Demarcation of the FUA Olomouc+ area

4.2 Definition of the time period, during which the suburbanization takes place

In the second phase the period during which the population of the central city (Olomouc) was decreasing because of migration was defined. This period was delimited on the basis of the ČSÚ (Czech Statistical Office) data by means of the net migration (Fig. 7).

EMIGRANTS FROM OLOMOUC CITY
in 1991-2009

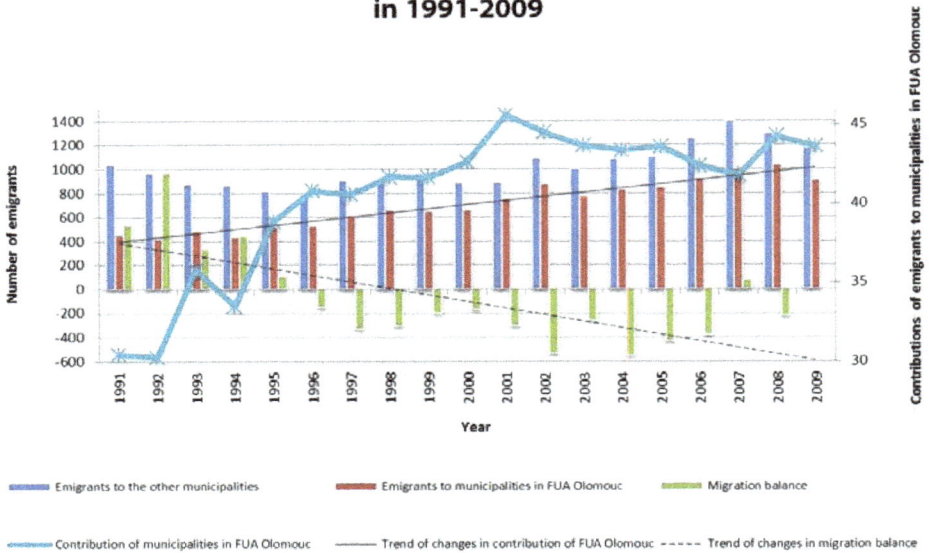

Fig. 7. Determination of the period of suburbanization

CHANGES IN NUMBER OF INHABITANTS IN OLOMOUC CITY AND FUA OLOMOUC
in 1991-2008

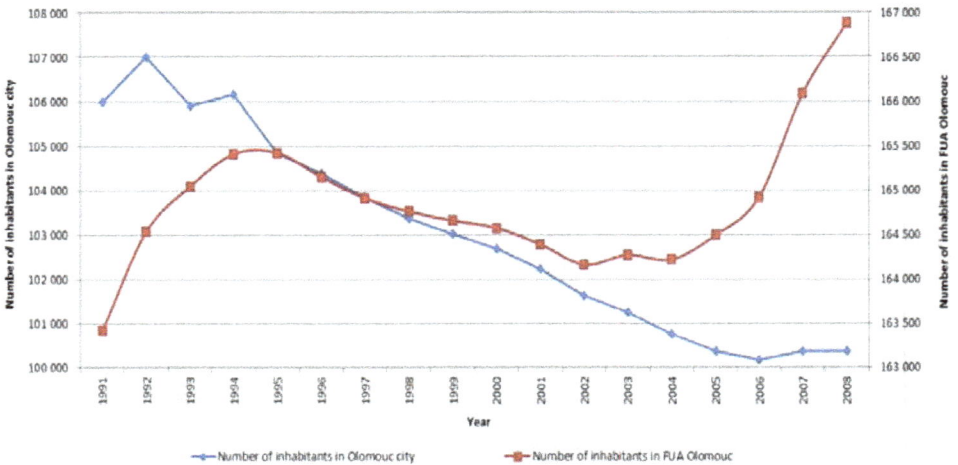

Fig. 8. Comparison of the population of the city of Olomouc and the FUA Olomouc+

As the development trend clearly shows, since 1991 the net migration has had a steady downward tendency and since 1996 till 2008 it had a negative value (within the range of -150 to -550) with the exception of 2007 when the net migration reached positive values (+78). The Cumulative net migration from the period between the years 1996 and 2008 is -3746. As the figure 8 clearly shows it is a period during which the city of Olomouc was losing inhabitants whereas the whole region (FUA Olomouc+) was gaining them rapidly.

This period was designated as the time during which an emigration of inhabitants occurred and the process of suburbanization is very likely to occur. Therefore, in the other parts of the model the data from the years 1996–2008 are dealt with.

4.3 Determination of the intensity of the individual criteria

The criteria used to determine the intensity of suburbanization by means of the constructed model were chosen on the basis of the study of specialized literature introducing these criteria as the most significant for studying the urbanization processes. More specific selection was based on multi-criteria decision analysis.

The methods of geometrical mean, Fuller's rectangle and Saaty method of pair comparison were used (Saaty, 1983). All methods are commonly used in statistics and multi-criteria analysis but in area of geography and urbanism are not so known. All of them are based on comparison all criteria, Saaty method allows to set up not only difference between criteria but also level of these differences. These methods are well described in many papers or books about multi criteria evaluation (Saaty, 1983 or Keeney, 1992). Almost it is not possible to select the most suitable method of evaluation. That is why the results of three most significant methods were used and compared. The final result of selection criteria was based on Saaty method, criteria with value lower than 0,1 were eliminated and the only 4 criteria left. Those were evaluated and compared once again and final values on intensities were estimated.

Even though absolute values (number of constructed housing units or number of immigrants) advert to a possible process of suburbanization, they can not be used to determine the level of suburbanization. Whereas a construction of 20 family houses in a municipality with the population of 5000 can be almost insignificant, in a municipality with 200 inhabitants it suggests a far more accentuated process of urbanization. This is a reason all the selected criteria were converted into their relative values (per 1000 inhabitants) so that they could be compared and evaluated together. Because of the subsequent computer processing (weighting of the layers and their summarization) the resulting relative values of every particular criterion were reclassified to new values within the range of 0–10. In case of the housing development intensity and the migration intensity the original values were negative; therefore also the reclassified values are negative.

The criteria used to determine the level of suburbanization:

- Migration intensity (crude migration rate) – the average value of net migration per 1000 middle class inhabitants between the years 1996–2008,
- Intensity of increase in built-up area – the average value of increase in built-up area per area of the municipality,
- Intensity of house building – the average value of increase in number of constructed housing units per 1000 middle class inhabitants between the years 1996–2008,

- Intensity of commuting to Olomouc – the number of job and school commuters to Olomouc per 1000 middle class inhabitants between the years 1996–2008,

4.4 Determination of the Suburbanization Intensity

Computer model „Suburban Analyst"realizes calculation of all mentioned relative criteria and allows estimating intensity of suburbanization. The constructed computer model executes the following proposed method to determine the suburbanization intensity (Fig. 9).

To calculate the suburbanization intensity it was necessary to define the weight of the individual criteria first. The weights were estimated by multi-criteria analysis, based on Saaty method.

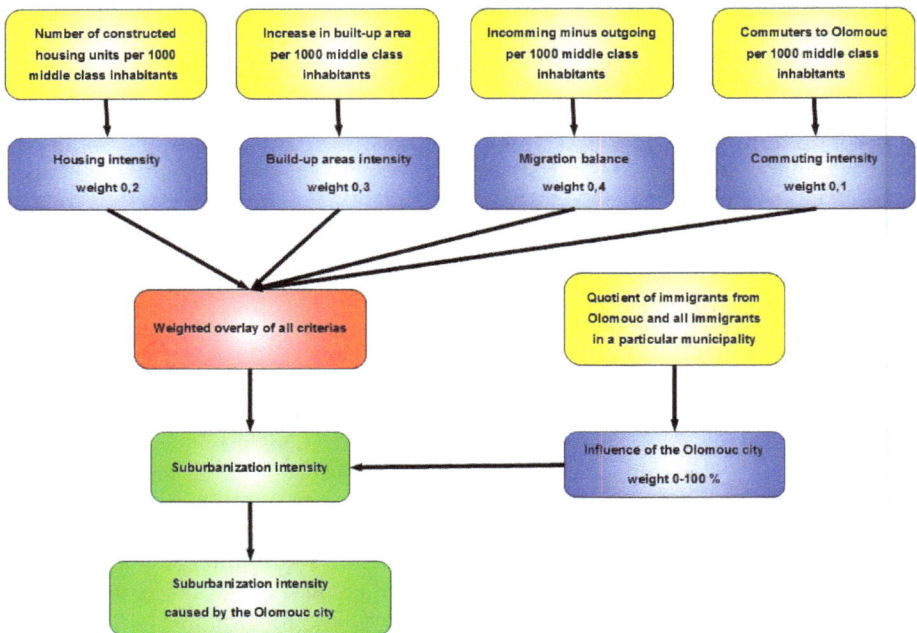

Fig. 9. The schematic method to determine the suburbanization intensity

The migration intensity with the weight of 0,4 was designated as the most significant criterion, which is usually the basic starting point of the process of suburbanization. Furthermore, the intensity of house building (0,2), the intensity of commuting to Olomouc (0,1) and the intensity of increase in built-up area (0,3) were designated. First these weights are used in the model to multiply all four reclassified intensities which are subsequently summarized. In this way the suburbanization intensity was calculated for every municipality and listed in a newly created attribute. Within the frame of the model it is possible to change the individual weights and carry out the whole calculation repeatedly.

4.5 Determination of the suburbanization intensity caused by the city of Olomouc

Suburbanization is characterized by the migration of inhabitants from a city to its suburbs. However, in a number of specialized publications suburbanization is also used to refer to the process of migration from other parts of a state to the suburbs of the monitored city. The constructed model carries out all calculations using data which encompass the influence of the whole state. Therefore, in the last step the final values of the suburbanization intensity were multiplied by the values of the level of influence of Olomouc. This level has been defined as a quotient of the cumulative number of immigrants from Olomouc and the cumulative number of all immigrants in a particular municipality in the period of 1996–2008. The resulting quotient is depicted in the Figure 7 together with the development trend; here it is fairly apparent that since 1996, when the city of Olomouc had a negative net migration, it has always assumed values higher than 40 %. The values of the quotient for the Olomouc+ FUA range from the minimum values of 30 % (the municipalities of Střeň, Olbramice, Slatinky, Štranov, Velký újezd) to the maximum values of 80 % (Bukovany, Skrbeň, Ústín, Tovéř, Hněvotín, Hlušovice).

5. The suburbanization of the Olomouc region

Ptáček (Ptáček et al., 2007) state that the development of the residential functions outside the city limits has been taking place gradually since 1997; this development concerned the localities within the accessibility zone of 10km (15–20 minutes). Furthermore, the authors claim that the localities undergoing the most intensive development during this period of time were Dolany, Slušovice, and Křelov. They demarcate the administrative boundary of the city of Olomouc as an area in which it is possible to identify various intensities of e.g. the process of suburbanization in 21 municipalities. House building in this area (both in the city and its suburbs) manifested itself most significantly in the second half of the 1990s.

On the basis of the results provided by the Suburban Analyst model it is possible to agree with the conclusions of these authors and elaborate on them. The results of the model are shown on the figure 10.

According to the results of model "Suburban Analyst" is possible to delimited 21 municipalities with intensity of suburbanization higher then average value (2,13) and 25 municipalities with intensity of suburbanization higher then median (1,9). When the influence of all municipalities in the Czech Republic is taken into account the intensity of suburbanization reaches its peak in Hlušovice (7,3), Bystrovany (7,2), Dolany (4,7), Bukovany (4,4), Tovéř (4,2), Samotišky (4,2), Velký Týnec (4,2) and Mrsklesy (4,1). Those municipalities are characteristic by the massive housing activities that can be easily documented by photos (fig. 12). These municipalities are followed by the municipalities of Křelov-Břuchotín, Hněvotín, Slatinky, Velký Újezd, Bohuňovice and Majetín.

When only the influence of Olomouc is taken into account the order of the municipalities stays almost the same; the only difference being lower final values: Hlušovice (5,3), Bystrovany (5), Bukovany (3,8), Tovéř (3,3), Samotišky (3), Dolany (2,9), Hněvotín (2,7), Křelov-Břuchotín (2,6), a Velký Týnec (2,5). The value of suburbanization intensity is higher than average value (1,23) in 23 municipalities and higher than median (1,9) in 26 municipalities. Correlation coefficient was calculated for each municipality and its average value (0,96) shows the strong cohesion between both values of suburbanization.

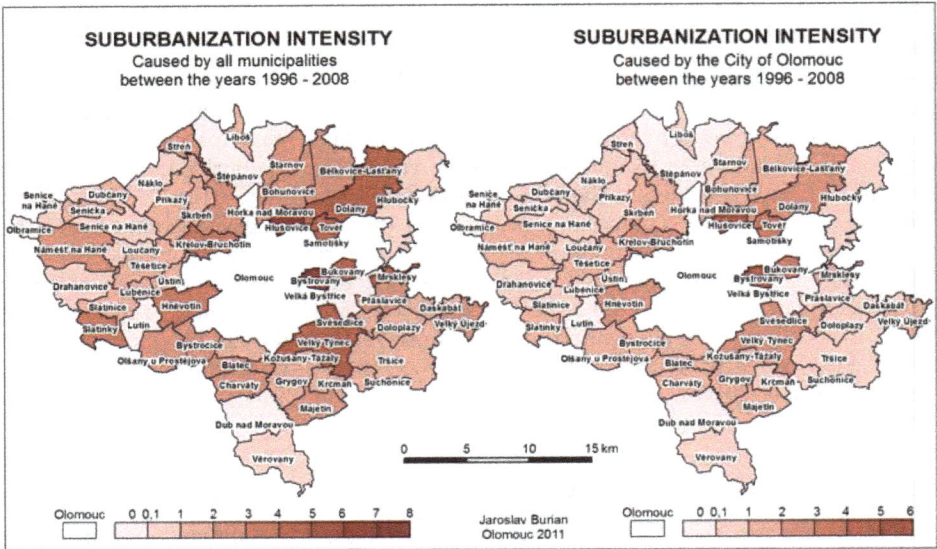

Fig. 10. Visualization of the model's results

The only exception can be seen in some specific municipalities. These municipalities are, rather naturally, in the close vicinity of Olomouc, whereas in the previous case (suburbanization influenced by all municipalities in ČR) some of the municipalities with considerable suburbanization intensity are more distant (e.g. Slatinky, Majetín or Velký Újezd). These are the localities attractiveness of which does not influence only Olomouc. The municipality which stands out the most from the results achieved is Hlušovice – probably the most typical example of a satellite town, population of which more than doubled between the years 1996 and 2008 (Fig. 11).

Ptáček and Szczyrba (Ptáček & Szczyrba, 2004) state that the most pronounced house building in the suburbs of the city is concentrated in the north and northeast; peripheral parts of the city are being developed mostly in the south and southeast. However, currently the municipalities in the northeast of the city are beginning to reach their capacity and their potential for further development is insignificant. This fact is responded to by the increased intensity of house building in other directions (e.g. Hněvotín in the west or Majetín and Charváty in the south of the city of Olomouc).

The model results can be evaluated with respect to the valid municipality plans. In municipalities with higher value of the suburbanization intensity are all planned areas for housing used and new houses are or have been built. In these municipalities the new plans are not prepared or no next development is suggested (Hlušovice, Bukovany, Bystrovany). This situation, described as relative saturation leads to increasing intensity of housing in various parts of Olomouc region. The new houses are built in municipalities that have enough areas for housing suggested by the municipality plan (Hněvotín, Bystročice, Majetín and Charváty).

Fig. 11. Example of suburbanization observation based on aerial photographs

All results of all analysis were evaluated by terrain survey. All areas of new housing in studied municipalities were documented by photos (fig. 12) and also all coordinates were stored by using GPS. For selected municipalities, examples of urban development caused by suburbanization were created by using aerial photographs from different ages (fig. 11).

From the above mentioned findings it follows that the process of the development of new residential localities in the suburban zone of Olomouc has just started (Ptáček & Szczyrba, 2004). In the following period, which is till 2008, more pronounced manifestations of this process took place in a number of other municipalities and with different intensity too. So far there are no statistical data for the period since 2008; however, on the basis of the development trend of suburbanization in the individual municipalities one can assume that this process will go on. According to relative saturation is possible to expect process of suburbanization in municipalities that serve in its municipality plans enough areas for housing. As a good example can be mentioned Tršice (local part Hostkovice) where 40 new houses are going to build in 2011.

Fig. 12. Residential suburbanization in municipalities of FUA Olomouc +

6. Conclusion

Suburbanization and its manifestations are being studied and elaborated on from various perspectives by number of specialists. Quantification of this phenomenon is not so frequent an approach; however, in the GIS environment its solution is not only possible but also relatively easy. The constructed computer model and the subsequent results can serve as an example of a possible utilization of GIS for quantification and objectification of the study of the urbanization processes The Suburban Analyst model employs aggregated statistical data which are considered to be optimal for the study of suburbanization by most of the specialists.

The model uses multi-criteria analysis and weighted overlaying of vector layers to determine final values of suburbanization intensity. Selected criteria were chosen by multi-criteria evaluation an can be considered as a objective. Pre-selection of criteria was based on various references describe urban processes. The whole process was consulted with experts – urban planners and geographers from Office of urban planning of Olomouc city and from Office of urban planning of Regional District of Olomouc city

Created model Suburban Analyst and calculated values of suburbanization intensity are one of the first approaches to quantification of urban processes by using statistical data and GIS software. According to the terrain survey and strong cohesion of the results with new housing can be the results of the model considered as relevant.

With regard to a slightly different understanding of the term suburbanization (an influence of immigrants from a central city only or an influence of immigrants from all municipalities in a given state) there are two types of results or resulting intensities. Therefore, it is necessary to provide them with an appropriate comment.

Simulation and modeling of the urbanization processes can serve as a useful approach to the understanding of the consequences of the contemporary politics of regional planning, thus providing a possibility to predict development of a region and its impact on sustainable growth.

The analysis of urban processes by estimation of intensity of suburbanization can show significant approach to study of contemporary policy of urban planning and trends of citizen movement. The results can help to predict possible future development of the Olomouc region and to predict impact of decision system of Czech urban planning.

7. Acknowledgment

This article is a part of the "Research of the citizen movement between urban and suburban space of the Olomouc region" project (number of project PrF_2010_14) with the aid of the Inner Grant Competition of the Palacký University in Olomouc.

8. References

Batty, M. & Densham, P. J. (1996). Decision support, GIS, and urban planning, In: Sistema terra, 15. 08. 2011, Available from:
< http://www.acturban.org/biennial/doc_planners/decission_gis_planning.htm>
Berg, L. van den; Drewet, R. ; Klaasen, L. H. ; Rossi, A.; Vijverberg, C. H. T. (1982). *A Study of Growth and Decline*, Pergamon Press, Oxford, New York, Totonto, Sydney, Paris, Frankfurt
Brail, R. K. & Klosterman, R. E. (2001). *Planning Support Systems*, ESRI Press, ISBN 978-15-894-8011-7 , Redlands, USA
Burian, J. (2008). GIS analytical tools for planning and management of urban proceses, *Proceedings of GIS Ostrava 2008*, pp. 13, ISBN 987-80-254-1340-1, Ostrava, Czech Republic, January, 2008
Burian, J. & Šťávová, Z. (2009). Kartografické a geoinformatické chyby v územních plánech. *Geografie - Sborník ČGS*, Vol. 114, No. 3, (2009), pp. (179-191), ISSN 1212-0014
Burian, J. et al. (2010). *Development of Olomouc city in 1930 – 2009: based on analysis of functional areas.* Palacký University in Olomouc, ISBN 978-80-244-2698-3, Olomouc, Czech Republic
Burian, J.; Miřijovský, J.; Macková, M. (2011b). Suburbanizace Olomouce, *Urbanismus a územní rozvoj*, Vol. 14, (in print), ISSN 1212-0855
Clarke, K. C. & Gaydos, L. J. (1998). Loose-coupling a cellular automaton model and GIS. Geographical informatik science, 7, (12), 699-714.
Dobesova, Z. & Krivka, T. (2011), Walkability index in the urban planning: A case study in Olomouc city, *Urban Planning*, InTech, ISBN 979-953-307-412-1
Frey, W. H. & Zimmer, Z. (2001). Defining the City, In: *Handbook of Urban Studies,* Paddison, R. et al., pp. 14-36, Sage, ISBN 978-08-039-7695-5, London, Great Britain

Geyer, H. S. & Kontuly, T. M. (1993). A Theoretical Foundation for the Concept of Differential Urbanization, In: *Differential Urbanization: Integrating Spatial Models*, Geyer, H. S. & Kontuly, T. M., pp. 290-308, ISBN 978-03-406-6285-4, Arnold, London, Great Britain

Grill, S. ; Vorel, J. ; Maier, K. ; Čtyroký, J. ; Drda, F. (2008). Simulation and assessement model of urban development, *Proceedings of GIS Ostrava 2008*, pp. 11, ISBN 987-80-254-1340-1, Ostrava, Czech Republic, January, 2008

Heisig, J.; Burian, J.; Miřijovský, J. (2011). Změny intenzity osobní automobilové dopravy a vliv na prostorovou diferenciaci suburbanizace, *Perner's Contacts*. Vol. 6, No. 1, (April 20011), pp. (431-450), ISSN 1801-674X

Hlásný, T. (2007). *Geografické informačné systémy – Priestorové analýzy*, Zephyros&Národné lesnické centrum – Lesnický výzkumný ústav Zvolen, ISBN 978-80-8093-029-5, Banská Bystrica, Slovakia

Johnson, R. J. ; Gregory, D. ; Pratt, G. ; Watts, M. (2000). *The Dictionary of Human Geography*, Blackwell Publishers, ISBN 978-06-312-0561-6, Oxford, Great Britain

Karabegovic, A.; Avdagic, Z.; Ponjavic, M. (2006). Applications of Fuzzy Logic in Geographic Information Systems for Multiple Criteria Decision Making, *Proceedings of CORP 2006 - 11th International Conference on Urban Planning & Regional Development in the Information Society*, pp. 235-244, ISBN 978-39-502-1390-4, Vienna, Austria, May 2006

Keeney, R. L. (1992). *Value focused thinking: A path to creative decisionmaking*, Harvard University Press., ISBN 978-06-749-3198-5, Cambridge, England

Klostermann, R. E. (1999). What-If? Collaborative Planning Support System. *Environment and Planning B: Planning and Design*, Vol. 26, No. 3, (1999), pp. (393-408), ISSN 0265-8135

Kumar, V. R. ; Satya, A. V.; Sinha, P. K. (2006). Urban Planning with Free and Open Source Geographic Information system. *Geological Survey of India, Southern Region, Hyderabad*, No. 1, (2006), pp. (393-408)

Laurini, R. (2001). *Information Systems for Urban Planning*, CRC Press, ISBN 978-07-484-0963-1, London and New York

Maantay, J. & Ziegler, J. (2007). *GIS for the Urban Environment*, ESRI Press, ISBN 978-15-894-8082-7, Redlands, USA

Maier, K. ; Drda, F. ; Mulíček, O. ; Sýkora. L. (2007). Dopravní dostupnost funkčních městských regionů a urbanizovaných zón v České republice. *Urbanismus a územní rozvoj*, Vol. 10, No. 3, (May 2007), pp. (75-80), ISSN 1212-0855

Mulíček, O. & Olšová, I. (2002). Město Brno a důsledky různých forem urbanizace. *Urbanismus a územní rozvoj*, Vol. 5, No. 6, (December 2002), pp. (17-20), ISSN 1212-0855

Ouředníček, M. (2002). Suburbanizace v kontextu urbanizačního procesu, In: *Suburbanizace a její sociální, ekonomické a ekologické důsledky*, Sýkora, L., pp. 39-54, ISBN 978-80-901-9149-5, Ústav pro ekopolitiku, Praha, Czech Republic

Ouředníček, M. (2000). Teorie stádií vývoje měst a diferenciální urbanizace, *Sborník ČGS*, Vol. 105, No. 4, (2000), pp. (361-369), ISSN 1212-0014

Ouředníček, M. et al. (2008). Suburbanizace, In: Suburbanizace, 20. 7. 2011, Available from:

<http://www.suburbanizace.cz>

Ouředníček, M. & Temelová, J. (2008). Současná česká suburbanizace a její důsledky. *Veřejná správa*, Vol. 11, No. 4, pp. (1-4), ISSN 1213-6581

Pászto, V. ; Tuček, P.; Marek, L.; Kuprová, L.; Burian, J. (2010) Statistical inferences - visualization possibilities and fuzzy approach computing, In: *Advances in Geoinformation Technologies 2010*, Horák, J., pp. (163-178), ISBN 978-80-248-2357-7, VŠB – Technical University of Ostrava, Ostrava, Czech Republic

Pechanec, V. (2006). *Nástroje pro podporu rozhodování v GIS*, Palacký University in Olomouc, ISBN 80-244-1553-4, Olomouc, Czech Republic

Pechanec, V.; Burian, J.; Kilianová, H.; Šťávová, Z.(2009). Analysis and prediction of flood hazards in urban planning, *Proceedings of Cartography and Geoinformatics for Early Warning and Emergency Management: Towards Better Solutions*, pp. 493-500, ISBN 978-80-210-4796-9, Prague, Czech Republic, January, 2009

Pechanec, V.; Burian, J.; Kilianová, H.; Němcová, Z. (2011a). Geospatial analysis of the spatial conflicts of flood hazard, *Moravian Geographical Reports*. Vol. 19, No. 1, (March 2011), pp. (11-19), ISSN 1210-8812

Pechanec, V.; Dobešová, Z.; Burian, J. (2011b). Neural networks and cellular automata in modelling land use changes, *Neural Network World*, Vol. 21, No. 4, (in print), pp. (1-27), ISSN 1210-0552

Ptáček, P. ; Szczyrba, Z. ; Fňukal, M. (2007). Proměny prostorové struktury města Olomouce s důrazem na rezidenční funkce, *Urbanismus a územní rozvoj*, Vol. 10, No. 2, (March 2007), pp. (19-26), ISSN 1212-0855

Ptáček, P. & Szczyrba, Z. (2004). Olomouc – profil města s identifikací problémů spojených se suburbanizací. In: *Krize měst z neregulovaného růstu*, Gremlica, T., pp. 23, Ústav pro ekopolitiku, Prague, Czech Republic

Pulselli R. M. & Ratti C. (2005). *Mobile Landscapes*, Equilbri , Il Mulino, Italy

Saaty, Th. L. (1983). Priority Setting in Complex Problems, *Proceedings of the Fifth International Conference on Multiple Criteria Decision Making*, pp. 326 – 336, Berlin/Heidelberg/New York, 1983

Schaal, P. (2004). Landscape planning and GIS-based data management. Proceedings of ESRI-International UC, pp. 8, San Diego, USA, 2004

Schaller, J. (2007). ArcGIS – ModelBuilder Applications for Regional and Development Planning in the Region of Munich (Bavaria), *Proceedings of 16. konference GIS ESRI a Leica Geosystems v ČR*, pp. 1, ISBN 978-80-254-0299-3, Praha, Czech Republic, January, 2007

Stuchlíková, M. (2009). Hradecko-pardubická aglomerace – socioekonomické podmínky, *Proceedings of seminar AUÚP*, pp. 58-62, ISBN 978-80-87318-03-4, Brno, Czech Republic, May 2009

Sýkora, L. (2003). Suburbanizace a její společenské důsledky, *Sociologický časopis*, Praha. Vol. 39, No. 2, (2003), pp. (217-233), ISSN 0038-0288

Vorel, J. (2006). *Informace o urbánním prostředí, jejich konceptualizace a komunikace*, Fakulta architektury ČVUT v Praze, Praha, Czech Republic

Voženílek, V. (2005). *Cartography for GIS – geovisualization and map communication*, Palacký University in Olomouc, ISBN 80-244-1047-8, Olomouc, Czech Republic

Yeh, A. G. & Li, X. (2002). A cellular automata model to simulate development density for urban planning, *Environment and Planning B: Planning and Design*, Vol. 29, No. 1, (2002), pp. (431-450), ISSN ISSN 0265-8135

Zwick, P. & Carr, M. (2007). *Smart Land-Use Analysis, The LUCIS Model*, ESRI Press, ISBN 978-1-58948-174-9 , Redlands, USA

9

Analyzing Neighbourhoods Suitable for Urban Renewal Programs with Autocorrelation Techniques

Beniamino Murgante[1], Maria Danese[2] and Giuseppe Las Casas[1]
[1]University of Basilicata
[2]Archaeological and monumental heritage institute, National Research Council
Italy

1. Introduction

Since the Industrial Revolution the main model of urban development has been based on the concept of urban expansion, where new parts are added to existing towns in order to satisfy the housing demand. While in pre-industrial cities, main activities influence a portion of space within or immediately next to the border of urban settlements, the rapid growth of cities in the industrial era represents the transition from an almost sustainable city to a city that takes advantage of the carrying capacity of neighbouring regions (Stren et al., 1992).

In order to maximize production efficiency in the new industrial era, new plans decomposed the city into distinct areas for different functions creating the principle of functional zoning.

The idea of city disappeared behind the imperatives of moving, residing, trading, producing, etc. (Alberti et al., 1994). The great pressure of housing demand has led to focus the attention on dwelling realization. But a city is not a collection of houses, it is home, a community where it is possible to realize urban functions (Salsano, 1998).

This approach generated cities with a low level of urban quality, primarily trying to provide a place to live to the great part of population migrated from rural areas to the city.

Neighbourhoods built in this period are characterised by environment deterioration, lack of open spaces, low availability of car parking and green areas, deficiency of street furniture.

Awareness of the inadequacy of traditional planning instruments marks the transition from an approach based on urban expansion to a phase based on urban transformation concept. The first and fundamental reason for this choice is the need to improve quality of neighbourhoods. Cities are encouraged to improve the built up, degraded or underused parts, rather than to expand. The second reason is to avoid new soil consumption, moreover it would be unimaginable to abandon degraded urban areas for building new neighbourhoods with a better quality.

It is also important to pay attention to heavily populated areas with a high number of unemployed and a low rate of school attendance, where risks of social conflicts are elevated, especially today, since great part of immigration from other countries is concentrated in these areas.

In order to find a solution to such problems a new type of planning instruments has been adopted: urban regeneration programs. Such programs, also called urban renewal in some countries, are based on the idea that the space where we will live in the next future is already built (Secchi, 1984).

Urban regeneration programs have contributed to encourage the transition from the concept of urban planning focused on town expansion to the qualitative transformation of the existing city in a decisive way. Targets of urban renewal programs can be summarized as increasing the quality of housing and open space, policy of social cohesion and balance, paying attention to migration phenomena and improvements of area reputation (Kleinhans, 2004).

Various experiences of urban regeneration policy from all over the world differ in contents and forms. Contents can be synthesized as functional integration, social mixing, increase of environmental quality, improvement of infrastructure systems. Several experiences have been implemented with community involvement, other forms considered the possibility to share a programme among many towns creating a form of city network to reach better results.

In order to revitalize degraded parts of towns, private sector in several cases is interested in investing money. Generally, it is rather easy to find a private investor interested in transforming a totally abandoned area, completely changing its possible uses (i.e. an old industry or station can be transformed in a trading centre with multi-screen cinema). In fact, such areas are very accessible, fairly central and can provide important services to the whole town. On the contrary, it is not that easy to find a private investor attracted by a degraded residential area to revitalize, without completely changing its possible uses. In the last case, it is important to reach an agreement between private investors and local authorities in order to use public and private funds.

Also in areas where private investors can have great interests, a public superintendence it is important, because the market cannot be the only possible solution in planning, real estate and environmental fields.

Public funding can be important even in areas where private companies have huge interests in order to guarantee housing policy, social and functional mixing. In this case, public intervention should not be considered as a dirigiste interference, but as an action for ensuring urban quality; moreover, in this period we are learning by the last global economic crisis that profit, market and liberalism are not the only possible solution.

An important goal of public action in city management is to balance imperfections and inefficiencies related to the market. The role of public can be fundamental in contrasting the rent produced by market failure, restoring suitable conditions of allocative efficiency and ensuring equitable treatment to private property.

The economic difficulties of local authorities and the increasing complexity of public investments require the involvement of private funding in services production and management. More recent forms of urban renewal programs concern negotiation of new relationships between public administration and private sector, using financial resources of private companies, pursuing the need to integrate public investments. There is an evolution from a public administration as a building contractor to a central government stimulating competition among local authorities.

The transition from indiscriminate financing to a competition among cities is not only in gaining new resources, but also in a wider strategy for a possible "urban success" based on uncommon and strategic services.

The traditional form of direct financing from public administration has been replaced or integrated by other forms of funding, such as project financing, tax incentives for investors, sponsorships, private investments with public guarantees to banks, complete private financing.

The main change is rooted on the shift from an approach based on a huge amount of public funding to a urban renewal through the market mechanism, reinserting areas into real estate market (Bonneville, 2005).

Methodological and operational innovation introduced in these experiences have encouraged and helped to develop attitude to integration and competition in local authorities. A new assessment system has been introduced in allocating public funds, overcoming the concept of indiscriminate financing and applying reward for quality design, innovation and community needs. The urban regeneration program has introduced also recomposition of socio-economic and financial programming on one hand and land-use planning and urban design on the other.

The designation of Urban Regeneration Program areas generates profound debates.

Considering that private investment is crucial, it is important to analyze in a detailed way areas to be transformed.

Generally a municipality proposes an area as suitable for a urban regeneration program, considering the edge of neighbourhoods established by bureaucrats. Socio-economic analysis can account for a huge amount of data related to the whole neighbourhood. Nevertheless, sometimes it is possible that only a part of a neighbourhood is interested by social degradation phenomena, because it could have been designed considering a social mix.

In these cases the indicator is diluted and it does not capture the phenomenon throughout its importance. Furthermore, it is possible that the more deteriorated area belongs to two different adjacent neighbourhoods: in this case the municipality will consider an area completely included in a single neighbourhood, even though it may show lesser problems.

Spatial statistics techniques can provide a huge support when choosing areas with high intervention priorities.

These methods allow more accurate analysis considering social data at a building scale. In this way areas will be determined considering the spatialization of very detailed data, and neighbourhood limits will be overcome.

Such an approach has been tested in two municipalities both located in Apulia (southern Italy).

Bari is a dynamic trading centre with important industrial activities and high immigration fluxes from Albania and north Africa.

Taranto was one of the main centres of Magna Grecia and has an important trading port, second in Italy for freight traffic.

2. An overview of spatial statistics techniques

The main aim of spatial analysis is a better understanding of spatial phenomena aggregations and their spatial relationships. Spatial statistical analyses are techniques using statistical methods in order to determine if data show the same behaviour of the statistical model.

Data are treated as random variables. The *events* are spatial occurrences of the considered phenomenon, while *points* are each other arbitrary locations. Each event has a set of attributes describing the nature of the event itself.

Intensity and *weight* are the most important attributes; the first one is a measure identifying event strength, the second is defined by the analyst who assigns a parameter in order to define if an event is more or less important according to some criteria.

Spatial statistics techniques can be grouped in three main categories: *Point Pattern Analysis, Spatially Continuous Data Analysis* and *Area Data Analysis* (Bailey and Gatrell, 1995).

The first group considers the distribution of point data in the space. They follow three different criteria:

- random distribution: the position of each point is independent on the others points;
- regular distribution: points have a uniform spatial distribution;
- clustered distribution: points are concentrated in clusters.

The second group takes into account spatial location and attributes associated to points, which represent discrete measures of a continuous phenomenon.

The third group analyzes aggregated data on the basis of Waldo Tobler's (1970) first law of geography: "Everything is related to everything else, but near things are more related than distant things".

These analyses aim to identify both relationships among variables and *spatial autocorrelation.* If some clusters are found in some regions and a positive spatial autocorrelation is verified during the analysis, it can describe an attraction among points. Negative spatial autocorrelation occurs when deep differences exist in their properties, despite closeness among events (Diggle, 1983; Boots and Getis, 1988; Brunsdon, 195).

It is impossible to define clusters of the same property in some areas, because a sort of repulsion occurs. Null autocorrelation arises when no effects are surveyed in locations and properties.

It can be defined as the case in which events have a random distribution over the study area (O'Sullivan and Unwin, 2002). Essentially, autocorrelation concept is complementary to independence: events of a distribution can be independent if any kind of spatial relationship exists among them.

Spatial distribution can be affected by two factors:

- first order effect, when it depends on the number of events located in one region;
- second order effect, when it depends on the interaction among events.

2.1 Kernel density estimation

Kernel density estimation (Fix and Hodges, 1951; Rosenblatt, 1956, Silverman, 1986) is a point pattern analysis technique, where input data are point themes and outputs are grids. While simple density computes the number of events included in a cell grid considering intensity as an attribute, *kernel density* takes into account a mobile three-dimensional surface which visits each point.

The output grid classifies the event L_i according to its distance from the point L, which is the centre of the ellipse generated from the intersection between the surface and the plane containing the events (Gatrell et al., 1996).

The *influence function* defines the influence of a point on its neighbourhood. The sum of the *influence functions* of each point can be calculated by means of the *density function*, defined by:

$$\lambda(L) = \sum_{i=1}^{n} k\left(\frac{L - L_i}{\tau}\right) \tag{1}$$

where:

- λ is the distribution intensity of points, measured in L;
- L_i is the event i;
- K is the kernel function;
- τ is the bandwidth.

The first factor influencing density values is bandwidth: if τ is too big, then λ value is closer to simple density; if τ is too small, then the surface does not capture the phenomenon. The second factor influencing density values is cell size as in every grid analysis.

2.2 Nearest neighbor distance

Nearest neighbour distance is a second order property of point pattern analysis and describes the event distribution measuring the distance. This technique analyzes the distance of a point from the nearest source. The events to analyze can be points or cells. The distance between the events is normally defined by the following function:

$$d\left(L_i, L_j\right) = \sqrt{\left(x_i - x_j\right)^2 + \left(y_i - y_j\right)^2} \tag{2}$$

If $d_{min}(L_i)$ is the nearest neighbour distance for an L_i event, it is possible to consider the mean nearest neighbour distance defined by Clark and Evans (1954) as:

$$\overline{d}_{min} = \frac{\sum_{i=1}^{n} d_{min}\left(L_i\right)}{n} \tag{3}$$

2.3 Moran index

Moran index (Moran, 1948) allows to transform a simple correlation into a spatial one. This index takes into account the number of events occurring in a certain zone and their intensity. It is a measure of the first order property and can be defined by the following equation:

$$I = \frac{N\Sigma_i\Sigma_j w_{ij}\left(X_i - \overline{X}\right)\left(X_j - \overline{X}\right)}{\left(\Sigma_i\Sigma_j w_{ij}\right)\left(X_i - \overline{X}\right)^2} \tag{4}$$

where:

- N is the number of events;
- X_i and X_j are intensity values in the points i and j (with i≠j), respectively;
- $\overline{X}$ is the average of variables;
- $N\Sigma_i\Sigma_j w_{ij}\left(X_i - \overline{X}\right)\left(X_j - \overline{X}\right)$ is the covariance multiplied by an element of the weight matrix. If X_i and X_j are both either higher or lower than the mean, this term will be positive, if the two terms are in opposite positions compared to the mean the product will be negative;
- w_{ij} is an element of the weight matrix which depends on the contiguity of events. This matrix is strictly connected to the adjacency matrix.

There are two methods to determine w_{ij}: the "Inverse Distance" and the "Fixed Distance Band". In the first method, weights vary according to inverse relation to the distance among events $W_{ij}=d^z_{ij}$ where z is a number smaller then 0.

The second method defines a critical distance beyond which two events will never be adjacent. If the areas to which i and j belong are contiguous, w_{ij} will be equal to 1, otherwise w_{ij} will be equal to 0. *Moran index* I ranges between -1 and 1. If the term is high, autocorrelation is positive, otherwise it is negative.

2.4 G function by Getis and Ord

The G function by Getis and Ord (1992), takes into account disaggregated measures of autocorrelation, considering the similitude or the difference of some zones. This index measures the number of events with homogenous features included within a distance d, located for each distribution event.

This distance represents the extension within which clusters are produced for particularly high or low intensity values. Getis and Ord's function is represented by the following equation:

$$G_i(d) = \frac{\sum_{i=1}^{n} w_i(d)\, x_i - \overset{-}{x_i} \sum_{i=1}^{n} w_i(d)}{S(i)\sqrt{\left[(N-1)\sum_{i=1}^{n} w_i(d) - \left(\sum_{i=1}^{n} w_i(d)\right)^2\right] \Big/ N-2}} \tag{5}$$

3. The case study

By the middle of nineteenth century, Great Britain was a highly urbanized country, more than thirty percent of the population lived in cities with the working masses concentrated around factories, located mostly in the biggest city; on the contrary in Italy this phenomenon occurred about fifty years later with a smaller dimension.

Compared to other European countries Italian Industrial Revolution can be considered the Giolittian Era, at the beginnings of the last century, characterized by a transition from agricultural to industrial economy. Consequently, the significant urbanization, occurred much earlier in other European countries, in Italy took place during this period and only in certain areas.

This phenomenon produced a growth rate of urbanized areas only in big towns, mainly concentrated in the northern part of the Country. The major part of migration phenomenon from rural zones to urban areas occurred after the second world war, a period also characterized by a high birth rate.

This too sudden growth coupled with not exceptional quality of plans has caused a significant lack of standards, not so much from a quantitative point of view as a lack of urban quality.

In order to verify the possibility to apply these techniques, Bari and Taranto municipalities have been chosen for this research since they show very different social contexts. Both towns are located close to the "heel" of Italy, which historically promoted its trading tradition.

Bari is one of the more developed areas of southern Italy for industrial and tertiary sectors. The location characterizes this area as a sort of gate for migratory fluxes. This municipality has 325.052 inhabitants distributed over 116,20 km².

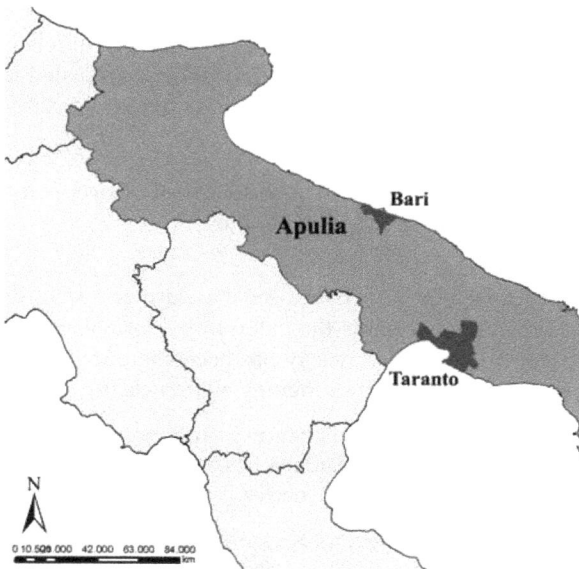

Fig. 1. Study area location

Taranto has 191.810 inhabitants distributed over 209,64 km² and is the second Italian trading port for freight traffic, mainly connected with Asia. At the same time Taranto has important industries in the fields of iron, steel and oil refinery. These activities have produced a lot of health and environmental problems. Taranto is one of the most polluted cities in Western Europe due to industrial emissions.

The localization of these activities during the '60s generated a great dwelling demand, complied with the construction of very intensive neighbourhoods. Such a disordered growth realized without a master-plan, produced urbanization in areas largely disconnected and without continuity. In order to have a more complete results municipalities surrounding Taranto have been considered in the application.

According to equation (1) all buildings have been represented as points, in order to apply Point Pattern Analysis techniques. The main difference with classical statistical approaches is the possibility to locate each event L_i in the space, by its coordinates (x_i, y_i), in an unambiguous way. An L_i event (equation 6) is a function of its position and attributes characterizing it and quantifying its intensity:

$$L_i = \left(x_i, y_i, A_1, A_2, ..., A_n \right) \tag{6}$$

The following attributes have been considered in order to calculate kernel density:

- dependency ratio is considered an indicator of economic and social significance. The numerator is composed of people, youth and elderly, who, because of age, cannot be considered economically independent and the denominator by the population older than 15 and younger than 64, who should provide for their livelihood. This index is important in urban regeneration programs because it can determine how many people can provide to building maintenance by themselves;
- foreign population per 100 residents. Normally foreign number is considered as capability of attractiveness, but in southern Italy, where concealed labour rate is 22,8% and unemployed rate is 20%, immigration phenomena can be considered a threat and not an opportunity;
- unemployment rate;
- percent of population which had never been to school or dropped out school without successfully completing primary school programs;
- number of people per room in flats occupied by residents.

All attribute values of various indicators have been standardized according to the following equation: $Z=(X- \mu)/\sigma$, where X represents the value to be normalized, μ is the mean value and σ is the standard deviation. Kernel density has been computed for each attribute and a grid synthetic index has been achieved by summing all kernels (fig.2).

A key factor of kernel density estimation is bandwidth dimension. In order to determine a suitable bandwidth, nearest neighbour distance has been applied. In this case bandwidth dimension is 128 metres, while cell size is 10 metres.

There are several kinds of kernel functions K (see equation 1) Gaussian, Triangular (Burt and Barber, 1996), Quartic and Epanechnikov (1969); in this case the Epanechnicov's kernel has been applied.

Figure 2 highlights two important issues. Highest values (black areas) represent the concentration of several negative social indicators. As supposed in the introduction, this kind of measure has all the limits of the edge of urban regeneration areas based on neighbourhood boundary.

The two details on the right part of figure 2 show as areas with high values are located on both parts of neighbourhood boundary and zones which need more urgent interventions are situated between the white line inside the yellow oval.

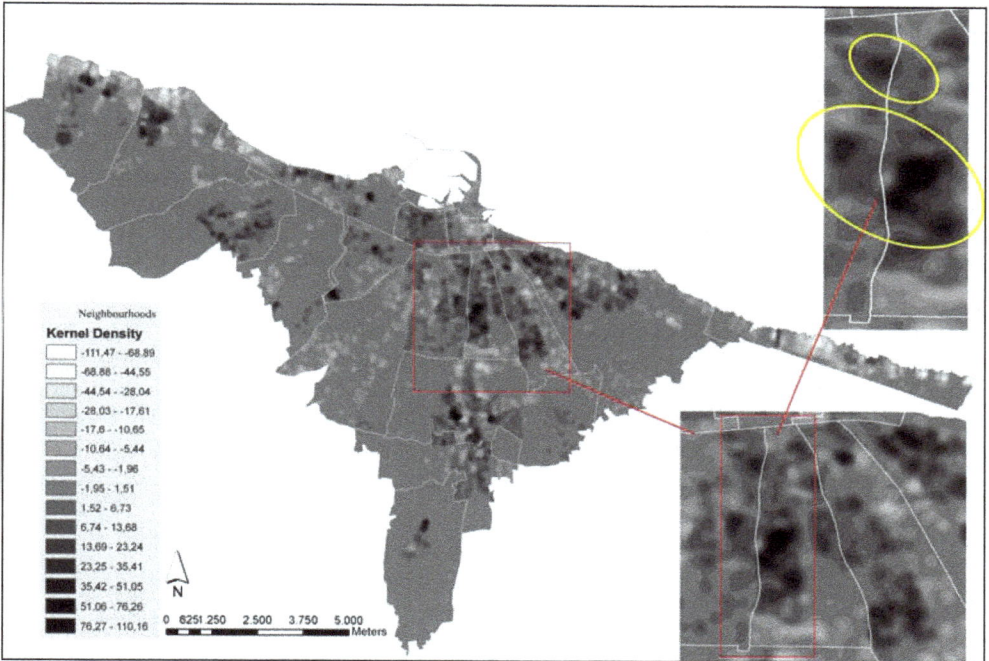

Fig. 2. Synthetic index of Kernel Density Estimation of social indicator

Moran index is a global measure of spatial autocorrelation which analyzes if and at which degree events are spatially correlated. This index does not give any information about event location. Considering that attributes are connected to buildings, rural areas can produce a sort of mitigation effect. Also, Moran index can be considered as a measure highlighting more concentrated problems determining policy priorities, in the case of a low global autocorrelation degree.

Table 1 shows that education, with very high Moran index in Bari and elevated value in Taranto, could be a great priority. In urban renewal programs a more diffused school network should be considered. It is impossible to know in which neighbourhood it is better to build a new school only using Moran index. In order to determine the exact location of these problems, further analyses are needed. It is important to adopt local autocorrelation measures. Getis and Ord's function is a suitable index to determine where several phenomena are concentrated and consequently where policies should be applied.

Indicator	Moran's I (Bari)	Distance band (Bari)	Moran's I (Taranto)	Distance band (Taranto)
Dependency ratio	0,17	40	0,39	40
Unemployment rate	0,28	30	0.39	50
Foreign population per 100 residents	0,12	20	0.57	40
Population which had never been to school or dropped out school without successfully completing primary school program	0,71	20	0.51	40
Number of people per room in flats occupied by residents	0,04	50	0.29	60

Table 1. Moran Index with related distance band for each indicator

Figure 3 shows where high autocorrelation of people with low educational level is located. It is important to pay attention to the class with very low autocorrelation in the external part of the towns. A low autocorrelation of people with a low educational level can be considered as a medium level of autocorrelation of people with a good educational level. This class is more related with urban sprawl phenomena.

Considering two or more variables at the same time it is possible to achieve other interpretations. Foreigners and unemployed (see figure 4) are not connected with urban sprawl phenomena and these variables are not related. As further remarks, areas with the strongest autocorrelation of immigration are far from zones with high concentration of unemployed. This means that no spatial correlation exists between immigration and unemployment.

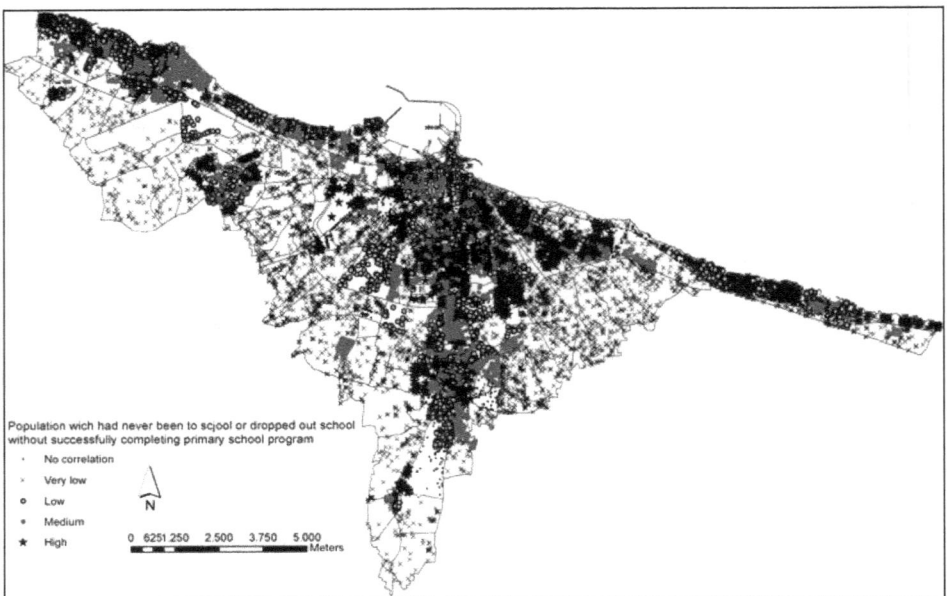

Fig. 3. Spatial autocorrelation of primary school graduation rates

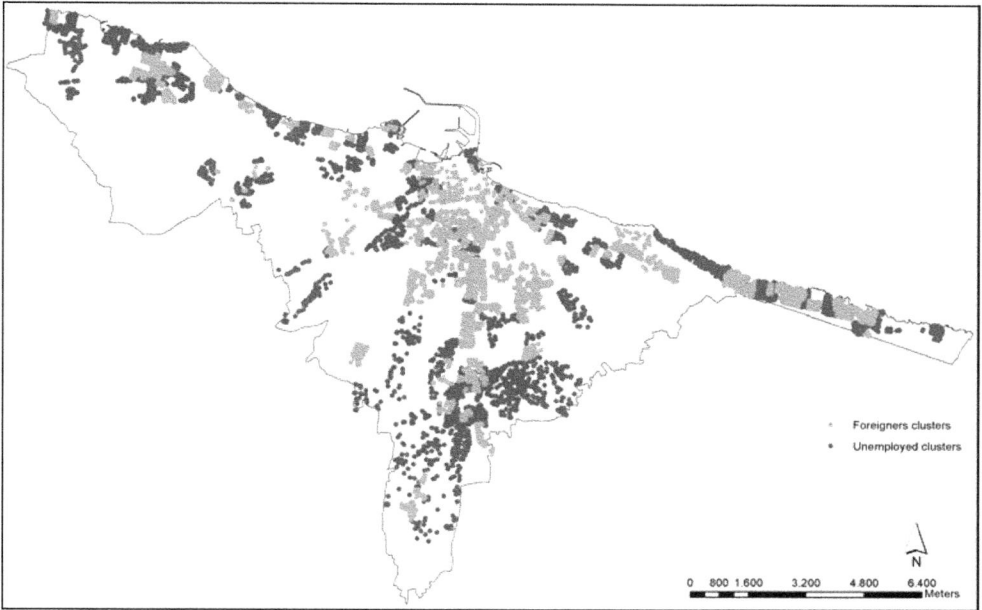

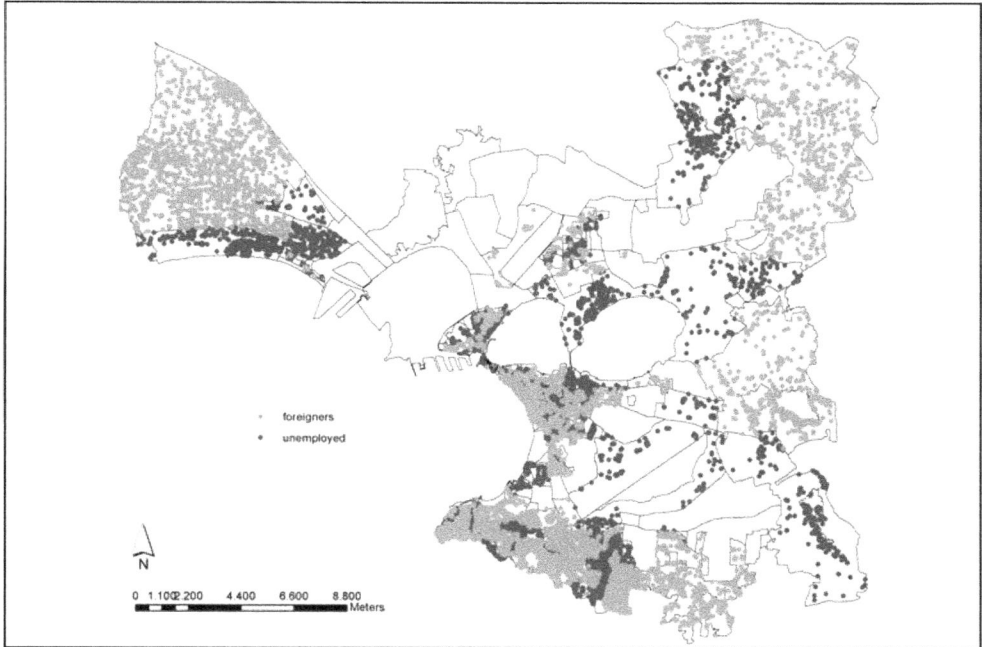

Fig. 4. Comparison of medium and high values of spatial autocorrelation of unemployment rate and foreign population per 100 residents

4. Conclusions

In order to pursue sustainability, urban renewal programmes can play a fundamental role. According to Giaoutzi and Nijkamp (1995) the concept of sustainability can be considered in a wider meaning, adding social and economical dimension to environmental aspects (Goodland, 1994; Capello et al., 1998; Murgante et al., 2011). For instance the concept of compact city can be considered sustainable under three points of view:

1. **environmental**: urban sprawl is one of the hugest environmental threats;
2. **social**: urban sprawl obliges people to travel many hours per day, leading to a total absence of social and neighbourhood relationships;
3. **economical**: compact cities produce agglomeration advantage in localizing services and activities and in realizing interventions and infrastructures (Jacobs, 1969; Nijkamp and Perrels, 1994).

If, on one side, urban sprawl represents a sort of antithesis of the city, on the other side the phenomenon has not deprived the city of urban functions (Nijkamp 2011). If urban sprawl enlarges the space where locating residential functions, the old town represents the barycentric and dense place where urban externalities have been produced for social and economical activities.

While computer technologies mainly oriented to graphics, multimedia, and representation have been adopted in supporting the planning process, quantitative methods and more particularly spatial statistics have not been completely used in urban analysis, despite one of the fathers of town planning, Patrick Geddes, defined an important sequence for planning: survey – analysis – plan (Geddes 1904, quoted in Faludi 1987).

One of the frequent accusations which is addressed to policy makers is to fail in understanding the complexity of the city and the real needs of degraded areas (Hull, 2001).

Analyzing urban renewal policies and programs adopted, there is a shift from the first experiences based on emphasizing built environment to some experiences based on citizen involvement in program choices (Carmon, 1999).

This increase in urban renewal quality lacks in defining areas needing more urgent interventions and policies fit to improve neighborhood conditions.

A great support can be represented by the use of spatial autocorrelation methods. These techniques have been adopted in many fields, from epidemics localization (Gatrell et al., 1996), to analysis of damage scenarios (Danese et al., 2008; Danese et al., 2009), to studies on spreading of city services (Borruso and Schoier, 2004; Borruso, 2003; Borruso and Porceddu, 2009; Porta et al. 2009; Produit et al. 2010), to assessment of Fire severity (Lanorte et al., 2011), to physical planning processes (Murgante et al., 2007; Murgante and Danese, 2011).

Such methods can determine areas with high priority of intervention in a more detailed way, increasing efficiency and effectiveness of investments in urban regeneration programs.

The experience explained in this paper is based on simple census data at the urban level and the lack of more specific data limits the number of analyses. In order to develop more interesting considerations, it would be opportune to integrate these indicators during the preliminary phase of urban renewal programs.

5. References

Alberti, M., Solera, G., & Tsetsi, V. (1994). La città sostenibile. Milano, Italy: FrancoAngeli.

Bailey T. C., Gatrell A. C. (1995). Interactive spatial data analysis. Prentice Hall.

Bonneville M. (2005). The ambiguity of urban renewal in France: Between continuity and rupture. Journal of Housing and the Built Environment (2005) 20: 229–242. Springer.

Boots B.N., Getis, A. (1988). Point Pattern Analysis. Sage Publications, Newbury Park.

Borruso G (2003) Network density and the delimitation of urban areas. Transactions in GIS 7:177–191.

Borruso G., Schoier G. (2004). Density Analysis on Large Geographical Databases. Search for an Index of Centrality of Services at Urban Scale. In Gavrilova M. L., Gervasi O., Kumar V., Laganà A., Mun Y. and Tan K. J. (ed), Computational Science and its Applications, Lecture Note in Computer Science LNCS 3044 - Springer-Verlag, Berlin.

Borruso G., Porceddu A (2009). A tale of two cities: density analysis of CBD on two midsize urban areas in northeastern Italy. Geocomputation & Urban Planning 176:37–56.

Brunsdon, C. (1995). Estimating Probability Surfaces for Geographical Points Data: An Adaptive Kernel Algorithm. Computers and Geosciences 21(7), 877–894.

Burt, J.E., Barber., G.M. (1996). Elementary Statistics for Geographers, 2nd edn. The Guilford Press, New York.

Capello, R., Nijkamp, P., and Pepping, G. (1998). Sustainable Cities and Energy Policy, Springer-Verlag, Berlin.

Carmon N., (1999). Three generations of urban renewal policies: analysis and policy implications. Geoforum 30 145-158. Elsevier Science.

Clark, P.J., Evans, F.C. (1994). Distance to Nearest Neighbour as a Measure of Spatial Relationships in Populations. Ecology 35, 445–453.

Danese, M., Lazzari, M., Murgante B. (2008). Kernel Density Estimation Methods for a Geostatistical Approach in Seismic Risk Analysis: the Case Study of Potenza Hilltop Town (southern Italy). Lecture Notes in Computer Science vol. 5072, pp. 415-427. Springer-Verlag, Berlin. ISSN: 0302-9743 doi: 10.1007/978-3-540-69839-5_31.

Danese M., Lazzari M., Murgante B. (2009). "Geostatistics in Historical Macroseismic Data Analysis" Transactions on Computational Sciences VI, LNCS Vol. 5730, pp. 324-341, Springer-Verlag, Berlin ISSN: 1611-3349, doi:10.1007/978-3-642-10649-1_19.

Diggle, P.J. (1983). The Statistical Analysis of Point Patterns. Academic Press, London.

Epanechnikov, V.A. (1969). Nonparametric Estimation of a Multivariate Probability Density. Theory of Probability and Its Applications 14, 153–158.

Faludi, A. (1987): A Decision-Centred View of Environmental Planning. Oxford: Pergamon.

Fix, E., Hodges, J.L. (1951). Discriminatory Analysis. Nonparametric Discrimination: Consistency Properties, reprinted by Silverman, B.W. and Jones, M.C. (1989) An Important Contribution to Nonparametric Discriminant Analysis and Density Estimation: Commentary on Fix and Hodges (1951). International Statistical Review 57: 233–247.

Gatrell A. C., Bailey T. C., Diggle P. J., Rowlingson B. S., (1996). Spatial point pattern analysis and its application in geographical epidemiology. Transaction of institute of British Geographer, NS 21 256–274 1996. Royal Geographical Society.

Getis, A. and Ord, J. K., (1992). The analysis of spatial association by use of distance statistics. Geo-graphical Analysis, 24, 189-206.

Giaoutzi, M., and Nijkamp, P. (1995). Decision Support Systems for Sustainable Regional Development, Avebury, Aldershot, UK.

Goodland, R. (1994). Environmental Sustainability and the Power Sector, Impact Assessment, vol. 12, pp. 275-299.

Hull A., (2001). Neighbourhood renewal: A toolkit for regeneration. GeoJournal 51: 301–310. Kluwer Academic Publishers.

Jacobs, J. (1969). The Economy of Cities, Penguin, London.

Kleinhans, R.J. (2004). Social implications of housing diversification in urban renewal: A review of recent literature. Journal of Housing and the Built Environment 19: 367–390, 2004. Kluwer Academic Publishers.

Lanorte, A., Danese, M., Lasaponara, R., Murgante, B. (2011). Multiscale mapping of burn area and severity using multisensor satellite data and spatial autocorrelation analysis. International Journal of Applied Earth Observation and Geoinformation, Elsevier, doi:10.1016/j.jag.2011.09.005.

Moran, P. (1948). The interpretation of statistical maps. Journal of the Royal Statistical Society, n.10.

Murgante B. Danese M. (2011). "Urban versus Rural: the decrease of agricultural areas and the development of urban zones analyzed with spatial statistics" Special Issue on "Environmental and agricultural data processing for water and territory management" International Journal of Agricultural and Environmental Information Systems (IJAEIS) volume 2(2) pp. 16–28 IGI Global, ISSN 1947-3192, DOI: 10.4018/jaeis.2011070102.

Murgante B., Borruso G., Lapucci A. (2011) "Sustainable Development: concepts and methods for its application in urban and environmental planning" In Murgante B., Borruso G., Lapucci A. (eds.) "Geocomputation, Sustainability and Environmental Planning" Studies in Computational Intelligence, Vol. 348. Springer-Verlag, Berlin DOI: 10.1007/978-3-642-19733-8_1.

Murgante B., (2005). Le vicende urbanistiche di Potenza. EditricErmes, Potenza.

Murgante B., Las Casas G., Danese M., (2007), The periurban city: geostatistical methods for its definition. In Rumor M., Coors V., Fendel E. M., Zlatanova S. (Ed), Urban and Regional Data Management, Taylor and Francis, London.

Nijkamp, P., and Perrels, A. (1994). Sustainable Cities in Europe, Earthscan, London.

Nijkamp, P., (2011). The role of evaluation in supporting a human sustainable development: a cosmonomic perspective, Regional Science Inquiry Journal, Vol. III, (1), 2011, pp13-22.

O'Sullivan D., Unwin D., (2002). Geographic Information Analysis. John Wiley & Sons.

Porta S, Strano E, Iacoviello V, Messora R, Latora V, Cardillo A, Wang F, Scellato S (2009) Street centrality and densities of retail and services in Bologna. Environment and Planning B: Planning and Design 36:450–466.

Produit T. ,Lachance-Bernard N., Strano E., Porta S. and Joost S. (2010). A Network BasedKernel Density Estimator Applied to Barcelona Economic Activities, Lecture Notes in Computer Science LNCS vol. 6016, pp. 32 – 45. Springer-Verlag, Berlin.

Rosenblatt, M. (1956). Remarks on Some Nonparametric Estimates of a Density Function. Ann Math Statist 27, 832–837.

Salsano, E. (1998). Fondamenti di urbanistica. Bari, Italy: Laterza.

Secchi B. (1984). Le condizioni sono cambiate, Casabella: Architettura come modificazione. n.498/9. Electa Periodici.

Silverman, B.W. (1986). Density Estimation for Statistics and Data Analysis, Chapman and Hall, London.

Stren, R., White, R. and Whitney, J.(editors) (1992). Sustainable Cities: Urbanization and the Environment in International Perspective. Westview Press, Boulder, Colorado and Oxford.

Tobler, W. R., (1970). A Computer Model Simulating Urban Growth in the Detroit Region. Economic Geography, 46: 234-240.

10

Statistical Analysis of Environmental Quality Indices in an Urban Street Network

Androniki Tsouchlaraki, Georgios Achilleos and Vasiliki Mantadaki
Technical University of Crete
Greece

1. Introduction

Nowadays, 80% of Europeans live in urban areas, facing a continuous degrading of the quality of the urban environment. The development of economy has reached high levels during the last decades, followed by an increase in the mobility of people around the city, the excessive use of passenger cars, and the concession of a substantial part of the city's vital space to the development of the public transportation network (Vlastos & Siolas 1994; Pitsiava, 1991; Raccioppi et al., 2002; Siddhartha, 1999; Dekoster & Schollaert, 1999).

The consequences of this policy and of the citizens' attitude are apparent, especially in the central city areas, which constitute a focal point of the transportation network, because of the density of the population in them and the diversity of their functions. Furthermore, it is observed that increasingly frequent occurrences such as conditions of traffic congestion, parking problems, accidents, noise, environmental pollution, functions, and visual obstructions of the central arterial routes, lower the citizens' quality of life and consist the major causes of a tendency to decentralize, thus creating the need for further expansion (Aravantinos, 1999; Masters, 1991; Vlastos & Polyzos, 1999; Porta & Latora 2007; Ferrarini et al., 2001).

It is therefore considered essential to design a geodatabase in a GIS, where environmental quality indices are developed. This geodatabase enables us to provide a specific and detailed description and analysis of the environmental indices of the urban roads. It can also become a convenient tool in the hands of specialists and of representatives of the state authorities, by facilitating their decision making on a number of issues concerning measures and interventions in the urban web. At the same time it will enable continuous observation and updating of the urban environment. Local communities which lack such a tool can attempt only occasional interventions which are difficult to evaluate as far as their effectiveness is concerned (Tsouchlaraki et al., 2009; Tsouchlaraki & Zoaki, 2008; Yildiz, 2008).

Such a geodatabase though, is useless if the data that are kept within it, are uncertain and do not contribute significantly towards the research of the environmental quality of urban streets. It is therefore considered critical to statistically analyse these data.

This chapter discusses the effort to research various indices and to statistically analyse the descriptive part of the geodatabase. Using methods of descriptive, inferential and multivariate statistics, it is attempted to examine the indices, to research their correlations, to define their degree of importance, as well as to evaluate and to classify the environmental quality of the urban streets.

2. The research

2.1 The research outline

In this context, the Department of Environmental Engineering of the Technical University of Crete attempted to build a methodology for the assessment of the environmental quality of urban streets. The aim was to create a useful tool to assist in decision making regarding measures and action within the urban web. Towards this direction, a pure research project, titled "Development of indices for recording the environmental quality in cities' streets using GIS – GPS" developed 124 qualitative and quantitative indices describing the environmental quality of the urban street web (Abdulaal, 2007; Aisopos, 2003; Aravantinos, 1997; Beckman, 2001; De Jong, 2008; Gehl, 2002; Kartalis, 1999; Lymberopoulos, 2003; Mendes, 2005; Nicol & Wong, 2005; Nikolopoulou, 2004; Tsouchlaraki & Zoaki, 2008; Yannas, 2001; Zivas, 2003).

These indices have been categorized into eight definable groups: Urban Planning and Architecture, Construction Materials, Road Infrastructure, Traffic, Real Estate Use, Pollution, Climate, Other. The indices within these groups are further categorized into 20 sub-categories (Tables 1, 2).

a.	Urban Planning And Architectural	1.	Urban planning indices
		2.	Architectural indices
		3.	Road geometry
b.	Construction Materials	4.	Road surface materials
		5.	Paving materials
c.	Road Equipment	6.	Street equipment
		7.	Street facilities
d.	Road Traffic	8.	Static road traffic indices
		9.	Dynamic road traffic indices
e.	Land Uses	10.	Commercial uses
		11.	Services
		12.	Communal spaces
		13.	Other uses
f.	Pollution	14.	Air pollution indices
		15.	Noise pollution indices
		16.	Visual pollution indices
g.	Climatic	17.	Climatic indices
h.	Other	18.	Economic indices
		19.	Hygiene indices
		20.	Other indices

Table 1. Categories and Sub-categories of the Selected Indices (left)

The Municipality of Chania is the largest in population in the Prefecture of Chania, with 53373 inhabitants, (2001 population census) and has an expanse of approximately 7000 hectares. This one-division municipality borders with the municipalities of Nea Kydonia,

Akrotiri, Souda, and Eleftheriou Venizelou. Figure 1 (Figure 1) is a map which depicts the urban complex of Chania.

The municipality is divided into five urban units which diversify in terms of their building permit, building density, building system, height of buildings, their distance from the city centre and the different uses of land. There are further diversifications as to the morphology of the ground, and the geometrical, functional, financial and traffic characteristics of the roads.In order to record the indices which contribute to the environmental quality of roads, especially those of the urban complex of the municipality of Chania, a part of the city of approximately 500 hectares is selected, including parts from sections I, II, III, IV (Figure 1). Of these, some are part of the city centre (sec. I, II), others border the city centre (sec. III), while sec. IV represents a more remote part of the city.

a.   Urban Planning And Architectural Indices
1.   Urban Planning Indices
Building Permit Limit
Plot Coverage Percentage
Building System
Entrance Gardens
Length Of Building Blocks
Maximum Permissible Height Of Buildings
Existing Minimum Floor Number
Existing Maximum Floor Number
Surface Area Of Building Blocks
Arcades (In Parallel Or Across Streets)
2.   Architectural Indices
Level Of Enclosure
Number Of Listed Buildings
Hypsometric Distribution Of Buildings
First Year Of Building Construction
Newest Year Of Building Construction
Average Year Of Construction
Subterranean Domiciles / Semi-Subterranean Domiciles
Building Shape
Building Color
Façade Details
3.   Road Geometry
Pavement Width
Road Width
Road Orientation
Road Gradient
Multi-Level Footpath

Table 2. First Category (URBAN PLANNING AND ARCHITECTURAL INDICES) and its Three sub-categories (right)

The task of conducting research on the entire municipality of Chania was not feasible given the budget and time limits of this particular research project. It was additionally recommendable to conduct a pilot scheme, to make observations, to reach certain conclusions and based on them, to reorganise the investigation and improve it where it is necessary, and if feasible, to re-conduct the research on a larger part of the city, or throughout the city as well.

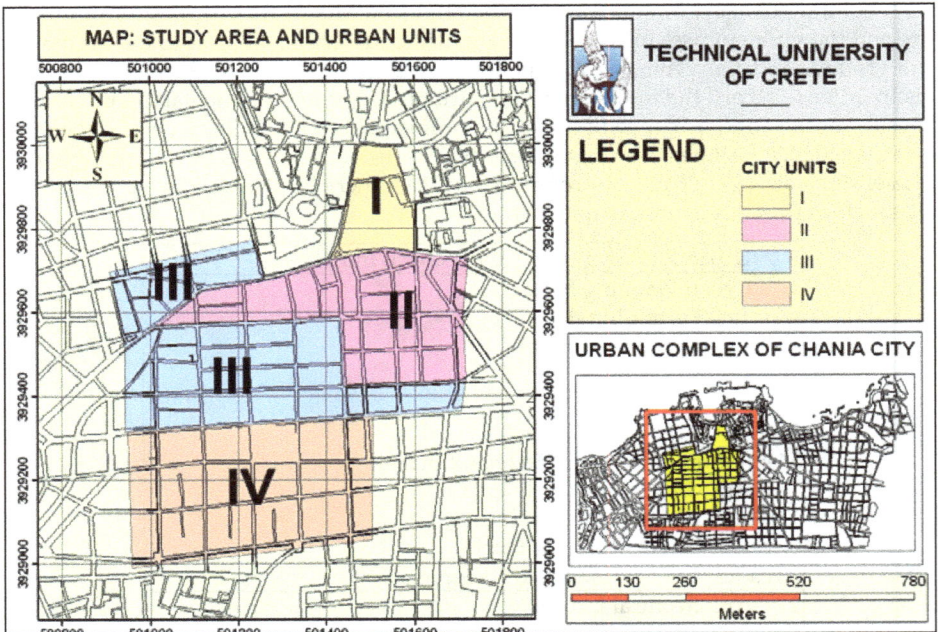

Fig. 1. Urban Complex of Chania, Research area and the four of the five Urban Units (right)

2.2 Data collection

The collection of data and the measurement of the values of indices of the selected area were conducted through public bodies, while the collection was based on already existing research and mainly on site recording (Tsouchlaraki et al., 2009). At the same time analytical photographical shootings were taken for every junction, facing all directions.

The collection of most indices was carried out via on site observation. Such indices were the archives, the existence of underground basements, semi underground / lower ground residences, informative signs, the existence of water draining grids, roadside handrails for pedestrians and so on. Other indices were digits of numbers, for example the number of preserved buildings in a street, the minimum and maximum number of floors of a building, the number of bus stops, etc. As for other indices such as the width of a street, they are estimated by providing a range of fluctuation, namely between 1 and 1.5 meters. Other indices define the material that was used either on the road surface or for layering the pavements. There were also indices that took the values of satisfactory, average and poor, for example when characterizing the appearance of a building.

Some of the indices were obtained through several governmental bodies. The building permit limit, the plot coverage percentage and the building system for example were provided by the Urban Planning Office of Chania. Noise levels, were obtained by the department of traffic planning of the municipality, while the commerciality index was obtained by the Tax and Revenue Office of Chania.

Apart from acquiring indices with the aforementioned methods – on site observation or through governmental bodies – other indices were obtained through a digital map of a software program. This digital map contained the urban streets layer and the building blocks layer and it was managed through the AutoCAD software which was very flexible to to the measurements. Such indices were the square surface of building blocks, the width of streets, the orientation of streets and the length of Building Blocks (B.B).

Measurements of some indices were practically impossible either because the access to some of the files of certain public offices was impaired (i.e. unknown date of building construction) or because there was a lack of relevant research, which would lead us to insufficient data.

The indices with no data were not taken into consideration in the evaluation of the streets. Those indices were the oldest, the most recent and the average year of building construction, those included in the category of dynamic traffic indices, and the ones which represent the concentration of COX , NOX, SO2, HC, of Ozone, of Suspended Particles and Dust. In addition, the indices of the average annual temperature, of wind speed, rainfall and absolute humidity, the indices of the value of land and buildings and of the population of building blocks as well.

In addition to the data collected in the ways described above, a survey was conducted on 192 inhabitants from all city areas. The answers were given either in writing or via interviews. The aim of the survey was to record the public's opinion on issues concerning the convenience, aesthetics and attractiveness of the roads, the traffic, the public transportation, pedestrian safety, drivers' safety, etc (Aravantinos, 1997). One of the basic questions that were included in the questionnaire was whether there was an index that had not been included and it should be taken into consideration. According to the outcomes, no new indices resulted.

2.3 The geodatabase

A database plays a major role in a GIS while it directly affects its cost. A database is the foundation of the use of a GIS enabling its user to do programming, develop an application, analyze and derive secondary data, which will in turn help in decision making (Arctur & Zeiler, 2004; Koutsopoulos, 2002). The distance between two junctions was selected to be the spatial unit, as in some cases the entire length of a street is heterogeneous and too large to be examined as an entity.

The junctions of the streets were photographed for the entire study area (Figure 2). Through these photographs one can make observations on the area, cross-check data of the base, possibly collect further data of interest, thus having a combined overall visual perception of the data. In this way, the photographic file provides evidence, reference, explanation and most likely a mechanism of cross-checking the database.

Three spatial information sources were used:

- A topographic map of the research area of a scale 1:200 (which results from field measurements in combination with an existing background map of a large scale). This map contained two layers, one representing the geometry of the urban street network and one representing the geometry of the urban blocks.

- A map of the city of Chania (a section within the administrative boundaries of the Chania Municipality) of a scale 1:2.000. This map contained two layers of information, the geometry of the urban street network (not as accurate as the information in the first map) and the second one with the administrative boundaries.
- Axes and major intersections of the road network of the research area. This information was in digital form and it was digitized from the first map (the topographic map).

Fig. 2. Typical examples of photo shootings

This spatial information contained in these layers, was in shapefile format within the ArcGIS system in an ESRI geodatabase. The geometry of the urban street network data was stored as a linear layer while the geometry of the urban blocks data was stored as a polygon layer. The axes of the road network (divided into straight line sections, from junction to junction) were also stored as a linear layer and the intersections of the road network were stored as a point layer.

The geodatabase contained these layers which were the spatial information of the research. The axes of the road network, as it has already been mentioned, are the spatial units of the research. All the values of the indices that were gathered were the non spatial information of the research. This non spatial information refers directly to the spatial units of the research and thus, it does not need a complicated geodatabase structure to be stored. A simple relational geodatabase structure was used and all the non spatial data were connected to the spatial units through a unique spatial unit code. The aim of the research was to provide a significant group of indices, well defined, in order to define the environmental quality of the urban streets, therefore the geodatabase best structure finding was not one of the research goals. If the results of this research are to be expanded in the whole area of the city of Chania or if they are to be applied in an another large city, then the geodatabase structure is important and special attention should be paid in order to optimize its performance.

The values of indices vary. Others are quantitative (i.e. building permit limit, height of buildings, area of building blocks, coverage percentage etc), while others are qualitative as for example telephone booth existence (yes or no), pavement surface condition (satisfactory – medium - poor) (Abdullal, 2007; Tsouchlaraki & Zoaki, 2008; Bata & Obrsalova, 2009; Button, 2002).

2.4 Methodology of evaluating the environmental quality

The research on the evaluation of the environmental quality of the streets was based on two ways of approach:

- Evaluation on a theoretical level
- Evaluation on a practical level

On the first approach of a theoretical basis, the minimum and maximum theoretical sum of values was calculated for each and every of the eight categories of indices. The minimum theoretical sum of a category may be 0 in the case where all the indices included have zero value (which means that all recorded indices are "of poor environmental quality), whereas the maximum theoretical sum is calculated by adding up all the indices included in that category. In case where all indices are 1, (which means that all recorded indices were of "satisfactory environmental quality"), the maximum theoretical sum would be 10 for a category including 10 indices.

In the case of approaching the environmental quality of streets on a practical basis, the actual sum is recorded every time. Subsequently, the minimum sum is obtained by the street with the lowest sum of indices while the maximum sum is obtained by the street with the highest sum of indices. An example of the criteria of the traffic indices is presented in Table 3 (Table 3). The classification of the indices into the two main categories (satisfactory or poor environmental quality of a street) was empirical and it was based on the opinion of experts in issues pertaining Urban Planning. The classification of these indices was based on a scale of 0 to 1, where 0 was given to the streets of poor environmental quality and 1 to the ones of satisfactory environmental quality.

Environmental Quality	Satisfactory	Poor
Indices: TRAFFIC INDICES		
Traffic Lanes in each direction	One-way streets and pedestrianised walkways	Two-way streets
Parking Scheme	Controlled Parking	Free parking
Existence of parking meters	yes	no
Pedestrians Crossings	yes	no
Ramps	yes	no
Traffic Lights	yes	no
Existence of Bus Routes	yes	no
Existence of Bus Stops	yes	no
Existence of cycling lanes	yes	no
Specially Paved Lanes for people with vision difficulties	yes	no
Special Routes for Handicapped People	yes	no
Reserved Parking areas	yes	no
Prohibited parking spaces (building entrances, garage entrances etc)	yes	no

Table 3. Evaluation Criteria for the Environmental Quality of Streets – Traffic Indices

It should be noted at this point that the classification was carried out with those indices for which recorded data were available. Principally, an assumption is made at this phase, and that is "all indices carry the same weight and consequently their importance is equal in the corresponding sums, whether they are theoretical or actual".

	Percentage of Streets %			Percentage of Streets %		
	FAIR	AVER	POOR	FAIR	AVER	POOR
INDICES	THEORETICAL			ACTUAL		
Urban Planning - Architecture	0.00	79.00	21.00	28.00	50.00	22.00
Construction Materials	91.00	1.20	7.80	91.00	1.20	7.80
Road Equipment	1.00	52.00	47.00	4.00	80.00	16.00
Traffic	0.00	6.00	94.00	8.00	21.00	71.00
Land Uses	15.00	70.00	15.00	15.00	70.00	15.00
Pollution	20.00	48.00	32.00	26.00	42.00	43.00
Climatic	42.00	1.00	57.00	42.00	1.00	57.00
Other	15.00	70.00	15.00	15.00	70.00	15.00

Table 4. Evaluation Results

The minimum and maximum sums – both theoretical and actual – are classified into three categories of environmental quality (poor, average and satisfactory). This is achieved by dividing the range of values between the minimum and the maximum sum into three equal sections, and defining at the same time the range of values for each of these sections. The classification of all the streets of the study area follows, into the three categories of environmental quality (poor, average, satisfactory) both on a theoretical and on a practical basis (Table 4, Diagram 1).

As defined by the theoretical approach, the evaluated streets are comparable to streets of any other areas, which can also be evaluated with the same method, which means that the evaluation is absolute. On the contrary, the evaluation of streets on a practical basis defines that the evaluated streets are only comparable among them, namely to the ones found within the same study area, which means that the evaluation is relative.

The first approach will help us intervene in an area in order to improve the streets which do not meet the environmental quality criteria, whereas the second approach will help us intervene in the study area in order to improve the streets which are in a disadvantaged condition compared to other streets in the same area.

All the previous results, concerning the environmental quality of the urban streets, were mapped through the ArcGIS system, using the geodatabase of the research. These maps represent the environmental evaluation of these streets, as this was defined through the research (Figures 3, 4, 5, 6).

After analyzing the urban and architectural indices, it can be concluded that area III is in a dire need of intervention, compared to the other areas, as there are indications of lowered environmental quality. The inspection of the exterior colour of the buildings, the maintenance of preserved buildings and the construction of pavements of a larger width are the courses of action that are recommended.

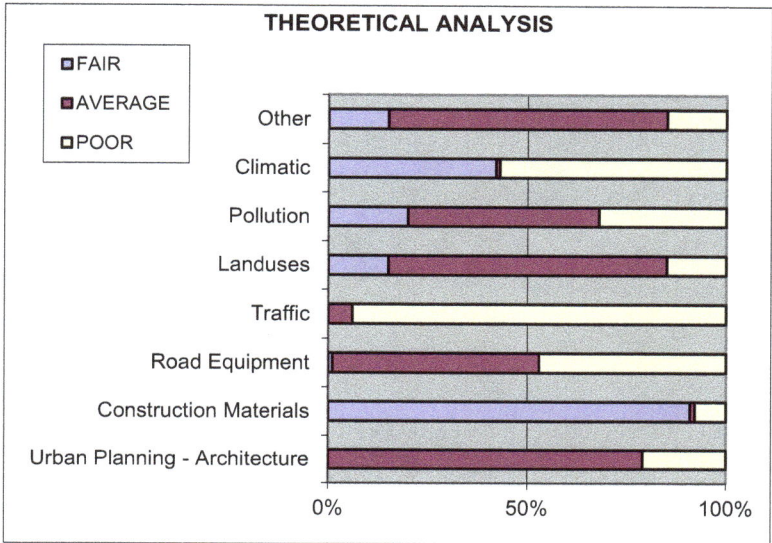

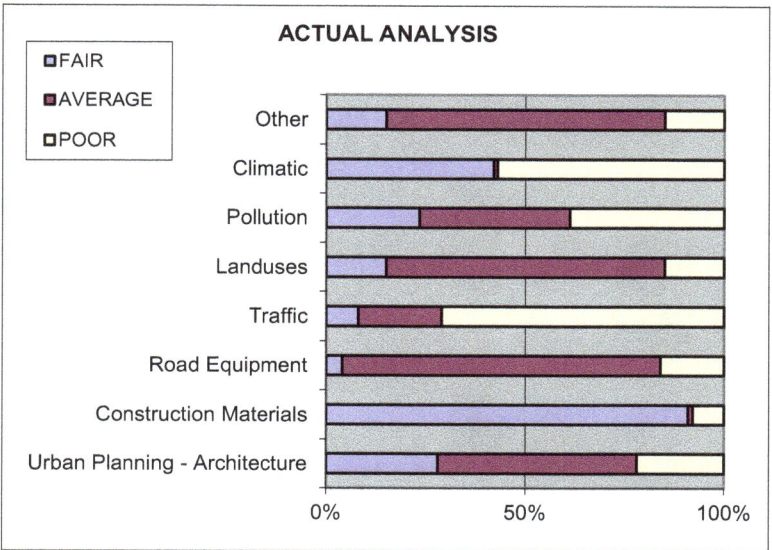

Diagrams 1. Evaluation Results

The environmental quality of the streets of the study area is considered satisfactory as far as its construction material indices are concerned, except for area III, where there is room for improvement and change. For these problematic sections of the streets, the use of friendly layering materials for the road surface and for the pavements, as well as their maintenance are recommended.

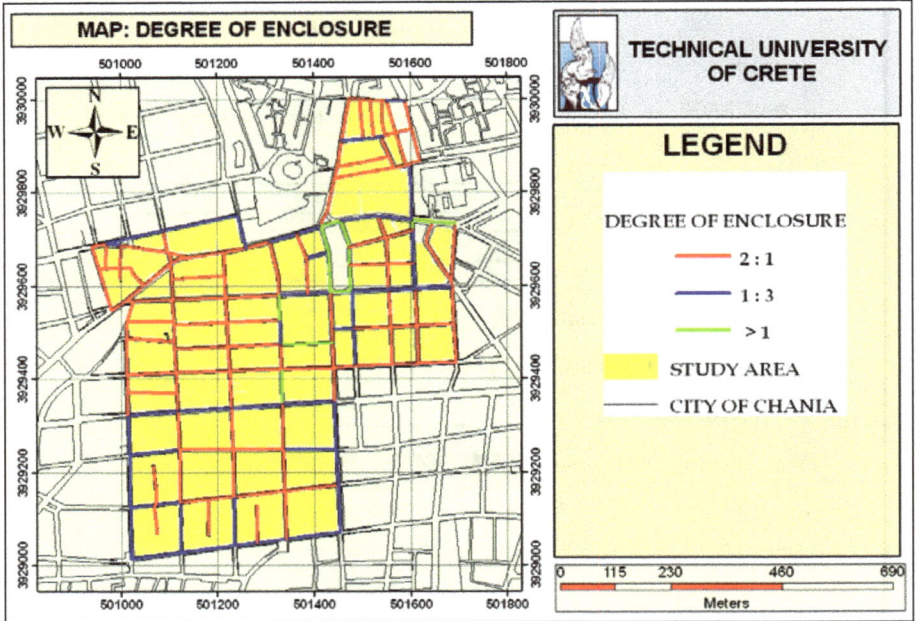

Fig. 3. Map – Degree of Enclosure

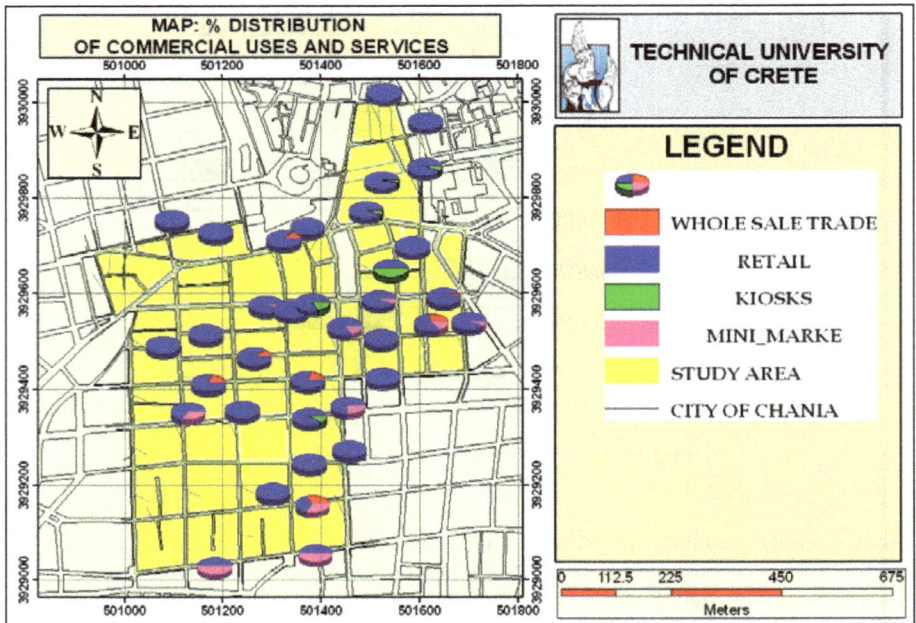

Fig. 4. Map – Percent Distribution of Commercial Uses and Services

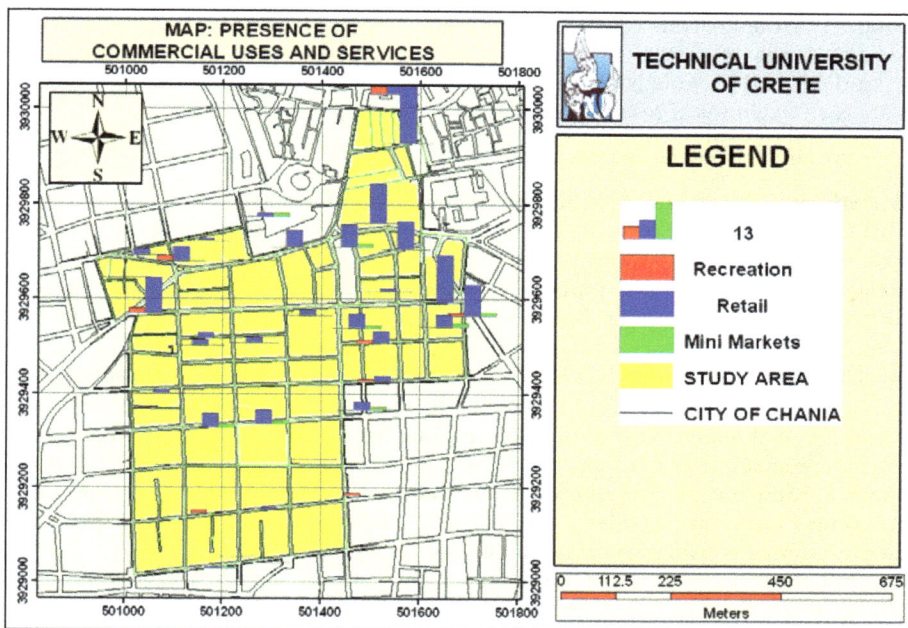

Fig. 5. Map – Commercial Uses and Services

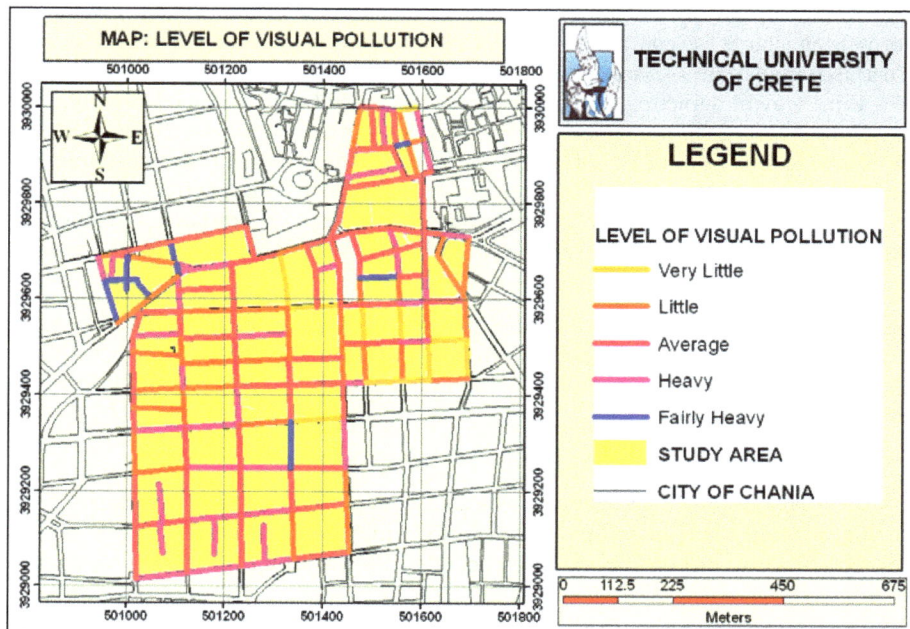

Fig. 6. Map – Level of Visual Pollution

The product of the analysis can be apparent and understandable provided that different mediums / forms of results presentation are used, and should be supervisory, easily understandable and provide with the capacity to compare and contrast between the results. Maps are such mediums of results presentation.

More specifically, the results enables one to:

- **Directly map the spatial distribution of the indices of the base**
  This form of mapping presentation enables researchers to observe the indices they are interested in and form a view on their spatial distribution. Observing this distribution leads to identifying the homogenous zones that are created and the spatial differentiations that appear. Examples of indices that can be depicted in this simple form are the index of enclosure level, of the presence of advertising signs, of cleanliness level, the index of building permit limit, and so on. Some of these indices appear further on. One can be either convergent with the area, or obtain data besides the environmental quality indices in order to draw useful conclusions about the picture of the indices in the research area.
- **Have a comparative inspection of the categories of the indices in the form of diagrams (histograms, pies, graphs etc), distributed in the area within the administrative boundaries that are being examined**
  The indices that are to be included into a group to be mapped must be comparable among them. For example, one can choose to chart the indices group concerning the commercial activity, some of them being the index of retail and wholesale transfers, of department stores, etc. In the event of indices being charted in the form of histograms then the comparison is full (indices directly comparable between them), while in the form of pie-charts the comparison is relative and on a per cent basis.
- **To search for certain sections of the area which adhere to certain criteria, spatial or not**
  The third way of depicting analysis products on a result map, is the creation of maps of multiple criteria search (spatial or not). This type of mapping reflects the logic of GIS since a search is conducted either on the spatial database or on the descriptive (non spatial) one. The search for parts of the research area which meet certain criteria (simple or complex) , helps the researchers to locate any occurrences – or lack of them - or even foresee any eventualities. The criteria could be so complex that it would be practically impossible to process without the use of a GIS.

The search for parts of the research area which meet certain criteria (simple or complex) , helps the researchers to locate any occurrences –or lack of them- or even foresee any eventualities. The criteria could be so complex that it would be practically impossible to process without the use of a GIS.

Concerning this research, examples of such searches could be:

- Which streets display high commercial activity but restricted pavement width
- In which streets the road and pavement surface quality is very low.
- The streets on which despite their restricted width, parking is permitted on both sides of it

In an effort to further research these various indices statistical analysis has been implemented on the descriptive part of the geodatabase. Using methods of descriptive, inferential and multivariate statistics, it was attempted to reexamine the indices, research their correlations, define their degree of importance, as well as evaluate and classify the environmental quality of the roads.

The geodatabase that is designed enables one to further examine the factors which are more important to affect the environmental quality of streets while it also provides the potential for their modeling. This evaluation process can be a useful tool in the hands of authorities in order to prioritize their interventions in the urban web, aiming at the improvement of the environment.

3. Methodology of the statistical analysis

Statistical analysis is implemented in an effort to define the indices affecting the environmental quality of urban streets. Through processes of descriptive and inferential statistics, the indices are examined, the cross-correlations that may exist between them are researched and those with the greatest importance are determined. Statistical classification methods are, also, used to identify homogeneous areas in a selected part of the city of Chania, which are, then, evaluated according to the level of their environmental quality.

Statistically non significant indices are identified regarding the data from the selected study area. These include the indices with zero or near zero variances, i.e. whose values are the same for all or nearly all the reviewed road sections. Therefore, they have a negligible effect in the evaluation of environmental quality of the examined roads. These indices are not taken into consideration in the subsequent statistical processes (inferential and multivariate statistics).

Inferential statistics analysis and evaluation of the sum of results leads to the conclusion that most of the reviewed indices present significant correlations with, at least, another index, while several of them present strong-significant correlations. Of these, the Building permit limit and Commerciality coefficient are the most important ones, since they have the greater number of strong correlations while, at the same time, they present independency in the context of affecting without being affected by the rest of the indices. Along with the Length and the Square footage of the city blocks, Width, Orientation and Elevation of the Roads, they form a group of inherent road characteristics which hardly allow any changes.

In order to evaluate the level of environmental quality of the resulting groups of roads, at first the profile of each is identified, based on the frequency distributions of the same binary variables of each of the three groups. Then, the presence of each road characteristic is defined as either positive or negative.

SPSS software (Statistical Package for Social Sciences) has been used for the processing and analysis of data. Urban streets environmental quality indices represent the data variables while the road sections represent the data cases. Note that the data measurement level varies. There are either scale (e.g. Road width, measured in meters), ordinal (e.g. Traffic flow direction values: 1. one way, 2. two way or 3. pedestrian way) or nominal variables (e.g. Paving material values: 1. asphalt pavement, 2. stone tiles, 3. concrete tiles, 4. concrete pavement).

3.1 Descriptive statistics

As a first step, in order to statistically research the environmental indices, methods of descriptive statistics were used. In specific, indices were organised and presented using frequency distributions (frequency tables and charts), measures of central tendency (mean, median and mode) as well as measures of dispersion (standard deviation and variance). Contingency tables and the respective clustered or stacked bar charts were used for the combined presentation of pairs of variables.

3.2 Inferential statistics

Next step was the evaluation of the interrelations between the indices through inferential statistics methods. Purpose was to highlight the interactions and the degree of correlation. This allowed the identification of those which are most influential.

In detail, the relation between two indices at a time was studied. Depending on the nature of the indices, different parametric test methods were used. These are the Independent-Samples T test, ANOVA (Analysis of Variance) and Pearson correlation. For the categorical variables, the nonparametric Chi-Square test was used. In cases were the requirements for the parametric tests were not met, the respective non-parametric tests of Mann-Whitney U, Kruskal-Wallis H and Spearman correlation were used. For all statistical tests performed, a 0.05 level of significance, otherwise alpha level, was chosen.

3.3 Cluster analysis

Multivariate classification statistical analysis was the final step in the statistical investigation of the indices. In specific, Hierarchical Cluster Analysis was implemented for all indices which were or could be transformed into binary variables. The values of '1' and '2' of these variables indicate the presence and absence, respectively, of the road characteristic each index is connected with. In the process of cluster analysis, agglomerative method was used. The aim of the Hierarchical Cluster Analysis was the grouping of cases, as well as grouping of variables.

Grouping of cases was made in order to create groups of roads with common environmental quality indices, i.e. their road characteristics. Ultimate purpose was these homogeneous groups to be evaluated according to the level of their environmental quality and define the good, average or poor environmental quality of each. Based on this rationale, it was decided that the resulting number of clusters should be three. As far as cluster method and dissimilarity measure are concerned, Between-groups linkage and Squared Euclidean distance, respectively, were used.

On the other hand, grouping of variables aimed in creating groups of related indices. In this case, a dendrogram was created in order to determine the number of clusters. Squared Euclidean distance was used as a dissimilarity measure and Ward's method as a cluster method. The reason Ward's method was chosen is because it tends to create clusters of approximately same size, in contrary with the, equally good, Between-groups linkage method.

The level of environmental quality and the profiles of the three groups define the needs for remedial action for each. In this manner, Cluster 1 requires actions in almost every aspect related to the quality indices, while Clusters 2 and 3 only require partial improvements.

The differentiation of the three groups formed, makes sense. Cluster 1 refers to road characteristics met, primarily, in parts of city with dense construction, i.e. large built space coefficients (>1), commercial areas, public services, means of advertisement and promotion (such as signs and tents) etc. Cluster 2 is related with road characteristics met in the city centre, with plenty of commercial uses, since it includes indices referring to commercial related characteristics, squares, road furniture, traffic control infrastructure, atmospheric – visual pollution, sidewalk parking etc. Finally, Cluster 3 refers to road characteristics relevant to residential areas, which ensure a certain level of quality of life within the urban environment. These include indices such as no-parking areas, gardens, trees, good lighting and air quality etc. Overall, indices have been divided according to the different uses of the urban environment; Commercial for Cluster 2, Residential for Cluster 3. Cluster 1 combines commercial use and dense construction. These come as a verification of the importance of the Building permit limit and Commerciality coefficient identified through inferential statistics.

The selection of statistical analysis for the study of the environmental quality indices of the urban roads proves to be effective. At first level, allows for the identification of the indices worth considering. At second level, it picks out the most significant ones of the lot. Finally, it leads to the distinction between areas with different road characteristics, and therefore different level of environmental quality. The level of quality was also registered.

Summing up, statistical analysis results in significant conclusions regarding the relationship and the importance of the environmental indices, as well as the distribution and assessment of the environmental quality of the selected study area streets.

4. Results and discussion

4.1 Findings of descriptive statistics

Statistical analysis is implemented in an effort to define the indices affecting significantly the environmental quality of the urban streets.

Further to the presentation of environmental quality indices' distribution, through descriptive statistics, statistically non significant indices were identified regarding the selected study area. These include the indices with zero or near zero variances, i.e. whose values are the same for all or nearly all the reviewed road sections. Therefore, they have a negligible effect in the evaluation of environmental quality of the examined roads. This is why these indices were not taken into consideration in the subsequent statistical processes (inferential and multivariate statistics).

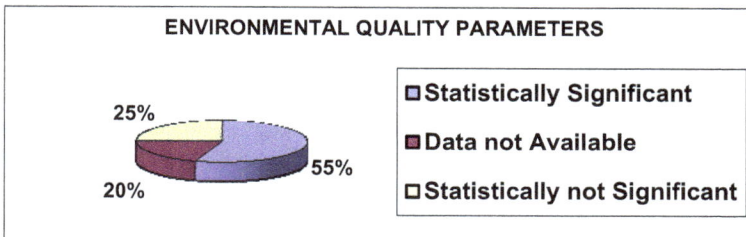

Diagram 2. Findings of descriptive statistics

As shown in Diagram 2 (Diagram 2), from the sum of the reviewed indices, statistically non significant accounted for 25%, while 55% represented statistically significant environmental indices which, through statistical analysis, can assist in deriving useful conclusions for the total of road sections and their environmental quality. Note that the remaining 20% represented indices for which no information was made available during data collection.

It is interesting to compare three frequency distribution diagrams of the indices Length of Building Block, Surface of Building Block and the index of Commerciality (Diagrams 3, 4, 5). If we look carefully, we can see that these frequency distribution diagrams present a clustering of three groups. The two first frequency distribution diagrams represent the geometry of the building blocks and therefore it is expectable to present a correlation. The third frequency distribution diagram, presents a relation with the other two (has a three group clustering). Probably, the areas have their commerciality index values based on the geometry of the building blocks and thus, commerciality index presents a similar distribution with the geometry indices.

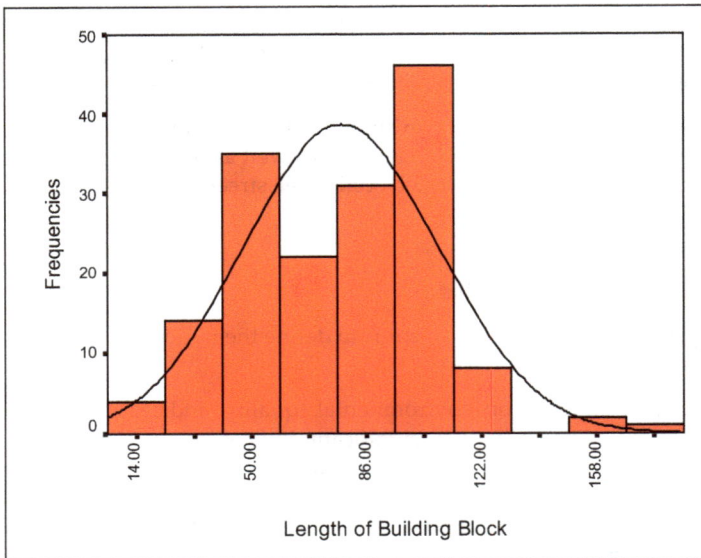

Diagram 3. Frequency Distribution Diagram of Index "Length of building blocks"

4.2 Findings of inferential statistics

The application of the inferential statistics analysis and the evaluation of the sum of results led to the conclusion that 98.5% of the reviewed indices present significant correlations with, at least, another index, while 69% present strong-significant correlations. Strong correlations are the ones for which the measures of association (Phi coefficient or Cramer's V) equal or exceed 0.4 (Bata & Obrsalova, 2009) or the correlation coefficients equal or exceed 0.6 (Button, 2002). Of these, the Building permit limit and Commerciality coefficient are the most important ones, since they have the greater number of strong correlations while, at the same time, they present independency in the context of affecting without being affected by the rest of the indices. Along with the length and the square footage of the city blocks, width, orientation and elevation of the roads, they form a group of inherent road characteristics which hardly allow any changes.

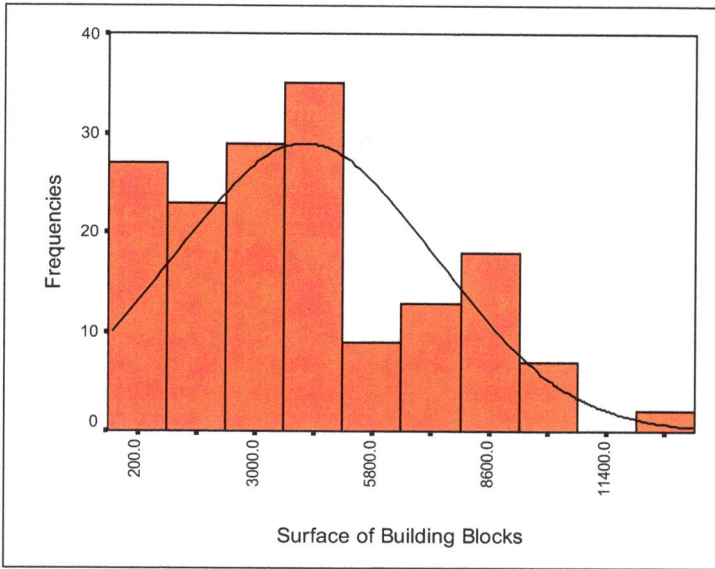

Diagram 4. Frequency Distribution Diagram of Index "Surface area of building blocks"

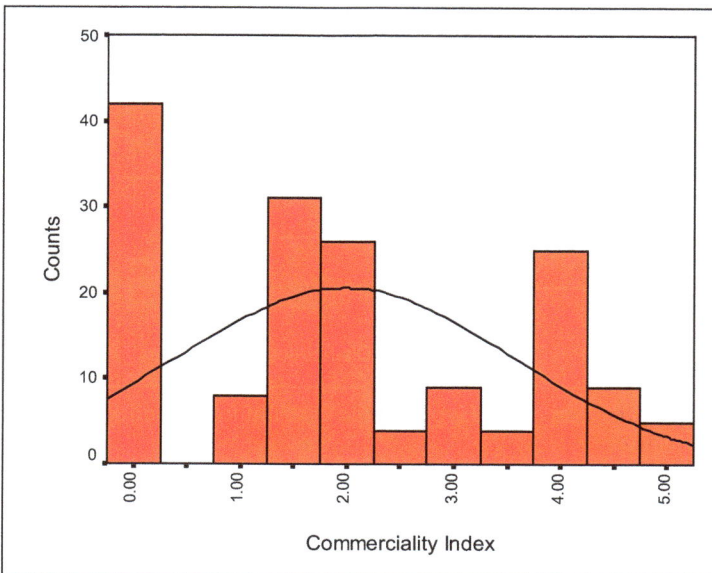

Diagram 5. Frequency Distribution Diagram of Index "Commerciality"

Comparing two of the indices, the Length of Building Block from the Urban Planning group and the Hypsometric Distribution of Buildings from the Architectural Group, we can see some interesting remarks (Diagram 6, Table 5). It can be seen from the boxplot diagram that the three groups of data for the Hypsometric Distribution of Buildings Index (Aligned, Irregular, With Gaps) present a different mean value for the Length of Building Block Index. The Hypsometric Distribution of Buildings, values "With Gaps" are met at high values of the Length of Building Blocks Index (with a mean value of Length equal to 92.5m), while the values "Aligned" are met at low values of the Length of Building Blocks Index (with a mean value of 50.2m). This observation can be statistically tested and certified.

These differences can be explained if we take into account that that the high values for the Length of Building Blocks Index can be found in urban streets which are usually away from the city centre and therefore, there are usually empty spaces in those streets (spaces that have not yet been built).

4.3 Findings of cluster analysis

Regarding the classification of the cases, i.e. the road sections, into 3 clusters (Figure 7), Cluster Membership results were mapped in order to facilitate their overview (Figure 8).

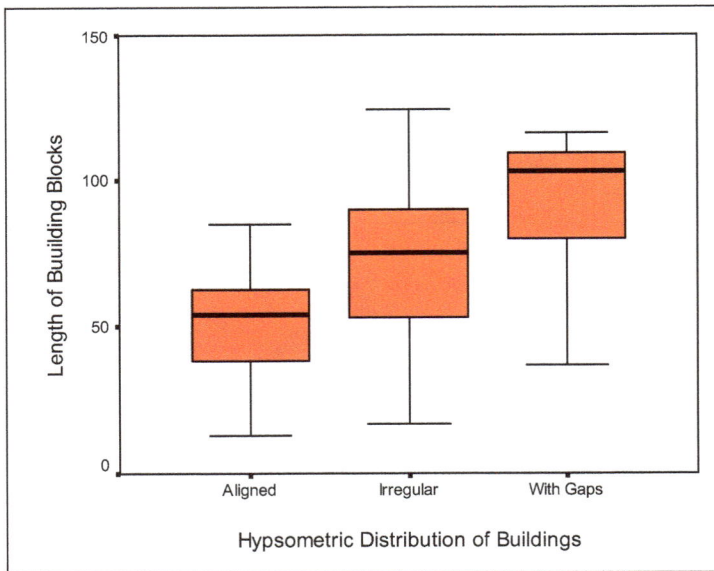

Diagram 6. Boxplot Diagram of the Hypsometric Distribution of Buildings and the Length of Building Blocks Indices

In order to evaluate the level of environmental quality of the resulting groups of roads, at first the profile of each was identified, based on the frequency distributions of the same binary variables of each of the three groups. Then, the presence of each road characteristic was defined as either positive or negative. The frequency of presence (or absence) of each index in each group was characterized by the symbols illustrated in Table 6 (Table 6).

For example, regarding the Recycle bins index, whose presence is considered positive, it appeared to 16% of Cluster 1 road sections, hence Cluster 1, according to Table 1 is characterized with the symbol ☹, 67% of Cluster 2 road sections, hence the symbol ☺/☺ and 32% of Cluster 3 road sections, hence the symbol ☹/☺. The same procedure was followed for each of the indices and ultimately, the total "score" for each group was calculated.

Length of Building (m)								
					95% Confidence Interval for Mean			
	N	Mean	Std. Dev.	Std. Error	Lower Bound	Upper Bound	MIN	MAX
Aligned	15	50.20	21.69	5.60	38.18	62.22	13	85
Irregular	93	74.00	30.92	3.20	67.63	80.37	17	181
With Gaps	55	92.53	22.84	3,08	86.35	98.70	31	116
Total	163	78.06	30.16	2.36	73.40	82.73	13	181

Table 5. Comparison of the Hypsometric Distribution of Buildings and the Length of Building Blocks Indices

Cluster 3 showed the best environmental quality of roads, scoring 36 ☺, 14 ☺ and 30 ☹. Cluster 2 comes second with a score of 32 ☺, 22 ☺ and 31 ☹ while Cluster 1 is the one with the lowest level of environmental quality, scoring a merely 8 ☺, 23 ☺ και 38 ☹.

Frequency of appearance of each index in each group	Symbol	
	Positive impact index	Negative impact index
0 %	☹ ☹	☺ ☺
0% - 20%	☹	☺
20% - 40%	☹/☺	☺/☺
40% - 60%	☺	☺
60% - 80%	☺/☺	☹/☺
80% - 100%	☺	☹
100%	☺ ☺	☹ ☹

Table 6. Indices' frequencies and symbols

The level of environmental quality and the profiles of the three groups define the needs for remedial action for each. In this manner, Cluster 1 requires actions in almost every aspect related to the quality indices, while Clusters 2 and 3 only require partial improvements.

As for the grouping of the variables, the cluster analysis results are presented in the form of a dendrogram, having emerged from the procedure (Figure 7).

One can notice that, at first level, nine groups are formed which, in return, form three larger clusters. Their differentiation makes sense.

First cluster refers to road characteristics met, primarily, in parts of city with dense construction, i.e. large built space coefficients (>1), commercial areas, public services, means of advertisement and promotion (such as signs and tents) etc.

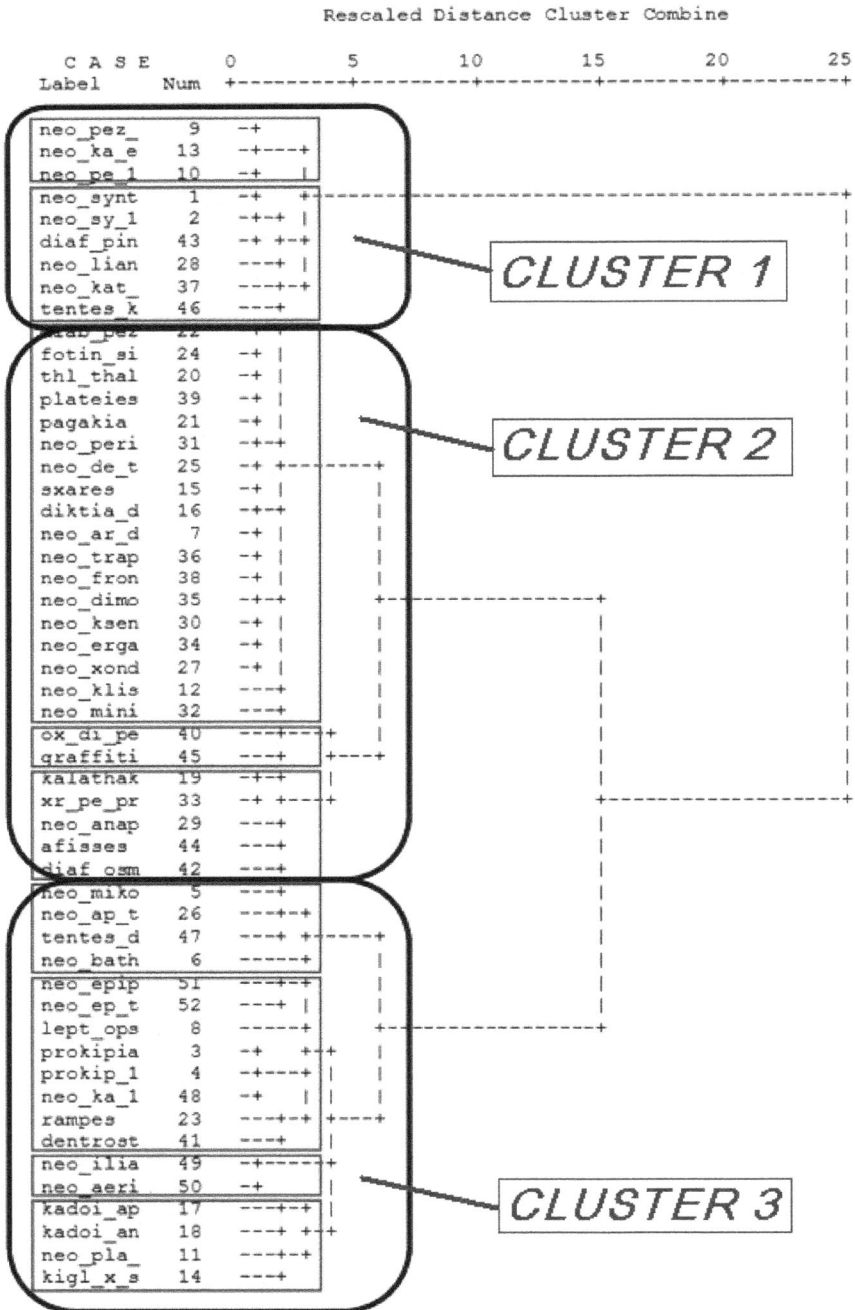

Fig. 7. Dendrogram using Ward's method

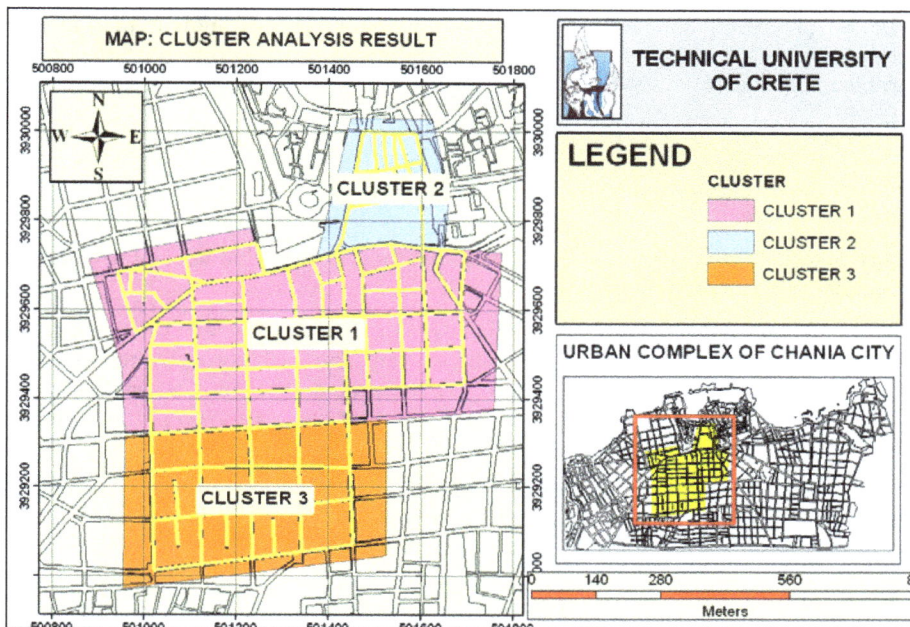

Fig. 8. Visual presentation of cluster analysis results regarding groups of cases

Second cluster is related with road characteristics met in the city centre, with plenty of commercial uses, since it includes indices referring to commercial related characteristics, squares, road furniture, traffic control infrastructure, atmospheric – visual pollution, sidewalk parking etc.

Finally, Cluster 3 refers to road characteristics relevant to residential areas, which ensure a certain level of quality of life within the urban environment. These include indices such as no-parking areas, gardens, trees, good lighting and air quality etc.

Overall, indices have been divided according to the different uses of the urban environment; Commercial for Cluster 2, Residential for Cluster 3. Cluster 1 combines commercial use and dense construction.

These come as a verification of the importance of the Building permit limit and Commerciality coefficient identified through inferential statistics.

5. Conclusions

The aim of this research was to design and develop a mechanism (a geodatabase within a GIS system) which would include all the elements relevant to the quality of the urban roads, so that the local administration bodies are facilitated in handling the urban environment and in their decision making.

This database was designed and could serve as a pilot research for the local administration offices. The 124 indices aimed to cover the whole spectrum of the environmental factors. The potential for processing and analyzing is endless; mapping, classification of areas according to their homogeneity, environmental evaluation based on multiple criteria, and the diachronic updating and observing of the urban environment.

One of the problems encountered was that for some of the indices, no data was obtained. This is a problem that occurs frequently throughout Greece and is owed to the fact that every municipality have conducted their own projects, which means that the conclusive data is quite diverse.

This particular database though, includes fields for those indices that haven't been recorded so far, so that when more data becomes available, the base can be updated.

Another obstacle was that in some cases the available data were in relation to the city as a whole (i.e climatic indices, temperature, rainfall, etc). The variations throughout the city though, are unknown, which is why it is recommended that new research should be conducted, based on recording every individual street, or finding an accepted technique for interpolating these generalised spatial data on the spatial unit level.

Yet another problem is the specialization of the people who take part in on site recordings of data. Certain indices -such as colour variety, harmonious mixture of colours- should ideally be carried out by an architect. In such cases, the research team should consist of researchers from all related sciences.

The Geo-Database that was designed enables one to further examine the factors which affect the environmental quality of streets while it also provides the potential for their modelling.

In a subsequent research program we attempt the examination of the recorded factors in order to establish which ones are the most important and to compile them hierarchically according to their significance in urban planning as well as explore their modelling potential.

This evaluation process can be a useful tool in the hands of authorities in order to prioritize their interventions in the urban web, aiming at the improvement of the environment.

The selection of statistical analysis for the research of the environmental quality indices of the urban roads, proved effective.

At first level, it allowed for the identification of the indices worth considering.

At a second level, it picked out the most significant ones of the lot.

Finally, it led to the distinction between areas with different road characteristics, and therefore different level of environmental quality. The level of quality was also registered.

Summing up, statistical analysis resulted in significant conclusions regarding the relationship and the importance of the environmental indices, as well as the distribution and assessment of the environmental quality of the selected study area streets.

6. References

Abdulaal W. A., (2007), Developing a Generic GIS Inspection Application for Saudi Municipalities, *10th International Conference on Computers in Urban Planning and Urban Management*, Iguassu Falls, Brazil, 2007.

Aisopos G., (2003), The Greek public space. The aesthetics of cities and the policy of intervention – contribution to the regeneration of the urban space, *Unification of the Archaeological Sites of Athens S.A.*, Athens, Greece, October 13-14, 2003 (in Greek).

Aravantinos A., (1997), *Urban Planning*, Symmetria, Athens (in Greek).

Aravantinos A., (1999), Urban uses of land and consequential environmental impact, , In: *Environmental impact planning and assessment methods*, (Part B, Chapter 1), Planning of cities and environmental impact, Hellenic Open University, Patras (in Greek).

Arctur D., Zeiler M., (2004), *Designing Geodatabases, Case Studies in GIS Data Modeling*, ESRI Press, Redlands – California, USA.

Bata R., Obrsalova I., (2009), Sustainable Environment Indicators and Possibilities of their Aggregation by Means of Petri Nets, *Proceedings of the 7th WSEAS International Conference on ENVIRONMENT, ECOSYSTEMS and DEVELOPMENT (EED '09)*, Puerto de la Cruz, Tenerife, Canary Islands, Spain, December 14-16, 2009.

Beckman E.P., (2001), Principles and methods for planning open areas and means for improving the micro-climate of structured environments. In: *Bio-climatic design of buildings and the environment, Restoration problems on the small scale of the urban space, Bio-climatic design of open areas*, (Chapter 3), Hellenic Open University, Patras, Greece (in Greek).

Button K., (2002), City Management and Urban Environmental Indicators, *Ecological Economics*, Vol. 40, (2002), pp. 217–233.

De Jong T., (2008), An urban designers' road hierarchy, *Proceedings of the WSEAS International Conference on URBAN PLANNING and TRANSPORTATION (UPT'07)*, Heraklion, Crete Island, Greece, July 22-24, 2008.

Dekoster J., Schollaert U., (1999), *Cycling: the way ahead for towns and cities*, Office for Official Publications of the European Communities, Luxembourg.

Ferrarini A., Bodini A., Becchi M., (2001), Environmental quality and sustainability in the province of Reggio Emilia (Italy): using multi - criteria analysis to assess and compare municipal performance, *Journal of Environmental Management*, Vol. 63, (2001), pp. 117–131.

Gehl I., (2002), *Public spaces and public life*, Adelaide, Australia.

Kartalis K., (1999), *Meteorology. Introduction to the natural and anthropogenic environment*, Hellenic Open University, Patras (in Greek).

Koutsopoulos K., (2002), *Geographic Information Systems and Spatial Analysis*, Papasotiriou Publications, Athens.

Lyberopoulos, E.L., (2003), *Study for integrating the bicycle in Heraklion, Crete*, (MSc dissertation), Hellenic Open University, Athens (in Greek).

Masters G.M., (1991), *Introducing Environmental Engineering & Science*, Prentice Hall International Editions.

Mendes M. A., (2005), Urban Environmental Management, *Proceedings of the Conference Challenges in Asia*, Institute for Global Environmental Strategies (IGES), Japan, 2005.

Nichol J., Wong M. S., (2005), Modeling urban environmental quality in a tropical city, *Landscape and Urban Planning*, Vol. 73, (2005), pp. 49–58.

Nikolopoulou M., (2004), *Planning of open urban areas with bio-climatic criteria*, RUROS KAPE (in Greek).

Pitsiava – Latinopoulou M., (1991), *Road traffic control*, Aristotle University of Thessaloniki Publication Office, Thessaloniki (in Greek).

Porta S., Latora V., (2007), Correlating Street Centrality and Land Uses: An Evidence-Based Support for the Multiple Centrality Assessment of City Spaces, *10th International Conference on Computers in Urban Planning and Urban Management*, Iguassu Falls, Brazil, 2007.

Racioppi F., Dora C., Krech R. & Von Ehrenstein O., (2002), *A physically active life through everyday transport*, Word Health Organization, Regional Office of Europe, Copenhagen, Denmark.

Siddhartha S., (1999), *Toward a typology of transportation-related urban design problems and solutions: case studies of small and medium sized cities in the eastern United States*, National Transportation Center, Morgan State University Baltimore, Maryland, USA.

Tsouchlaraki A., Zoaki E., (2008), Environmental Quality of Roads in Heraklion, Crete, *Proceedings of the WSEAS International Conference on URBAN PLANNING and TRANSPORTATION (UPT'07)*, Heraklion, Crete Island, Greece, July 22-24, 2008.

Tsouchlaraki A., Achilleos G., Nasioula Z., Nikolidakis A., (2009), Designing and Creating a Database for the Environmental Quality of Urban Roads, using GIS, *Proceedings of the 7th WSEAS International Conference on ENVIRONMENT, ECOSYSTEMS and DEVELOPMENT (EED '09)*, Puerto de la Cruz, Tenerife, Canary Islands, Spain, December 14-16, 2009.

Vlastos Th., Siolas A., (1994), The contribution of transport networks to the strategy for articulating and reuniting urban entities in view of restructuring the city. The case of western Athens, *Technika Chronika*, Vol. 14 (1994), pp. (in Greek).

Vlastos Th., Polyzos, I., (1999), Policies for the urban environment – the European experience, In: *Design, environmental impact and methods for their assessment, Urban planning and environmental impact*, (Part A, Chapter 2), Hellenic Open University, Patras, Greece (in Greek).

Yannas S., (2001), Bio-climatic principles of urban planning, In: *Environmental design of cities and open areas, Environmental technology*, (Chapter 4), Hellenic Open University, Patras, Greece (in Greek).

Yildiz P., (2008), Environmental Designing Parameters regarding Sustainable Tourism among Coastline Cities with Comparisons in Turkey, *Proceedings of the WSEAS International Conference CULTURAL HERITAGE AND TOURISM (CUHT'08)*, Heraklion, Crete Island, Greece, July 22-24, 2008.

Zivas D., (2003), *The public space of the city. Its formation, function and educative mission. The aesthetics of cities and the policy of intervention – contribution to the regeneration of the urban space*, Unification of the Archaeological Sites of Athens S.A., Athens, Greece (in Greek).

11

Walkability Index in the Urban Planning: A Case Study in Olomouc City

Zdena Dobesova and Tomas Krivka
Palacký University in Olomouc
Czech Republic

1. Introduction

There is the relationship between physical activity of people and the city environment in which they live. This relationship is researched by project IPEN (International Physical Activity and the Environment Network). IPEN is an international interdisciplinary project. It studies physical activity of citizens in the various structures of the town. There are many determinants that shape the citizen physical activity. The important factors are, for example, distances from home to shops and schools, density of streets, land use in town etc.

Project IPEN has developed a methodology for analyzing city environment according to the physical activity of people. The main result of this methodology is a Walkability index. This index consists of four partial indexes: Connectivity index, Entropy index, FAR index (floor area ratio) and Household density index. Final Walkability index is the sum of partial indexes (where the weight of the Connectivity index is two times bigger than the weight of the other indexes).

The Walkability index can be useful is in the stage of urban planning. High value of Walkability index means that a particular arrangement of the city supports physical activity of people. The low value of Walkability index means that people very often used cars in the everyday life which results into minimum physical activity. The enumeration of the Walkability index is possible for both existing structure of the town and for any proposed urban plan. It is, therefore, a good opportunity for urban planners to evaluate the proposed urban plan from the point of influence on the peoples´ physical activities. Urban planner can so increase the everyday physical activities of citizen by good arrangement of streets and high diversity of land use.

Nowadays, the urban plans are mostly in a digital form. Geographical information systems (GIS) are very often used when creating and maintaining the urban plans. New program tool for automatic calculation of Walkability index has been designed in the Czech Republic at Palacký University. The tool was made as a custom toolbox with tools for ArcGIS software version 9.3 (produced by company Esri). The custom toolbox consists of five scripts in Python language. It is an easy task to calculate Walkability index on the digital urban plan. The tools were tested on the case of Olomouc town structure.

The Walkability index is calculated according to IPEN methodology. The town is divided into several sub areas - small urban units of town. The final Walkability index is calculated for each sub area of town. First partial index is the Connectivity index. It is the number of the streets crossings per sub area. Second is the Entropy index, which represents the diversity in the land use (portion of commercial, residential area, …). Third index calculates FAR index (floor area ratio). This means the ratio of the actual area of commercial objects per area of whole commercial zone in land use category (e.g. the actual area of a shop is only a part of the commercial zone which also includes a car park and other resources). The Household density index represents the number of households per area category used for living and is obtained from the land use layer. Program also adds three new attributes to the table of the urban areas in GIS layers of urban plan.

Urban planners can automatically recalculate the Walkability index of any new version of their urban plan by the program. Finally, it is possible to see the influence of the various versions of urban plan on value of the Walkability index. This index can than be a significant criterion when choosing the most convenient urban plan.

2. Urban area and the mobility of people

Phases of urban development are related to the economic development of transportation. They are: (a) Initial capitalism phase; (b) Industrial development; and (c) Big traffic problem on the cities (Ávila, 2010). The support of walking and cycling in cities is a partial solution of the traffic problem.

The planners implemented formalist ideas, treating the "city as a machine" after World War II. The towns have increased number of inhabitants and their total areas. Mass industrialization during the 20th and 21st century led to car-centric development, where walking from one place to another is not feasible anymore (Saligaros, 2011). This situation started in the USA and Western Europe. Subsequently, rising of area of cities has appeared in Eastern Europe during last three decades.

The Congress for the New Urbanism (CNU, 1993) officially started the New Urbanism in the USA in 1993. The New Urbanist movement began as a human-scaled alternative to Modernist city planning: while the latter is based upon distances, spaces, and speeds that suit machines and the needs of industry, the former considers instead the very different needs of human beings. Among other things, New Urbanism promotes walkable communities (where people can live, work, and socialize without being totally dependent on cars), and nonrigid zoning that allows a mixture of work, industry, and housing, all done with well-proportioned buildings that borrow heavily from traditional forms and techniques (Saligaros, 2011).

Walking has three dimensions. The first is practical dimension. It is easy to get somewhere by a walk when the distance is short. The second impact is the social dimension. Meeting with other people increases social cohesion. Old cities (Latin tradition) had the central square. In fact, the evening walk around square has a social dimension. The third impact of walking is health impact. This last dimension is in the centre of research of IPEN project.

Jane Jacobs (1992) mentioned in her book that the streets and sidewalks are the most crucial part of city's safety, because that is how people travel. Not just policemen and cameras solve the criminality. There must be people looking after other people, in a mutually functioning city life in which people can coexist. This idea can belong to the social impact of walking and existence of sidewalks.

Having access to a pedestrian environment offers the possibility of a walk on foot that can be of any duration. A system of complex connected sidewalk geometry allows periodic activity around 15 min (e.g. a trip to a restaurant, shop or park) which are unfeasible in a car city. Some urban planners suggest next obligation. Government must build concrete sidewalks on both sides of all the new streets. Such thought could be especially useful for some of the cities in the USA where new roads are being built without sidewalk.

New phenomena "suburban sprawl" brought great numbers of isolated houses and isolated shopping malls at the edge of cities. This situation implicates necessity of a car when moving for a long commute to work, food shopping, getting children to school, accomplishing regular out-of-house chores, etc. Good idea of "shopping streets" in the central part of urban unit disappeared. People living in suburban cities have less physical activity than people living in the centre of town. Everyday commute to work, school and shopping require car when dwellers live at suburban area. On the other hand, more shops, restaurants and public areas are in the centre of the city. People living in the centre walk more in every day live than people living in the suburban parts.

Comparison of work travel and non-work travel from the point of time is presented by Zhang (2005). Activity-Travel Survey in Boston (1991) compares non-work travel and accessibility to entertaining, recreational, eat-out, and other social opportunities.

The mobility is nowadays a very diverse and complex issue, due to the tendency for a more disperse residential occupation and for more decentralized location of most commercial and service activities, as well as of different population mobility habits resulting from their daily situation. As a consequence, urban mobility has been ever more dependent on the private car and, in many cases, by the existence of inefficient and costly public transport systems, with obvious negative impacts at the environmental, social and economic levels for the society (Seco, Silva, 2010).

Sum of negative influences on dwellers calls for a change in urban planning. Better arrangement of houses, schools, food shops, parks in urban units is essential in order to improve the environment. "Innovation" requires a city environment that encourages a state of emotional and physical well being (Ward and Holtham, 2000).

Physical inactivity is one of the most important public health issues in the U.S. and internationally (Bronson et al., 2009). The physical activity is currently one of the most important factors in the area of public health. In general, it is a policy of various states to increase their citizens' motoric activity. It is, therefore, a subject of study of many researches all around the world to obtain information about the city structures and evaluate the impact on their inhabitants' physical activity. The primary assumption of these studies is that the environment greatly determines the amount of physical activity of its inhabitants (Abu-Omar & Rütten, 2008; Badland et al., 2008; Giles-Corti & Donovan, 2002). Moreover, it was found out that new city quarters designed with attention to the active transport of people really increases their physical activity (Saelens, Sallis, Black, & Chen, 2003). An urban area with positive influence on its residents' physical activity typically shows higher number of population, sufficient connectivity, higher value of Entropy index and FAR (Floor area ratio) index (Mitáš, et al. 2008).

2.1 Measurement of the physical walking activity

There are various approaches to measuring the physical walking activity of people. It is, for example, possible to perform a questionnaire survey (Bronson, 2009). Another option is to obtain the data from contact chips that are worn by a group of respondents for given time. It is also possible to monitor the number of people passing through a certain place using optical sensors. In addition, there is an option to find the amount of peoples' physical activity indirectly, by evaluating the arrangement of the city structures (the net of sidewalks and streets) knowing the type of the land use (residential, commercial, recreational). The last indirect method of measuring the physical activity of inhabitants using the GIS in urban units is the subject of this chapter and is described in detail in part 4.

Project IPEN used a small accelerometer ActiGraph (Fig. 1). Respondents wear this sensor with them for one week. ActiGraph measures a frequency of steps, a step distance, number of steps and the intensity of movement. The ActiGraph worn near hip or at waistline gives the best results of measurement. Data are then exported from the sensor for processing and calculation of physical activity. The average number of steps per day is used for evaluation of overall physical activity. Respondents also fill a form about their daily activity for the purpose of checking the ActiGraph measurement. The ActiGraph device was also used during the case study in Olomouc city. Subsequently, the results of the ActiGraph were compared with the output of the GIS (geographic information systems) evaluation.

Fig. 1. Accelerometer ActiGraph (Photo: T. Krivka)

Furthermore, literature describes interesting studies of physical activity measurement. Pikora (2002) described a special scan instrument – SPACES (Systematic Pedestrian and Cycling Environment Scan instrument). Scan collects data from a total of 1987 km of roads in metropolitan Perth, Western Australia. Additional city information was collected using desktop methods and GIS. The result from Perth was that neighborhood environmental factors correlated with walking near home.

Another method of acquiring the data about inhabitants' activity shows Hoedl (2010). The Bikeability and Walkability Evaluation Table (BiWET) was developed in 2007 at the University in Graz, Austria. The BiWET is designed for fast assessment of 15 predetermined characteristics of 10-m street segments. Inter-rater and Intra-rater reliability was calculated in 2008 based on the auditing data of 152 km of commuting routes of city dwellers in 2007. Research project in Graz developed and tested a simple and efficient audit tool to get a quick overview of built environment along a route or in a neighborhood. Result was that the mean auditing pace was 16.4 minutes/km.

3. Project IPEN and the Walkability index

IPEN is the International Physical Activity and the Environment Network (IPEN, 2011). It was launched by Professor Jim Sallis (USA), Dr Ilse DeBourdeaudhuij (Belgium) and Professor Neville Owen (Australia) at the International Congress of Behavioral Medicine in Mainz Germany in August 2004.

The network declares this aims (IPEN, 2011):

- increase communication and collaboration between researchers investigating, environmental correlates of physical activity,
- stimulate research in physical activity and the environment,
- recommend common methods and measures,
- support researchers through sharing of information, feedback, letters of support etc.
- bring together data from multiple countries for joint analyzes aid in the publication of data through papers, special journal issues, symposia etc.

Fig. 2. Logo of IPEN project

The Czech Republic also takes part in IPEN project together with 14 countries around the world. The main research department for IPEN project in the Czech Republic is the Center for Kinanthropology Research at Faculty of Physical Culture at Palacký University (CFKR, 2011).

IPEN project confirmed that the built environment – land use, structure of roads, number and location of shops, etc. has an influence on physical activity of people. Digital data about the urban environment are processed in GIS according to a recommended methodology. This methodology is described in step by step instructions. In 2011, Department of Geoinformatics at Faculty of Science at Palacký University developed a program for GIS that automatically processes all steps of spatial analysis in ArcGIS software.

The physical activity in a city environment, mainly in urban units of the city, is evaluated by Walkability index. Walkability index is calculated from four partial indexes. High value of the index means that the environment encourages dwellers to walk. Low value of Walkability index means that dwellers are forced to use cars and thus their physical activity is low in everyday life (much less than 10 000 steps).

The necessary input data for calculation of every index is a polygon geometry of urban units. The urban units are represented as a polygon feature layer with attribute table (name of urban unit, area of urban unit) in GIS. Urban units are set by Czech Statistical Office as a base unit for census. Apart from the data about the urban units, other specific data are taken into account when partial indexes are evaluated. Only vector data are necessary for calculation (no raster data). The scale of data is 1 : 10 000. All data were taken from Center for Kinanthropology Research.

3.1 Connectivity index

The first partial index that is an input for the resultant Walkability index is the Connectivity index. Connectivity is also called Intersection density. It is calculated from the number of intersections of roads per square kilometre of urban units. Input digital data is line geometry of roads in a city. Highways are excluded from the input data as they are not suitable for being walked along. Moreover, area of water bodies and rivers is subtracted from the area of urban units.

In order to calculate the index, each crossroad is assigned a Valence value. This value states the number of roads that meets at the particular crossroad. A "T" shaped crossroad has Valency value three, an "X" shaped four. An example is shown in the Fig. 3.

Fig. 3. Valence of intersection in the Olomouc centre

Any crossroads that are too close, that means closer than 15 m, are merged and treated as one crossroad. The distance of 15 m is recommended by the methodology of IPEN. These crossroads are then assigned higher Valence (Fig. 4).

High values of Connectivity index show that the assessed area is well interconnected. High connectivity is typical for the city centres where the conditions are not suitable for mass car transport as there are no fast road corridors and there are not enough parking spaces. In most cases, a respondent opts for walking rather than car or public transport. In addition, the distances are relatively short. In case of city centres, respondents can often be tracked within relatively small area.

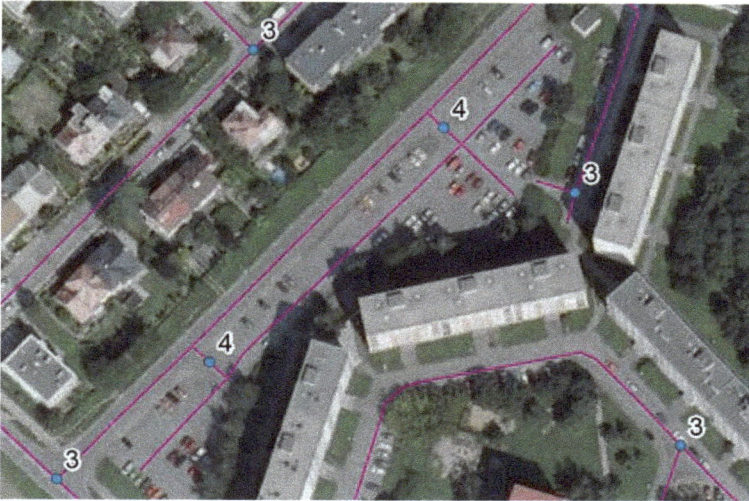

Fig. 4. Example of the merge intersection with Valence 4

The situation is, however, different for the areas of low connectivity, which are mostly those at the outskirts of the city. In these zones, dwellers usually do not work near their homes and thus have to travel longer distances, which cannot be done by active transport (on foot). People are then forced to use a car to perform most of their transport routines.

3.2 Entropy index (Shanon index)

The Entropy index represents how homogenous or heterogeneous the usage of a particular area is. The higher is the diversity of the land use, the higher is the Entropy index. The keystone of the Entropy index calculation is a high-quality land use polygon layer. The land use types are divided into following eight basic categories (Table 1).

Name of category	Coded value
Living	L
Commercial	C
Services	S
Industrial	I
Institutional	T
Recreational	R
Other	O
Water	W

Table 1. Type and coding of land-use according to IPEN methodology

In the USA, where the methodology was developed, each polygon (area) is expected to have exactly one type of usage. However, this assumption is not suitable for European (including the Czech) cities, where combined usage is common, especially at the city centres. We can, for example, find elements of living, commercial and institutional usage at the very same place. For that reason, at stage of analysis at the Centre of Kinantropological Research, substitutional codes were introduced to enable notation of combined use. For example, a combined use of an area which comprises of living, commercial and institutional types would be denoted by a string of three letters: LCT (living, commercial, institutional).

Evenness of distribution of building floor area of residential, commercial, and office development is the best for physical activity. If the value of entropy is high for a given district, we can expect that an inhabitant can carry out all common activities within a relatively small area. It is probable that he will be working, doing the shopping, seeking the entertainment, etc. within this district. Again in this diverse area (as far as the land use is concerned) he will have to travel some distances, yet as the distances will be short, he will probably walk. On the other hand, the situation would be different in a district with a low Entropy index. In such area, a residential or industrial type of land use prevails. People do not usually have the possibility to satisfy their needs (schooling, work, shopping, etc.) and are, therefore, forced to travel longer distances, which cannot be done on foot.

The term Entropy appears in the information theory; nevertheless, it is also very useful in mathematical statistics. Entropy is a number which determines how difficult it is to predict a value of a random quantity. It is obvious, that the prediction is more difficult if the distribution of the quantity's values is more even. Therefore, the entropy characterizes the evenness of the distribution of values or the extent of disorder in a system (Vajda, 2004).

For the needs of the IPEN project, the equation for calculating the entropy was modified into the following form:

$$H(S) = \frac{-\sum_{i=1}^{k}\left[(p_i)\cdot(\ln p_i)\right]}{\ln k} \tag{1}$$

where

H(S)   Entropy index (Shannon index)
p_i the area of a particular category of land use over the total area of all categories (within the scope of one district)
k   the number of land use categories in the particular district

3.3 FAR Index - Floor area ratio

The FAR index represents the ratio of the area of shop buildings to the whole area of land use category labelled as commercial. It is estimated that high index shows that the place has a significant percentage of smaller retail shops. Such district will surely be more attractive for walking then the others. When the FAR index has low value, it is probable, that there are more large shops and shopping malls with extensive car parks. Therefore, it is more convenient to use a car to go shopping. Moreover, bigger shopping will be done so that the customer does not have to go shopping again soon. A region with high FAR will exhibit the opposite – inhabitants will go shopping relatively often, mostly on foot rather than by car.

The necessary data for the calculation of FAR index are all areas of buildings with commercial usage (shops, restaurants, etc.). The input into the evaluation is the point geometry (layer) of addressed points to which the size of the area is attached.

3.4 Household density index

In order to calculate the household density, it is necessary for the polygon geometry to hold information about the number of households in every urban unit. The number of households is divided by the area intended for living in the urban unit. The second necessary input is again the land use type (as it was for the Entropy index). Only zones tagged as L (living) are taken into account.

The index reflects the form of living in an urban district. High value represents high density of households. Such values are typical for the city centres, where distances are suitable for walking.

3.5 Walkability index

The Walkability index is obtained by simply adding the partial indexes: Connectivity index, Entropy index, FAR index and Household density index (Equation 2). All of the partial indexes are calculated for individual layer of urban districts. A set of partial indexes is assigned to every row of the table of urban districts. The resultant value of Walkability index is saved with each of the urban district as well.

$$WAI = (2*con) + ent + far + hdens \qquad (2)$$

where

WAI	Walkability index
con	the standard value of the Connectivity index
ent	the standard value of the Entropy index (Shannon index)
far	the standard value of the FAR (Floor area ratio)
hdens	the standard value of the Household density

The resultant value of Walkability index should indicate the physical activity of people in the particular district. The higher value of the Walkability index indicates the more probable it is that people will exercise more physical activity.

3.6 Utilization in urban planning

The Walkability index can be utilised by urban planners in the field of evaluation of the current state of a part of a city. Urban planners receive valuable information about the bad or good current situation in various parts of the city. Additionally, the index can be used to evaluate any newly proposed urban plans.

Moreover, the newly proposed plans can change the contemporary situation of already existing district. For example, new household zones or roads can arise in suburbanised areas. The Connectivity index changes whenever a new road is opened. A change in the Connectivity index can also be caused by a change as small as building a few short interconnecting streets or pavements in already existing urban zone. Another change can be caused by new recreational zones, industrial grounds, small shops and restaurants in current urban units. In these situations, it is possible to see the difference between the current index and the index recalculated for the proposed change. The value of the index can then serve as a clue when deciding for the final arrangement of the plan – for example the difference between a plan with a huge shopping mall can be compared to a plan that considers a lot of small shops scattered around a given area. As it is possible to see the probable effect of a change before the actual construction is realized, it is easier to eliminate or minimize unwanted effects of the new urban plan.

4. Program solution for the Walkability index calculation

Geographic Information Systems (GIS) are useful software for processing spatial data from the urban environment. GIS processes available spatial (graphic) data and census data in the urban units. Application ArcGIS 10 by Esri company was chosen for the data processing. The input format of all data was shapefile (SHP). Two input polygon feature classes are necessary – urban unit and land use. The third input feature class is lines of roads. The fourth input classes are points of shop with their area. The feature class of urban unit also has to contain the number of households.

There are two ways of how to process the input data according to the IPEN methodology. The first way is manual processing by interactive running of separate functions in ArcGIS. The functions are commonly called "tools" in ArcGIS. The second way is a construction of automatic program for batch processing of all data. The first way is very time consuming and requires a lot of manual work. The second way is better than first one. It fact, the second way is quicker and without operator errors.

The processing of a large amount of data can be run automatically by program extension designed especially for batch data processing. ArcGIS software offers a possibility to design the steps of data processing by data flow diagram in the graphic editor ModelBuilder. Program elements are represented by graphic elements, simply by boxes and arrows. The algorithm is simply constructed by drag-and-dropping the tools in graphic editor (Dobesova, 2011a). In some cases, the data flow diagram which is designed in ModelBuilder editor is not sufficient for all required tasks. In such cases, it is possible to automatically convert the data flow diagram into a Python language script and supplement the program code with other program constructions and commands (Dobesova, 2011b). Programming (scripting) language Python is very suitable for creating special program (script) extensions in ArcGIS. Language Python is supported by Esri and very well documented in manuals. Scripting in Python has several notable advantages. Script can call any of the tools from ArcCatalog and other methods from object Geoprocessor that are not directly accessible (Esri, 2009; 2010; Dobesova, 2011c).

A custom toolbox named "Walkability Index" was designed for the purpose of IPEN project and a case study in Olomouc at Department of Geoinformatics (Krivka, 2011). The custom toolbox can be added to ArcToolbox next to all other system toolboxes. Toolbox is situated at the bottom, according to the alphabetical order (Fig. 5).

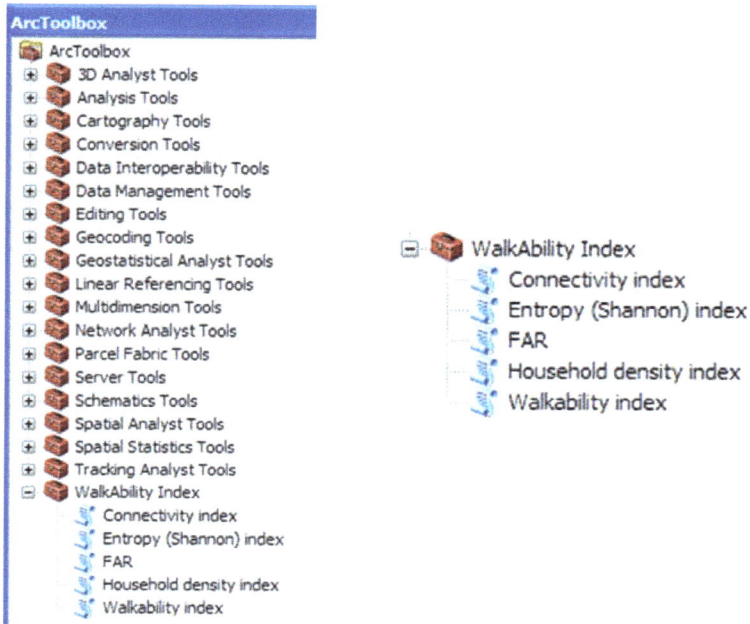

Fig. 5. Custom toolbox "Walkability Index" in ArcToolbox in ArcGIS (left), scripts (right)

The Walkability toolbox contains five scripts. Four scripts calculate the partial indexes and the fifth one calculates the final Walkability index. Each of the scripts can be run separately, simply by double-clicking its title. Scripts then launch an interface to allow the user to set the path and name of input feature class. The mandatory parameters have a green dot next to them. The explanations (help) are on the right side of the window (Fig. 6.). Help supports a utilization of the scripts by new users.

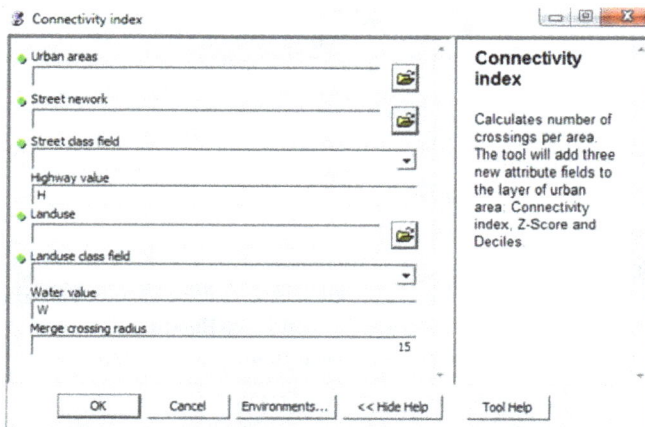

Fig. 6. Interface of a script for calculation of the Connectivity index

In our example, Python version 2.5.1 and ArcGIS version 9.3 were used. All scripts were also tested for ArcGIS version 10 and proved to be fully functional. Only the lowest licence of ArcView is necessary for running the scripts successfully. All functions, such as the subtraction of water areas from urban units, were solved directly in scripts.

The most difficult part of the script development was counting the value of Valence intersection in Connectivity index. For this purpose, a special algorithm was designed (Krivka, 2011). The idea was based on the number of crossing points with a temporal surrounding triangle (Fig. 7.).

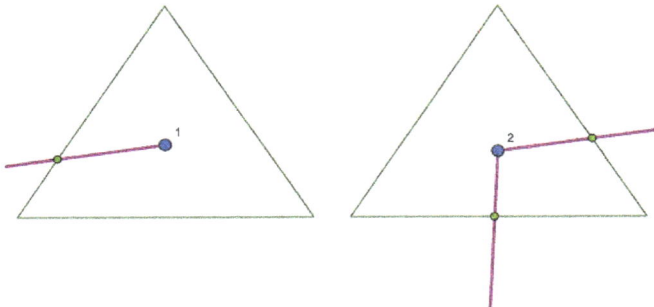

Fig. 7. Determination of the Valence value by the intersection of the surrounding triangle and the input streets

Three new attribute data are calculated for every index. The first is the absolute value of each index, the second value is the Z-score index and the third value is categorisation of Z-score to deciles (Krivka, 2011). The standard score (z-score) is quantified as a subtraction of the average from the absolute value. This difference is then divided by the standard deviation (Equation 3).

$$z = \frac{x - \mu}{\sigma} \tag{3}$$

where

z	Z-score
x	absolute value of index
μ	average
σ	standard deviation

Unexpected errors in scripts were solved by "try: except:" programming construction in Python language. Information messages about successfully running calculation of indexes were displayed to user in the ArcGIS result window. Result window also was used for errors messages in different colors of text. Additionally, a user receives a total time of run (Fig. 9).

Final toolbox is accessible from Esri web pages "Geoprocessing Model and Script Tool Gallery" (Esri, 2011).

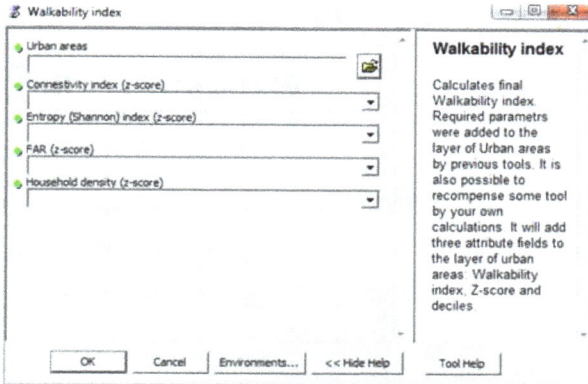

Fig. 8. Interface of the script for calculation of Walkability index – setting of the input layer and attributes

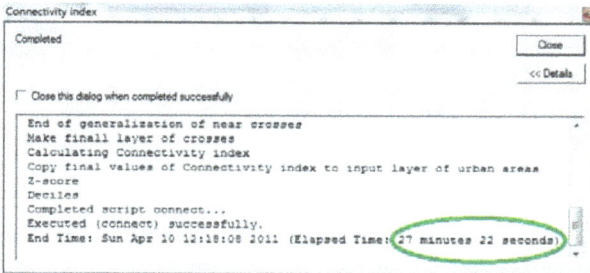

Fig. 9. A result window in ArcGIS for the Connectivity index

5. A case study in Olomouc city

Several regional analyses were conducted in Olomouc region and Olomouc city in the field of the citizen moving behaviour (Burian & Vozenilek, 2011). The evaluation of walkability in the city is also one of them. Custom toolbox "Walkability index" was tested on the data for Olomouc city.

Olomouc was divided into 37 urban units. Line geometry of roads was received from the municipal office. The number of households was taken from the national census data. Data about shops and area of shops was obtained by a terrain survey and by questionnaires. Fig. 10. depicts the results of calculated Walkability indexes in Olomouc. Computed values of Walkability indexes were divided into three levels: low, medium and high walkability. The source data for the map were taken from attribute "deciles of walkability" at attribute table.

The resultant values of the walkability in Olomouc corresponded with expectations. High Walkability index was measured in areas with high connectivity of street nets and household density. These areas are diverse in terms of land use and usually have a high percentage of small retail shops. The historical centre of Olomouc is located in the middle of the city and provides commercial objects as well as living area. Therefore, all urban units located in the city centre have high Walkability index.

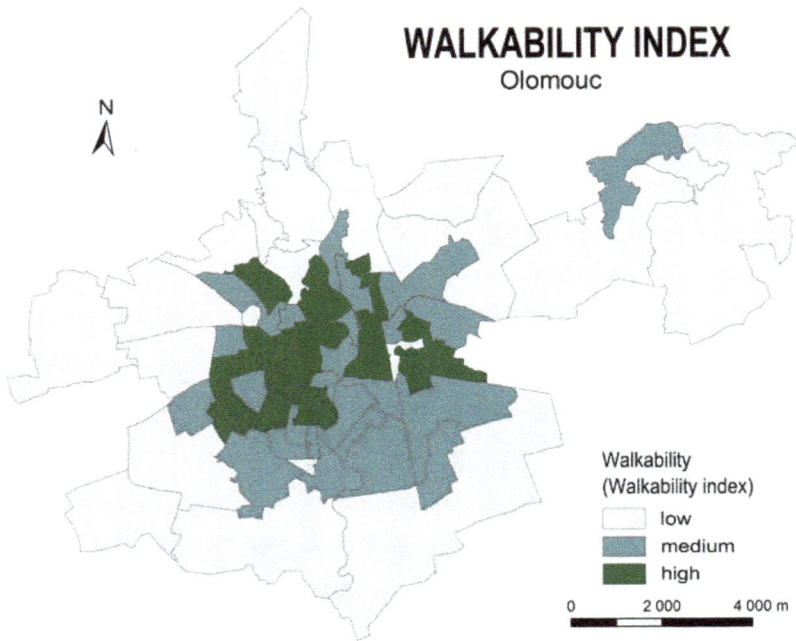

Fig. 10. Walkability index in the Olomouc urban units

At the same time, the Center for Kinanthropology Research carried out a measurement by ActiGraph. More than 350 respondents took part in the research.

For one week sensors were worn by respondents in Olomouc. However, this research was not done for all of the Olomouc parts. Grey urban units (Fig. 11) offered only small number of respondents (less than 5) and were not assessed. These units were excluded from the comparison. As the final step, the calculated values of Walkability index were compared with the data obtained by ActiGraph devices.

For the purpose of further statistical assessment, districts with notably high and low Walkability index (lower and upper three deciles) were picked. In these areas, the average number of steps of respondents were surveyed and analyzed.

It is interesting to compare the average numbers of steps with the computed values of Walkability index (Fig. 11). The highest numbers of steps were really obtained from the areas for which the calculated index was ranking high. Analogically, the lowest numbers came from the districts of the low Walkability index.

Nonetheless, the ActiGraph results from some of the researched districts did not correspond with the calculated values of walkability. This discrepancy could be cause by both insufficient number of respondents and their wrong selection. Furthermore, the methodology could be enhanced by taking the public transport and the cycling tracks into account. After a discussion with specialists, some improvements of the IPEN methodology could be introduced.

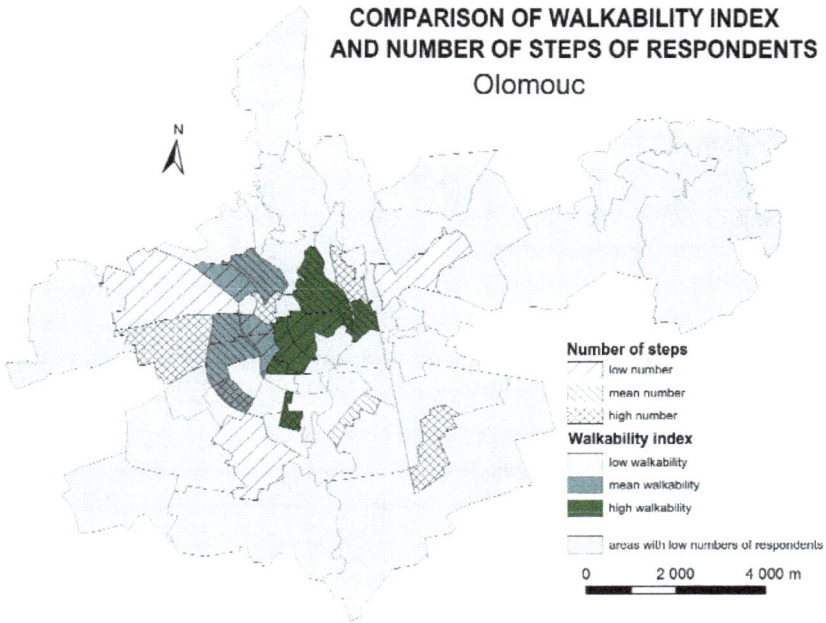

Fig. 11. Final comparison of Walkability index and the average number of steps of respondents

6. Conclusion

"Did you walk 10 000 steps today?" This is the questions that IPEN asks to its respondents. IPEN project proposes a method of evaluation of an urban environment using GIS. The method was transformed into a program, a "Walkability Toolbox" for ArcGIS software at Department of Geoinformatics. The advantage is that the program automatically calculates and examines the connection between the urban structure and the physical activity of dwellers. Program can be utilized in another city over the world.

Results of the case study in Olomouc also suggest a change in the IPEN method. The method, for example, does not consider the existence of public transport and its utilization by dwellers in everyday life. Dwellers can e.g. combine transport by walk to a bus and tram station. Moreover, the existence of cycling routes and bike transport is also not considered, even though cycling is also a form of active physical transport. This can be caused by the fact that (compared to, for example, Europe) the public transport and cycling is not so widespread in the USA where IPEN methodology arose. Nevertheless, calculation of the Walkability index brings significant results when evaluating an urban area.

Calculation of the Walkability index is notably useful for urban planners in the designing stage of a new urban plan. Innovative approach (walkability index) can be used for more precisely urban planning. They can automatically recalculate the Walkability index on a new version of urban plan by this program. Finally, it is possible to follow influence of the various version of urban plan on value of the Walkability index. The aim of urban planner is surly to create better city surrounding for healthy people.

7. Acknowledgment

This paper was supported by the Inner Grant Competition of the Palacký University in Olomouc, the project PrF_2010_14 "Research of the citizen movement between urban and suburban space of the Olomouc region".

8. References

Abu-Omar, K., & Rütten, A. (2008). Relation of leisure time, occupational, domestic, and commuting physical activity to health indicators in Europe. *Preventive Medicine*, 47(3), pp. 319–323

Ávila, G.M. (2010). A Contribution to Urban Transport System Analyses and Planning in Developing Countries, *Methods and Techniques in Urban Engineering*, Armando Carlos de Pina Filho and Aloisio Carlos de Pina (Ed.), ISBN: 978-953-307-096-4, InTech, Available from: http://www.intechopen.com/articles/show/title/a-contribution-to-urban-transport-system-analyses-and-planning-in-developing-countries

Badland, H. M., Duncan, M. J., & Mummery, W. K. (2008). Travel perceptions, behaviors, and environment by degree of urbanization. *Preventive Medicine*, 47(3), 265–269 p.

Brownson, R.C., Hoehner, C.M., Day, K., Forsyth, A., Sallis, J.F. (2009) Measuring the Built Environment for Physical Activity. State of the Science. *American Journal of Preventive Medicine*, 36 (4 SUPPL.), pp. S99-S123. doi: 10.1016/j.amepre.2009.01.005

Burian, J. & Vozenilek, V. (2011) Identification and analysis of urbanization and suburbanization in Olomouc region – possibilities of GIS analytical tools, *Urban Planning*, InTech, ISBN 979-953-307-412-1

CFRK. Center for Kinanthropology Research (2011) Available from: http://www.cfrk.eu

CNU. Congress for the New Urbanism. Available from: http://www.cnu.org/

Dobesova, Z. (2011a). Visual programming language in geographic information systems, Recent Researches in Applied Informatics, Proceedings of the 2nd International Conference on Applied Informatics and Computing Theory, AICT '11, Prague, WSEAS Press, pp. 276-280, ISBN 978-1-61804-034-3

Dobesova, Z. (2011b). Programming Language Python for Data Processing, *Proceedings of 2nd International Conference on Electrical and Control Engineering ICECE 2011*, Yichang, China, Volume 6, Institute of Electrical and Electronic Engineers (IEEE), pp. 4866-4869, ISBN 978-1-4244-8163-7

Dobesova, Z. (2011c). Automatic generation of digital elevation models using Python scripts, *Conference Proceedings SGEM 2011, 11th International Multidisciplinary Scientific*

GeoConference. STEF92 Technology Ltd., Sofia, Bulgaria, ISSN 1314-2704, pp. 599-604

Esri (2011). *Geoprocessing Model and Script Tool Gallery*, Available from: http://resources.arcgis.com/gallery/file/geoprocessing

Esri (2010). *ArcGIS Desktop Help 10, What is ModelBuilder?*, Available from: <http://help.arcgis.com/en/arcgisdesktop/10.0/help/index.html#/What_is_Mo delBuilder/002w00000001000000/>.

Esri (2009). *Geoprocessor Programming Model ArcGIS 9.3*, Available from: http://webhelp.esri.com/arcgisdesktop/9.3/pdf/Geoprocessor_93.pdf

Giles-Corti, B., & Donovan, R. J. (2002). The relative influence of individual, social and physical environment determinants of physical activity. *Social Science & Medicine*, 54(12), 1793–1812 p.

Hoedl, S., Titze, S., Oja, P. The Bikeability and Walkability evaluation table: Reliability and application, *American Journal of Preventive Medicine*, Volume 39, Issue 5, November 2010, pp. 457-459

IPEN. *International Physical Activity and the Environment Network*. Available from: http://www.ipenproject.org/

Jacobs, J. (1992). *The Death and Life of Great American Cities*, Vintage Books. New York, 480 p. ISBN 9780679741954

Krivka, T, (2011). *Spatial assessment of physical activities in built-up area*, diploma thesis. Department of Geoinformatics, Faculty of Science, Palacký University, Olomouc, 58 p. , Czech Republic, Available from: http://www.geoinformatics.upol.cz/dprace/magisterske/krivka11/summary.html

Mitáš, J., Dygrýn, J., & Frömel, K. (2008). Utilization of GIS for monitoring of the physical activity indictor [Využití geografických informačních systémů při sledování ukazatelů pohybové activity]. *Czech kinantropology*, 12(4), Prague, pp. 21–29 (In Czech)

Pikora, T.J., Bull, F.C.L., Jamrozik, K., Knuiman, M., Giles-Corti, B., Donovan, R.J. (2002) Developing a reliable audit instrument to measure the physical environment for physical activity, *American Journal of Preventive Medicine*, 23 (3), pp. 187-194. DOI: 10.1016/S0749-3797(02)00498-1

Python Software Foundation. *What is Python?* Executive Summary. Python documentation. Available from: http://http://www.python.org/doc/essays/blurb/

Saelens, B. E., Sallis, J. F., Black, J. B., & Chen, D. (2003). Neighborhood-based differences in physical activity: An environment scale evaluation. *American Journal of Public Health*, 93(9), pp. 1552–1558

Saligaros N.A. et al. (2011) *P2P Urbanism*. Umbau-Verlag, Solingen, Germany. 116 p. Available from: http://zeta.math.utsa.edu/~yxk833/P2PURBANISM.pdf

Seco A. & Silva A. (2010). Efficient Solutions for Urban Mobility - Policies, Strategies and Measures, *Methods and Techniques in Urban Engineering*, A. C. de Pina Filho and A. C. de Pina (Ed.), ISBN: 978-953-307-096-4, InTech, Available from: http://www.intechopen.com/articles/show/title/efficient-solutions-for-urban-mobility-policies-strategies-and-measures

Vajda, I. (2003*). Theory of information*. [Teorie informace]. Publishing house CTU, Prague, 109 p. (In Czech)

Ward, V. & Holtham C. (2000). *The Role of Private and Public Spaces in Knowledge Management*, Available from:
    <http://spark.spanner.org/documents/Public_Spaces_in_KM.pdf>.
Zhang, M. (2005). Exploring the relationship between urban form and nonwork travel through time use analysis, *Landscape and Urban Planning* 73 (2-3), pp. 244-261

Part 3

Practical Applications of Spatial Planning

12

Post-Industrial Land Transformation – An Approach to Sociocultural Aspects as Catalysts for Urban Redevelopment

Luís Loures[1] and Jon Burley[2]
[1]CIEO - Research Centre for Spatial and Organizational Dynamics
University of Algarve, Faro
[2]Landscape Architecture Program; School of Planning, Design, and Construction
Michigan State University, East Lansing
[1]Portugal
[2]USA

1. Introduction

The inheritance of idled, derelict and frequently abandoned post-industrial structures and sites we found nowadays in our landscapes is, arguably the result of human current and former land uses. One way or another, the present situation, enabled by technological innovation and structural economic change, is somehow based in human (ab)use of limited resources. As mentioned by Krinke, (2001, p.126) *"as the world moved from agriculture to industry, a mechanist view of the universe began to supplant the idea of an organic nature. A desire for "progress" and faith in technology implied that the earth was a place to extract resources and its "complementary" idea: that the earth could absorb anything humankind asked of it"*. However, the environmental and social consequences of such point of view enabled not only changes in society's values but also a different view, according to which the former production and consumption patterns were no longer acceptable.

As landscapes became economically disadvantaged, environmentally degraded and socially distressed, several planners, designers and developers started to react to decline, both by looking for answers to the social and economic problems caused by former activities (Secchi, 2007) and by developing new methods and frameworks to transform them. In this scenario, it became increasingly acknowledged that previously developed land (*e.g.* post-industrial landscapes) constitute an undervalued asset towards urban redevelopment. This idea is supported by the six key challenges for producing a sustainable built environment presented by the European Council for Construction Research, Development and Innovation (2001): urban sprawl; redeveloping industrial sites; regenerating brownfield sites; sustainable construction; green space, and regenerating distressed neighborhoods. Somehow, all these challenges may be directly or indirectly connected with post-industrial land transformation processes.

For this reason, all over the world, several regions and countries have begun to embrace the notion that post-industrial landscapes offer unique opportunities to the creation of renewed landscapes, viewing their value to society in a broad sense, recognizing that more than ecological and environmental reclamation opportunities those sites embodied alternative social, cultural and economic values (Doick *et al.* 2006). In fact, development of creative cultural and recreational amenities and the improvement of the image of the city through landscape transformations is increasingly acknowledge (Beriatos and Gospodini, 2004).

The fact that derelict landscapes, originally viewed as threats, became increasingly recognized as opportunities, not only because of their location, proximity with infrastructure, uniqueness in form and configuration, but also because they became often the only lands available for development in urban areas Meyer (2000), enabled the emergence of new approaches and perspectives towards landscape, especially previously developed and abandoned ones, as it is the strategies for designing with drosscapes presented by Berger (2006). Though at the beginning the practices and approaches towards post-industrial land transformation were primarily site-specific and driven mainly by economic motivations, undervaluing the importance of a contextual approach in achieving sustainable redevelopment, now they tend to be more inclusive and holistic, providing directions on how ecological restoration, cultural preservation, economic development and public needs and interest should be met.

As Adam (1998, p.55) observes, we moved "from single and dualistic approaches and abstract, functional perspectives to knowledge that emphasizes inclusiveness, connectivity, and implication". These new perspectives and approaches, besides addressing issues at multiple scales and across diverse areas of concern, acknowledged that benefits could arise from incorporating existing and remnant patterns of development into land transformation projects, suggesting that the resolution of the natural and culture conflict, evident in previous approaches which focused either land restoration or cultural preservation alone, might influence both design perspectives and processes (Ekman, 2004 and Tymoff, 2001). Regardless the used approach, planning and design options should maximize the reuse of previously developed land, using methods and principles which enable landscape's redefinition through community-based interdisciplinary actions, integrating multifunctional longer-term solutions that consider social-cultural aspect at the same level as economic, environmental and aesthetic ones.

This idea is increasingly recognized by practitioners and academics as it will be proved throughout this exposition, considering both current state of the art and two case studies that will be addressed further in this research, which represent successful examples of industrial heritage protection and public participation and involvement in post-industrial landscape redevelopment.

2. Socio-cultural aspects as catalysts for post-industrial redevelopment

It is generally acknowledged that the different dimensions of sustainable development are not always equally prioritized by policy makers and designers within the sustainability discourse (Colantonio, 2007). However, in an increasing demanding society, designers have increasingly recognized the importance of social and cultural factors in sustainable landscape redevelopment. Citizens have the right to live in aesthetic pleasant and functional landscapes (Lamas, 2004). The access to quality natural and built environments constitutes a social right that should be the foundation of architectural intervention (Lamas, 2004). Planning and design projects will be unsuccessful, if the proposed landscape fails to earn enough interest and care from the society. This interest and care towards a certain landscape depend mainly on two factors: how does a certain landscape fulfill their needs and desires, and how people experience and use them. However, as stressed out by Forman (2002, p.85) *"most landscape designers are still inspired by and primarily focused on aesthetics; society's other major objectives are secondary for them"*.

Regarding landscape redevelopment, designers need to understand that more than aesthetic pleasant and iconographic, landscapes need to be though in terms of function, purpose, and intrinsic values, considering as far as possible public will and needs, given that, as stressed out by Andresen (2005), public space belongs to the public. Indeed, as mentioned by Secchi (2007, p.10) *"it is important in an era dominated by the rhetoric of uncertainty to pay attention to visions that help people reflect on different possibilities and opportunities. Discussing these hypothetical scenarios with the public seemed the only valuable strategy to us"*. Thus, if society goals and concerns are to be incorporated, and design and planning proposals are to have legitimacy with those directly and/or indirectly affected by the plans, public participation need to be carried out from the outset of the process and throughout implementation.

Planning and design processes need to be site-specific ensuring that public aspirations are effectively addressed and proposals are thus appropriate for the site and that industrial heritage is safeguarded. In this sense industrial heritage preservation represent together with community participation major elements in post-industrial land transformation projects (CLARINET, 2002). In this scenario, recognizing the complexity of post-industrial landscapes, industrial heritage and public participation constitute crucial elements to the success of post-industrial land redevelopment projects.

2.1 Industrial heritage

Concepts as heritage and cultural heritage have clearly meant different things to different groups of scholars and the public interested in reclaiming traditions – and landscapes – presented as part of shared, remembered pasts (Carr, 2003). In 1949, the Statute of the Council of Europe, adopted in London, stressed out that in order to achieve a greater unity it is imperious to safeguard cultural heritage, facilitating economic and social progress. The years ahead proved that the conservation and safeguard of the natural and cultural environment (people's common heritage) was one of the major issues confronting society. The concept of heritage is normally divided in two groups: one cultural and another natural (ICOMOS, 2008). However, most of what is today protected or celebrated as patrimony has been chosen within industrialized societies as pre-industrial or non-qindustrial, as older, more rare, beautiful, spiritual, and/or traditional, though it is increasingly acknowledge that what is known as heritage can no longer be merely equated with monuments built before the eighteenth century. 'Recent' patrimony may no longer be considered, *a priori*, of lower value than old ones. It is essential to emphasise the idea that the history of the city and consequently our history do not stop in the eighteenth century (Custódio, 1993). The analysis of society and this understanding of cultural heritage lead to a social perception of the kind of place people wish to live in, and to the realisation that the whole city is "our heritage", regardless of whether it is large, small, historic, industrial, old or new (Fadigas, 2007; and Storelli, 2003).

These circumstances coupled with the destruction of relevant evidences from the industrial architecture during the middle of the twentieth century (Kuhl, 2004), with the reaction against the Urban Renewal policies of the 1950's and 60's which not only decimated the historic cores of many industrial cities, but also failed miserably in achieving the social and economic goals it purported (Rea, 1991; and Appleyard, 1979), and with the increasing contestation regarding *"existing criteria for monument classification and restoration"* (Custodio, 1993 in Ribeiro, 1998, p.118), created a *momentum* to the emergence of the industrial heritage concept and consequently to the interest in its preservation. Significant efforts have been developed in order to define the meaning and the scope of industrial heritage, establishing chronologic parameters and performing several studies, with the objective to define what to preserve and why to preserve it. Since then, cities have been increasingly recognized as cultural entities that contain representations from the past, via the present, to the future, running through the entire cultural evolution of the *"city as object"* (Loures, 2008a). In this way intrinsic values need to be determined, meaning given to elements of the city, its importance identified and exceptional sites highlighted.

In this sense, it is important to recognise that heritage, regardless of being architectural, vernacular, industrial, etc., is an irreplaceable expression of the wealth and diversity of common culture. It is an "entity" shared by several people, which every country must show real solidarity in preserving. While the definitions of why to protect and how to protect are sometimes dissimilar, it is commonly recognized that the concept of industrial heritage is applicable to every type of industrial activity and to every material or immaterial element created by the industrial society (Berliet, 1985; and Green, 1985). The fact that historic areas are progressively coming under threat of new development (Strike, 2003, Loures, 2008c), and that the impact of new construction is noticed not only at nationally important sites, but also in local areas – where small changes can be very significant, diminishing landscape character and local distinctiveness – increased the need to develop new strategies and frameworks to protect and highlight our cultural heritage and consequently the sense of place (Montaner, 2001; and Aguilar, 1998). To tackle this urban/cultural problem, there is a primary basic assumption that should be followed: in order to maintain this heritage, it is necessary to consider, first, the reasons behind the development of certain industrial landscape, second, the relation of that landscape with its surroundings and, third, its meanings to citizens.

In this regard, and even if industrial heritage did not have a "formal" document regarding its protection until the creation of the Nizhny Tagil Charter in 2003, followed by the Monterrey Charter, some of the principles enounced in several other international charters and conferences, supported by the Council of Europe (COE), the International Council on Monuments and Sites (ICOMOS) and United Nations Educational, Scientific and Cultural Organization (UNESCO) included somehow the protection of industrial buildings and landscapes as it may be confirmed in figure 1.

As Mendes (1995) points out from 1978 to 1994, the UNESCO list of World Heritage Sites included twelve elements with industrial characteristics. Created in 1978, the list contains today 890 world heritage properties (689 cultural, 176 natural and 25 mixed) considered as having outstanding universal value by the World Heritage Committee (UNESCO, 2010), of which over 60 related to old industry (Fuchs, 2010). However, the buildings, sites and landscapes, which are not listed as World Heritage Sites, but that are recognised as industrial heritage, falling under the scope of the Nizhny Tagil Charter, still face inappropriate material and cultural appraisal and stereotyped ideas of industry, once the way in which they were designed do not satisfy the aesthetic, ecological, and functional requirements and standards (Alanen and Melnick, 2000). Appearance was and continues to be almost everything, given that the assessment of the industrial heritage is often anchored to visual values rather than to any other consideration of function or history (Smith, 1974).

1964	Venice Charter – International Charter for the Conservation and Restoration of Monuments and Sites, ICOMOS	States on article 1 that: *the concept of an historic monument embraces not only the single architectural work but also the urban or rural setting in which is found the evidence of a particular civilization, a significant development or an historic event. This applies not only to great works of art but also to more modest works of the past which have acquired cultural significance with the passing of time.*
1972	Paris Convention Concerning the Protection of the World Cultural and Natural Heritage, UNESCO	States on the 1st article of the DEFINITION OF THE CULTURAL AND NATURAL HERITAGE chapter that shall be considered as "cultural heritage": *Groups of buildings: groups of separate or connected buildings which, because of their architecture, their homogeneity or their place in the landscape, are of outstanding universal value from the point of view of history, art or science;* *Sites: works of man or the combined works of nature and man, and areas including archaeological sites which are of outstanding universal value from the historical, aesthetic, ethnological or anthropological point of view.*
1975	European Charter of the Architectural Heritage, COE	States on article 1 that: *the European architectural heritage consists not only of our most important monuments: it also includes the groups of lesser buildings in our old towns and characteristic villages in their natural or manmade settings.* *(…) Today it is recognized that entire groups of buildings, even if they do not include any example of outstanding merit, may have an atmosphere that gives them the quality of works of art, welding different periods and styles into a harmonious whole. Such groups should also be preserved. The architectural heritage is an expression of history and helps us to understand the relevance of the past to contemporary life.*
1976	Recommendation concerning the Safeguarding and Contemporary Role of Historic Areas, UNESCO	States on the 1st article of the DEFINITIONS chapter that: *(a) `Historic and architectural (including vernacular) areas' shall be taken to mean any groups of buildings, structures and open spaces including archaeological and paleontological sites, constituting human settlements in an urban or rural environment, the cohesion and value of which, from the archaeological, architectural, prehistoric, historic, aesthetic or sociocultural point of view are recognized.*
1980	The Burra Charter, ICOMOS	States on article 1 that: *Cultural significance means aesthetic, historic, scientific, or social value for past, present, or future generations.*
1987	Charter on the Conservation of Historic Towns and Urban Areas, ICOMOS	States in the 2nd article of the PRINCIPLES AND OBJECTIVES chapter that: *qualities to be preserved include the historic character of the town or urban area and all those material and spiritual elements that express this character, especially: a) Urban patterns as defined by lots and streets; b) Relationships between buildings and green and open spaces; c) the formal appearance, interior and exterior, of buildings as defined by scale, size, style, construction, materials, color and decoration; d) The relationship between the town or urban area and its surrounding setting, both natural and man-made; and e) The various functions that the town or urban area has acquired over time.*
2000	Charter of Krakow – Principles for Conservation and Restoration of built Heritages, International Conference on Conservation	States in the preamble that: *each community, by means of its collective memory and consciousness of its past, is responsible for the identification as well as the management of its heritage. This cannot be defined in a fixed way. One can only define the way in which the heritage may be identified. Plurality in society entails a great diversity in heritage concepts as conceived by the entire community. The monuments, as individual elements of this heritage, are bearers of values, which may change in time. This variability of the individual values of monuments constitutes each time the specificity of the heritage. From this process of change, each community develops an awareness and consciousness of the need to look after the individual built elements as bearers of their own common heritage values.*
2003	Nizhny Tagil Charter for the Industrial Heritage, ICOMOS	
2006	Monterrey Charter for Industrial heritage Conservation, ICOMOS	

Fig. 1. Information present in international charters and conferences regarding heritage and historic matters. Loures (2011) –

When analyzing and re-developing these landscapes, landscape architects, architects, designers and other planning professionals need to realise that post-industrial, typically part of ordinary or vernacular landscapes, incorporate the passage of time (Loures, 2008a, and 2008b; Loures and Panagopoulos, 2007a; and Panagopoulos and Loures, 2007b), representing multiple layers of time and cultural activity therefore being part of the identity of a people and a place. In this sense, these landscapes should be seen as assets, once as historic sites they enhance the possibilities of creative practice in preservation, design, and planning, given that they are often unique, as a result from the combination of natural landforms and buildings defining a particular place or region. These changes in perception contributed to increase the relevance of industrial landscapes and to highlight the need to study and protect the material and immaterial remains of our industry from a different perspective (Casella and Symonds, 2005; and Stratton and Trinder, 2000).

2.1.1 What future for the industrial heritage?

As it was mentioned before, the concept of Industrial heritage was only introduced in England in the middle of the twentieth century, during a period when several industrial buildings and landscapes were destroyed (Casado, 2009; and Kuhl, 2004). From rural to urban, and now to industrial, the concept of heritage is now larger than ever. This enlargement is not only thematic; it is also spatial, once its scope changed from the protection of a single monument to the protection of a whole landscape, or even a whole city (Neyret, 2004). The notion of heritage includes now, the landscape which has become a part of inhabitants' identity. Nonetheless, the theme of urban heritage is still surrounded by a conceptual ambiguity, based in a clear ambivalence between ideological speech and practical policies (Ferreira, 1998). Civil society and decision makers have become more aware and are paying greater attention not only to the "environmental and economical dimensions" of urban rehabilitation, but also to the "socio-cultural dimension", currently recognised as a powerful driving force for local redevelopment, given the challenge of maintaining local identity. The preservation of the industrial heritage constitutes an important cultural objective, not only because it enlarges the sense of community (Brandt *et al.* 2000, Burley and Loures, 2008), but also because it constitutes a sustainable approach, once it encourages the positive re-use of redundant buildings that are part of our industrial and commercial heritage. In this regard, several efforts have been made in order to define what should and should not be considered as industrial heritage. In this sense, before planning the redevelopment of post-industrial landscapes it is important to find the answer for two different questions: 'why' and 'how' to reclaim and protect the industrial landscape?

The answer to the 'why' is often clear. As it was mentioned before, industrial landscapes describe an important part of the history of a place, thus, constituting a testimony of cultural, social and economic conception and evolution which documents and interprets considerable values for urban heritage. Furthermore, the analysis and recovery of these landscapes constitute an opportunity that tends to be lost in time, considering the growing urban pressure that, especially in pleasant and valuable landscapes, had several times led to the disappearance of various industrial infrastructures, some with high heritage value and significant relevance.

The answer to the 'how' is relatively more complex, once, even if there are several possible answers to this question; each one includes generally several restrictions enabled by the search for profit maximization by private and public sectors. Although it is recognized that the economic and social dimensions of the redevelopment process cannot be dissociated from the environmental and cultural dimensions, and that the cultural heritage has become a key factor in improving people's surroundings, addressing issues of social cohesion and encouraging economic development, little has been done in order to rehabilitate industrial buildings and its surrounding landscapes which were often the catalysis to the creation of the urban settlement; and in addition to that, design professionals tend to highlight 'how' to manage cultural landscapes (redevelopment proposals, analysis, cultural landscape reports, heritage management plans, etc.) but not 'why' should we be concerned with historic sites and places or 'what' are the expectations and 'which' are the objectives we seek to accomplish by working with them.

2.2 Public participation

As already mentioned it is generally recognized by landscape architects, landscape ecologists, and sociologists, among others, that the social component plays a relevant role in urban planning and management activities, and that participation processes are linked both to landscape and strategic environmental valuation. The last decades have seen a rapid change in attitudes towards the environment, which reflects a greater environmental awareness amongst professionals as well as the general public (Ozguner and Kendle, 2006). There is a growing trend in government to conclude that the commitment and will of the population is a crucial element to the development of a sustainable city (Giddings et al. 2005), and that the redevelopment of derelict, abandoned or underutilized land can play a significant role in future planning activities (Loures and Panagopoulos, 2007b). For this reason it is often recognized that the social component plays a relevant role in planning and management activities (Loures et al. 2008; Faga, 2006; Christensen et al. 1996; and Grumbine, 1994). The need of introducing public participation into planning and management activities has been reinforced not only by designers, governments and private associations, but also in several international meetings and conventions. Examples include Rio Declaration on Environment and Development in 1992; the Aarhus Convention on Access to Information, Public Participation in Decision-making and Access to Justice in Environmental Matters in 1998; the recent Leipzig Charter on Sustainable European Cities in 2007, and the Cimeira de Lisboa in 2008, among others.

In this sense, as stressed out by Bellah (1998, in Potts and Harrill, 1998, p.16), "a good community is one in which there is argument, even conflict, about the meaning of the shared values and goals, and certainly about how they will be actualized in everyday life. Community is not about silent consensus; it is a form of intelligent, reflective life, in which there is indeed consensus, but where the consensus can be challenged and changed — often gradually, sometimes radically — over time". More than active citizens, in order to achieve sustainable development, cities need active involvement on the entire policy and decision-making process, which needs to be decentralized and as far as possible focused at the local level (Taylor, 2000 in Camagni et al. 1998; and Selman and Parker ,1997).

In fact, public participation has become increasingly more important, playing a relevant role in determining the way society will manage, protect and reclaim not only the natural but also the built environment. The recognition that the economic and social dimensions cannot be dissociated from the environmental and cultural ones, contributed to increased the relevance of public participation (Loures, 2008a and 2008c).

A wide range of methods have been established all over the world, including new ways of people interacting, new types of event, new services and new support frameworks. Governments look now to provide greater community input in the identification of needs and problems, and in the design and implementation of remedial and preventive solutions (Creighton, 2005; and Hartig *et al.* 1998). However, according to Faga (2006, p.xiii) it is still common in Europe, *"elite professionals enter competitions and propose designs (often very exciting designs) that are selected by a panel of experts (...) a similar process is inconceivable in the United States, where community participation has become a central element in deciding what will be built".*

2.2.1 The use of public participation in landscape redevelopment

Although public participation in planning, management and redevelopment of post-industrial landscapes has gained wide acceptance among private and public domains, in part motivated by the introduction of public participation in several international design competition (Fresh Kills Parkland, Duisburg Nord Landschaftspark, Westergasfabriek Culture Park, among others), though academic literature and research offers still limited understanding on how to accomplish it and what contributes to its success (Beierle and Konisky, 1999; and Davies, 2001).

As it is common among "concept definitions", the concept of public participation is not unanimous, once there are always different perspectives of understanding a specific concept. Still, public participation may be generally defined as a descriptive and exploratory method, which enables the observation and analysis of specific issues and phenomena, allowing the establishment of relations among variables (Triviños, 1995; and Gil, 1994). In opposition to an experiment (which according to Strauss and Corbin (1990) is a research conducted in a laboratory under controlled conditions), public participation instruments (*e.g.* surveys and questionnaires) are conducted in a real-life context, and can be descriptive (using standardized questionnaires for describing a specific phenomena) or analytical (using qualitative and quantitative methods to find relations among variables and explanations.

The fact that *"democracy is a work in progress"* (Creighton, 2005, p.1) have contributed to the evolving meaning of public participation over time. However, as mentioned by Duffy and Hutchinson (1997, p.351) concepts indicating different levels of public involvement, often associated with different styles of political decision-making (Table 1), with direct influence on the acceptance of the project *"such as participation, incorporation, empowerment, capacity building and consultation",* although having different meanings are often used as synonymous. Public participation is not a neutral concept. According to the World Bank (1992, p.22) definition public participation is a process that *"enables the public to influence the quality or volume of a service through some form of articulation of preferences or demand",* a definition that is closely linked to the concept of governance. In a more direct definition Beierle and Cayford (2002, p.6) defined public participation as *"any of several 'mechanisms' intentionally instituted to involve the lay public or their representatives in administrative decision-making".*

Fiorino (1996) characterize public participation as the involvement of people outside formal governmental decision-making processes. Nevertheless, there are still some authors (Britton, 1998; and Pateman, 1970) that defend that public participation is one of the components (together with public consultation) of what they consider to be 'public involvement'.

Political Style	Main Decision-maker	Options	Criteria	Scientific Instruments	Project Acceptability
Bureaucratic	Political responsibility	No explicit alternative	Not clear	Technical reports	Low
Technocratic	Specialist	Explicit multiple alternatives, determined according to scientific rationality by specialists	Economic or similar, expressed in monetary units Multiple: measured in different units and determined by the specialists	Cost-benefit analysis or cost-effectiveness analysis Multi-criteria analysis	Medium
Participatory	Actors and/or concerned - affected citizens	Multiple alternatives proposed by concerned - affected actors	Various: measured in different units and determined by the affected actors	Multi-criteria analysis Citizen participation tools	High

Table 1. Political Decision Styles - Loures (2011)

These approaches are not contradictory in their main principles, once they all comprise public activities directed at cooperation and team work, providing the authority with opinions and information about public will, needs and objectives. Public participation in landscape redevelopment and management can take several different forms (Faga, 2006; Creighton, 2005; and Beierle and Cayford, 2002): Public meetings, workshops, charettes, citizen juries, focus groups, internet, mail interviews, face to face interviews, etc. each of them legitimate *a priori*, and justified by the context in which the project takes place (Vasconcelos, 2001 and Hester and Blazej, 1997).

2.2.1.1 The role of participation in project acceptability

The relevance of the social acceptability of a specific project should never be underestimated. In the past, scientific and technological options having a negative environmental impact appeared to be inappropriate, not in terms of technical performance but for reasons of social acceptability (RESCUE, 2004). In recent years, due in part to a need to reduce social conflict and litigation, the planning paradigm has shifted to give the general public greater input in environmental decisions (Steelman, 2001; Dustin and Schneider, 1998; Fiorino, 1996; and Gunderson, 1995).

As (Beatley, 2004) mentions it is through ownership, commitment and the infusion of "local knowledge" in project development, unique places, genuinely native to the culture and environment, can be sustained. Still, designers have to be aware that different people have different ideas, perspectives, needs, and concerns (table 2), reason why the participation process as to be as inclusive as possible, considering the opinion of each and every single group related directly or indirectly with the project.

Environmentalists	Citizens	Law Makers	Farmers	Realtors | Developers	Businesses
Preservation	Good schools	Jobs	Investment	Profitability	Taxes
Water Quality	Quality of life	Industry	Right to farm	Affordability	Market
Energy	Health	Security	Equity	Community	Talent
Food Security	Community	Equity	Viability	Opportunities	Competitiveness
Biodiversity	Jobs	Welfare	Property rights		
Stewardship	Safety	Policy	Food security		
	Future for children				

Table 2. People's Concerns and expectations, Loures (2011)

The social acceptability of results in a decision-making process is linked to the way the different parts involved in the process perceive it: if they feel it is adequate and equal, they find it legitimate. For this reason, improving the social acceptability of specific design options during the process often results in higher legitimacy of the whole process, which in this way depends largely on how much people affected by the plan have been involved in it (Steiner, 2000). Considering post-industrial redevelopment projects, as they are often located in highly visible and accessible areas, public perception and support is essential to the long-term success of the project (Nassauer, 1997) and to enhance the social, economic and environmental benefits that they provide.

In order to ensure better organization and efficiency it is necessary to develop new forms of interaction between the social and the political sphere, enabling the creation of conditions for an active and participative citizenship. In past years, government development of large post-industrial landscape reclamation projects have increased on international, national, regional and local levels. Professionals involved are becoming more and more aware of the fact that specific local human and social factors need to be considered and introduced in the planning process of rehabilitation of industrial derelict sites. Public participation holds nowadays an essential position in the post-industrial regeneration process.

3. Applied theory – The impact of socio-cultural aspects in post-industrial redevelopment

Post-industrial redevelopment is a complex topic with many actors and stakeholders who often pursue contrasting aims in the development process. A socially well balanced planning process, assuring participation opportunities for all the affected parties, provides the necessary conditions for sustainability standards and is as such a prerequisite for each post-industrial reclamation project. To exemplify the relevance of using public participation and protecting industrial heritage in post-industrial landscape redevelopment, as it was mentioned before, this chapter will address two practical case studies (Duisburg Nord and Westergasfabriek) emphasizing the way these socio-cultural aspects influenced and catalyzed urban redevelopment.

3.1 Duisburg nord – From a blast furnace plant into a landscape park

Duisburg Nord Park represents only a small portion of the effort, which has been made in the Ruhr River watershed to reuse old industrial areas: the International Building Exhibition (IBA – from the German Internationale Bauausstellung). Based in the 1988 structural program created with the objective of redeveloping the highly contaminated former industrial and coal mining area in the Ruhr region (European Academy of the Urban Environment, 2001), the Emscher Landscape Park (1989-1999) was presented as one of the main projects of the IBA (Bothmann and Auer, 2009; Shaw, 2002; and von Haaren, 2002).

The motivation for redeveloping this densely-populated, derelict industrial region was mainly driven by the fact that the area was becoming increasingly deteriorated as a result of the ongoing cease and abandonment of several heavy industries, initiated 30 years ago, which left behind a legacy of high unemployment, scars of environmental contamination, and the haunting shadows of the gigantic steel plants (EPA, 2009a, Shaw, 2002; and Hough, 1995). However, the unity and character of this landscape (considering the cultural and historic value of the former industrial buildings), coupled with the huge environmental and economical costs associated to a possible demolition of the existing industrial complexes, enabled the creation of a redevelopment strategy, based not only on the preservation of the industrial heritage, but also on ecological, economic and social principles to protect, enhance and develop the old industrialized region (figure 2) (EPA, 2009a; Sustainable Cities, 2008; and Latz + Partners, 2007).

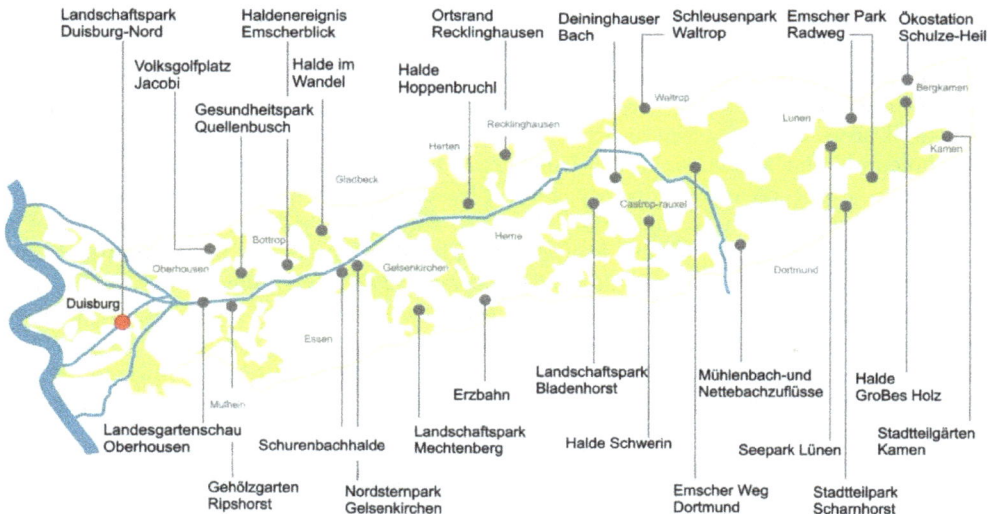

Fig. 2. Location of some of the most advertized projects developed in the Ruhr Valley during the International Building Exhibition (IBA). Loures (2011) – all rights reserved.

Considering these principles, the abandoned industrial landmarks of the region have been transformed to serve new recreational and leisure function, giving a greener and more sustainable image to the region while creating a more cohesive community with a sense of pride on the area's identity (Sustainable Cities, 2008). These facts, coupled with the impressive numbers involved in the overall process, brought the IBA project into international spotlight.

As mentioned by Shaw (2002, p.77), "one of the Exhibition's principle features was that restructuring should take a holistic view rather than simply trying to attract inward investment and jobs". While highlighting the memory and spirit of the place and protecting industrial heritage, new experiments for the future were promoted, enabling the creation of monuments which are at the same time historic and experimental (European Academy of the Urban Environment, 2001).

In this regard, Emscher Park constitutes the result of the combined effort of a multidisciplinary team of experts (architects, landscape architects, sociologists, engineers, among others) in order to achieve a set of pre-established goals: cleanup one of the most polluted areas from Europe; decontaminate and naturalize a fluvial network of approximately 350 kilometers; reuse former industrial buildings; develop several cultural and leisure routes, reorganizing rural areas and promoting the creation of cultural and artistic installations; and renew former worker's neighborhoods and develop a socio cultural network. As mentioned by Latz (2001, p.150) *"the Landscape Park Duisburg Nord is a key project of the IBA that reflects new ideas about landscape and nature"*.

Though the park was only completed recently, the proposal developed by Latz + Partner's constitutes an important legacy in the reclamation of derelict industrial sites in urban areas, not only as an individual case study but also as an element of the overall redevelopment strategy developed for the IBA Emscher Landscape Park. In fact, as new reclamation projects are looking to Park Duisburg Nord for inspiration it is evident that the way of looking at history, and at the world around us, is changing. By literally defining the park as a post-industrial landscape, Latz + Partner affected how people think not just about industrial areas but any place or space that helps to define a specific culture or cultural phenomenon. The attraction of the Duisburg-Nord Park lies in what Macaulay (1953) referred to as the pleasure of ruins, or the pleasure associated with exploring physical remains of the past. The combination of nature and industry enabled the creation of a landscape full and memories and feelings, considered as one of the most significant and noteworthy projects of the past decade (Nickerson, 2007; and Stilgenbauer, 2005). The relevance of the Duisburg Nord Landscape Park is evident not only in the high number of visitors (more than 500.000) it receives every year, but also in the ingenious way the program was merged with the industrial remnants. As mentioned by Vollmer and Berke (2006, p.60) the Duisburg Nord Landscape Park *"is not only a gigantic monument, but also an open-air museum, a free climbing and a scuba diving venue and an illuminated work of art"*.

The way the surrounding communities were involvement in the project using multiple public participation techniques, as a way of assessing divergent interests and assuring that the site was developed according to existing relations and effective needs from those who will use it the most, constituted also an important element of the project. However, the strategy envisioned by Peter Latz while considering the aforementioned objective moved a step further taking into account the application of a new vision of "re-cultivation" to deal with the derelict industrial landscapes, based on the search for the way in which new landscapes should seek their position within existing industrial dereliction, considering at the same level the spaces that are going to be changed and the ones that are going to be protected and highlighted as an integral part of our common industrial heritage (Latz, 2001).

In this regard, considering the fact that the site is a complex matrix of buildings and landscapes the designers' goal was to utilize the existing fragments of industry as layers that are recombined through the lens of park design (Krinke, 2001). In fact, as mentioned by Latz (1992) instead of creating a completely new landscape, the proposed approach attempts to celebrate the area's industrial past by integrating vegetation and industry, promoting sustainable development and maintaining the spirit of the place. Instead of tearing down the industrial buildings, the project integrates them, enhancing the past and creating a perfect symbiosis between the past, the present and the future landscape.

The design strategy developed to the park was based in the idea of interlacing the existing fragments into a new "landscape", integrating, shaping, developing and interlinking the existing patterns that were formed by the previous industrial use, while finding a new interpretation with a new syntax. In this industrial landscape nearly everything has been reused in some manner, playing with the distinctions between natural and artificial, while confusing our definition of "park" (Latz, 1992). This project highlights the interest in the "spirit of the place" rather than in the genius of the creator. Developed in layers, both spatially and historically, Landscape Park Duisburg-Nord represents the contemporary interest in exploring the site as a palimpsest. Landscape Park Duisburg-Nord combines human intervention and natural processes to create an environment that neither could have created alone (figures 3 and 4).

Fig. 3. Landscape Park Duisburg Nord, Elevated Paths. Loures (2011), all rights reserved.

Fig. 4. Ore bunker garden. Loures (2011), all rights reserved.

Considering the configuration of the former site the proposed strategy enabled the development of a program in which the old industrial structures were adapted to new cultural and corporate functions (Berrizbeitia, 2007; and Nickerson, 2007). In this regard, and bearing in mind the objective of developing a multifunctional park, Latz and Partner proposed a design based in a group of specific functional areas (Latz + Partner 2007b): the blast furnace park, the water park, the sinter park, the railway park, play points, the ore bunker gallery. Besides the abovementioned program and functional areas, the park is composed by several other former structures that have been converted to new functions as it is the case of the old central power station, and the blast and tapping buildings now housing an event area, the old administration building that was converted into a youth hostel, the loading area transformed into a multifunctional and leisure area (López, 2004; and Weilacher, 1999), and the wind mill which besides constituting a land art piece, is used as a water oxygenation system (Krauel, 2008). Additionally visitors can still find several conference spaces, a museum and visitor center, a restaurant and several other performance spaces. The space is the results of a very ambiguous design, which on one hand could be seen as an outdoor museum of the iron and steel industry, but on other hand constitutes a simple space that allows the development of several public activities associated to an ecologically sound outdoor environment (figure 5).

Fig. 5. Landscape Park Duisburg Nord, diagram of program and functional areas – Loures (2011), all rights reserved.

Besides the aforementioned elements which gave international recognition to this project, the approach used by Peter Latz's highlighted the importance of using a relevant theoretical basis in landscape design. The proposed design highlights the fact that even industrial wastelands can be filled with a new spirit and can be made worth living by keeping visible the spirit of the site and the characteristics that make it unique.

3.2 Westergasfabriek – Listening people's needs and desires

During the early nineteen sixties, with the discovery of the natural gas fields, The Netherlands initiated the process of changing over to natural gas, the old coal-gas production facilities started to shut down one by one, and the Westergasfabriek (West Gas Factory) gasworks, built in 1884, at the western edge of the inner city of Amsterdam was no exception, closing in 1967 (EPA, 2009b; and Koekebakker, 2003). After the end of the gas production, after approximately eighty years of activity, various uses (e.g. a tram depot, a train washing yard, etc.) emerged for the site, considering mainly its privileged location. However, the proposed options created resistance among local residents, who thought that the site should become a large park, an option supported, indeed, by a historic map dated from 1875, according to which a large park should be developed in this site (Koekebakker, 2003).

In 1981, the site was re-zoned as a recreation space, its proximity to the city centre and the existing historic structures gave it potential for cultural re-use (Landscape Institute, 2007). In this regard the landscape was not only required to be a green space for residents but also a location for open-air and cultural events. After the new function for this area was decided, park plans, building restoration and soil-cleaning operations were continually readjusted to each other (Bokern, 2006; and Koekebakker, 2003). Even if some buildings of the Westergasfabriek were still in use, by the municipal energy company, the district council launched an "appeal for ideas" in which everyone was invited to submit their thoughts/dreams for possible interim uses. From the 334 submitted entries, four plans were worked out in further detail: a Museum of Civil Engineering, an Amsterdam Center for the Arts, a Centre for Modern Music and the Rhizome Plan, which involved the use of the building by local residents and organizations (Bokern, 2006; and Koekebakker, 2003). The success of the interim use activities was so evident that the plan, which was supposed to be implemented for only one year, lasted for more than six, a period during which the site housed an endless list of events, including performers as the Canadian Circus – Cirque du Soleil, and festivals as the National Music Festival.

Nevertheless, in 1996 the district council approved the development plan for the Westergasfabriek, in which the biggest task was to design the new park. For this reason, following the developed plan, twelve landscape architects were invited to present a general proposal to the committee, which selected five of them to take part in the limited competition: Michael van Gessel, Adrian Geuze, Kathryn Gustafson, Edwin Santhagens, and Lodewijk Wiegersma (Koekebakker, 2003).

From the five selected designers the commission chosen the plan entitled "Changement" by Kathryn Gustafson, which using a simple layout proposed a park that guaranteed various experiences both in space and time, fulfilling the original intention to maintain the cultural activities in the park. The significance of this project is evident at three different levels: the first is connected to the initial perception by a variety of stakeholders, residents and city officials of the ongoing cultural, social and civic value of the site even in its former physical state; the second is related with the development of a consistent and creative vision for the site, robust but flexible over time, embracing stakeholders and local communities; and finally the one associated to physical, social and material qualities (Kirkwood, 2003 in Koekebakker, 2003, pp.5-6).

The uniqueness of the park is somehow evident in the combination of a very strong structure with a subtle detailing in which each place has a distinctive atmosphere. This was mainly achieved by using the vestiges of the partially dismantled industrial site layout, as the structure of the park (Gustafson and Porter, 2007). "*By representing such a clear example of the passage from conceptual design ideas to implemented built work, it stimulates both professional and public dialogue concerning the range of possibilities that may exist for such sites in the future around the world*" (Kirkwood, 2003 in Koekebakker, 2003, p.7).

As it was mentioned before, the title of the proposed redevelopment scheme by Kathryn Gustafson, "Changement", is symptomatic of the design strategy used in the park, suggesting a subtle balance between landscape and society, city and nature, and order and freedom (Landscape Institute, 2007).

Besides retaining 22 of the buildings of the power company's former gasworks (Gastil and Ryan, 2004), to which several different functions and activities had been attributed (e.g. restaurants, cafés, clubs, offices, galleries, a cinema, a kindergarten, a basketball club, etc.), the Park's program proposed by Kathryn Gustafson is very diverse responding both to site and context, and to people's needs and desires, leading to strong narrative interpretations often apparent in the use of memory and history in design (Spens, 2007) (figure 6).

In summary, the design, based in the concept of change and transformation, represents not only the transition from city to garden, to landscape, to nature (Bokern, 2006), but also the effort to build a resilient and adaptable park, according to the inputs acquired throughout the public participation process.

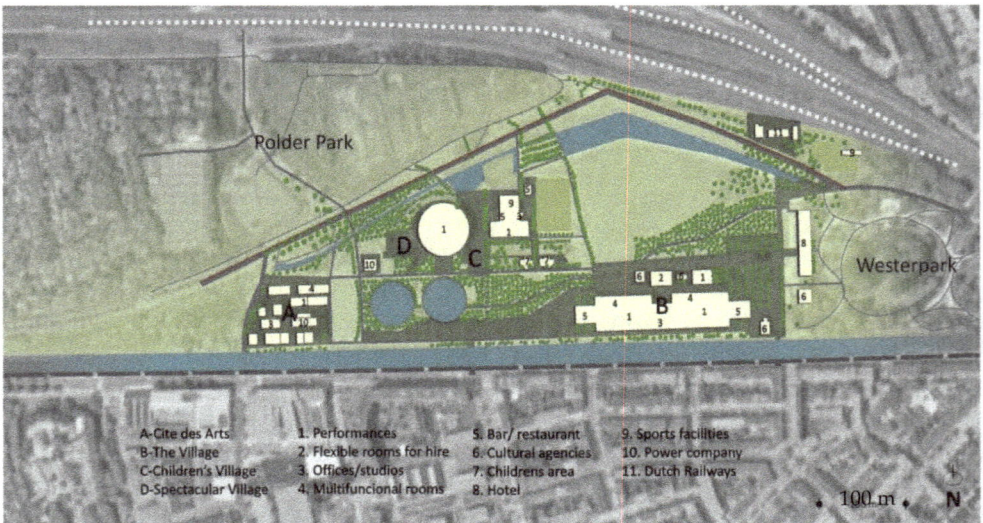

Fig. 6. Westergasfabriek Park functional areas proposed to the former buildings. Loures (2011), all rights reserved.

The Westergasfabriek project by folding historic surfaces, structures and places with emerging and progressive ideas in green open space, gives direction to other postindustrial communities in the need to protect and enhance the visceral qualities of modern cities in another step in their evolution (Kirkwood, 2003 in Koekebakker, 2003, p.6). However, one cannot disregard, on the one hand the fact that several industrial buildings of what has been the formerly Europe's largest gasometer, with high heritage value, were demolished, and on the other hand the crucial public role in preventing a plan, which proposed that a tabula rasa approach for the Westergasfabriek site, of being implemented. This fact highlights the relevance of the introduction of public participation in land transformation processes, not only in the design phase but from the outset of the process.

The use of new forms of cultural entrepreneurship, in which it is not the government that adapts the former buildings to their new function, but the private property developer, is also considered an important factor, given that in this way the selected activities require little or no public subside, forcing the users to take a relatively pragmatic and independent attitude, which in the particular case of the Westergasfabriek Park proved to promote freedom and dynamism (Gaventa, 2006; and Koekebakker, 2003). In conclusion one may say that the used open ended process enabled the creation of a dynamic place in which residents, tenants, politicians, designers, organizations and other partners were, are and will continue to be inspired by the project.

4. Concluding remarks

One of the problems that happen in post-industrial redevelopment projects is that sometimes the results do not match the original aspirations. Not only because some projects are just speculative, using "sustainability" and "heritage protection" as marketing labels, but also because public is often not a relevant part of the project. In recent years several Architects, Landscape Architects, Urban Planners and other planning specialists have built a number of outstanding iconic landscape reclamation designs that do not represent the community of which they are an integral part. These fail in what should be considered essential in a landscape reclamation project: connectivity to the place and to the society. In fact as mentioned by Loures and Panagopoulos (2010) greening is not enough, reason why socio-cultural aspects constitute essential drivers in post-industrial landscape redevelopment.

As it was confirmed on the addressed case studies the integration of public participation in the decision making process benefits both project quality and society. For this reason it is essential to develop specific frameworks according to which public participation can be introduced in the different planning phases. It is critical to shift the power paradigm in the urban planning process to allow residents to proactively envision and create public green spaces that would reflect the diversity of the society it represents. The use of public participation and the incorporation of human preferences and needs in post-industrial landscape reclamation is a safeguard to achieve success and to develop a sense of community.

However, it is essential to continue studying the city as an evolutionary 'object', looking at culture and heritage, and highlighting that the values and the history of the city do not end in the eighteenth century, they continue right to the twenty first century. And, as Dolores Hayden (2000) has written: *"cultural landscapes (including industrial ones) tell us who we are, far more effectively than most architecture or exhibits in museums ever can"*. For this reason, the redevelopment of post-industrial landscapes should be seen as part of larger, ongoing processes of architectural preservation and urban design, once, it is not confined to the most symbolic factories. It includes, also, all the additional elements and structures associated with the industrial activity. In this regard, it is imperious that politicians, developers, stakeholders and planning professionals understand that the maintenance of the urban layout is one of the most important features for the cultural identity of a city, and that the industrial landscape is an important part of it. A place is only a fragment of a cultural space, which was given consciously or subconsciously certain meanings during the course of its creation.

In this way, industrial preservation and reclamation becomes more than the celebration of the past, as important as that is; it becomes part of reconstructing the future. Thus, industrial heritage preservation that connects people, place, and history fosters a sense of place and the power for community renewal.

Industrial reclamation proposals should therefore be a part of an overall urban project - a local development strategy - which requires a broad, integrated approach comprising all urban policy areas and promoting the reconciliation of heritage conservation with social progress and sustainable economic development. The development of an increasingly multicultural urban society, emphasizes the need of rising the "socio-cultural dimension" of the city, where the rehabilitation of the industrial patrimony appears to be an essential contribution to the creation of a shared local identity and hence to the cohesion of the urban society. For this reason post-industrial landscapes should be viewed as a resource and its recovery as an opportunity to develop new multi-functional landscapes.

In summary it is possible to conclude that public participation and industrial heritage protection and preservation encourage awareness of "belonging to" a community, sharing common culture and creating identity. It improves community consciousness and responsibility while fostering a "collective sense". These are "feelings" of considerable importance in the development of new, satisfying and concerted post-industrial land transformation projects, fostering sustainability and urban development.

5. References

Adam, B. (1998). *Timescapes of Modernity*. Routledge, New York.

Aguilar, I. (1998). *Arquitectura industrial. Concepto, método y fuente*. Museu d' Etnologia de la Diputació de València, València.

Alanen, A. and Melnick, R. (2000). *Preserving Cultural Landscapes in America*. The Johns Hopkins University Press, Baltimore.

Andresen, T. (2005). A Obra vai começar, por favor estejam calados. *Público*, 23-03-2005.

Appleyard, D. (1979). *The Conservation of European Cities*. MIT Press, Cambridge.

Beatley, T. (2004). *Native to Nowhere: Sustaining Home and community in a global age*. Island Press, Washington DC.

Beierle, T. and Cayford, J. (2002). *Democracy in Practice. Public Participation in Environmental Decisions*. Resources for the future, Washington DC.

Beierle, T. and Konisky, D. (1999). *Public Participation in Environmental Planning in the Great Lakes Region*. Resources for the future, Washington DC.

Bellah, R. (1998). Community properly understood: A defense of 'democratic communitarianism. In: Etzioni, A. (Ed.), *The essential communitarian reader*. Rowman and Littlefield Publishers, New York. pp. 15-19.

Berger, A. (2006). *Drosscape: Wasting land in urban America*. Princeton Architectural Press, New York.

Beriatos, E. and Gospodini, A. (2004). Glocalizing urban landscapes: Athens and the 2004 Olympics. *Cities*, 21(3): 187-202.

Berliet, P. (1985). An approach to conservation of the industrial heritage: Marius Berliet Foundation. In: *The industrial heritage: What policies? Council of Europe Conference*. Lyons.

Berrizbeitia, A. (2007). Re-placing Process. In: Czerniak, J. and Hargreaves, G. (Eds.), *Large Parks*. Princeton Architectural Press, New York. pp. 175-197.

Bokern, A. (2006). Westergasfabriek Cultural Park. *TOPOS*, 56: 28-33.

Bothmann, F. and Auer, S. (2009). The New Emscher Valley – Reshaping an urban Landscape creates regional Identity. In: Schrenk, M., Popovich, V., Engelke, D. and Elisei, P. (Eds.), *Proceedings of the REAL CORP 2009 - 14th International Conference on Urban Planning, Regional Development and Information Society*, April 22-25, 2009, Sitges. pp. 907-909.

Brandt, J., Tress, B. and Tress, G. (Eds.), (2000). Multifunctional Landscapes: Interdisciplinary Approaches to Landscape Research and Management. *Material for the conference on "Multifunctional Landscapes"*. Centre for Landscape Research, 18-21 October 2000, Roskilde.

Britton, M. (1998). *An Evaluation of Public Involvement in Reclamation Decision Making at Three Metal Mines in British Columbia*. Master Thesis. The University of British Columbia, Vancouver.

Burley, J. and Loures, L. (2008). Conceptual Landscape Design Precedent: Four Historic Sites Revisited. In: Panagopoulos, T. and Burley, J. (Eds.), *New Aspects of Landscape Architecture Proceedings of the 1st WSEAS International Conference on Landscape Architecture*. Universidade do Algarve, June 11-13, 2008, Faro, Portugal. pp. 11-16.

Camagni, R., Capello, R. and Nijkamp, P. (1998). Towards Sustainable City Policy: An Economy-Environment Technology Nexus. *Ecological Economics*, 24: 103-118.

Carr, E. (2003). *Cultural Landscapes: Theory, Management, Design*. Graduate Seminar, Department of Landscape Architecture and Regional Planning. University of Massachusetts. Retrieved January 03, 2008, from http://www.umass.edu /history/ph/ph_word/Carrsyllabus.doc

Casado, I. (2009). *Breve historia de la protección del patrimonio industrial. Contribuciones a las Ciencias Sociales*. Retrieved November 06, 2009, from www.eumed.net/rev/cccss/06/icg4.htm.

Casella, E. and Symonds, J. (Eds.), (2005). *Industrial Archaeology: Future Directions*. Springer-Verlag, New York.

Christensen, L., Bartuska, A., Brown, J., Carpenter, S., D'Antonio, C., Francis, R. and Franklin, J. (1996). The report of the Ecological Society of America Commitee on the Scientific Basis for Ecosystem Management. *Ecological Applications*, 6(3): 665-691.

Colantonio, A. (2007). *Social Sustainability: An Exploratory Analysis of its Definition, Assessment Methods, Metrics and Tools*. Retrieved July 14, 2008, from

http://www.brookes.ac.uk/schools/be/oisd/sustai
    nable_communities/resources/SocialSustainability_Metrics_and_Tools.pdf
Contaminated Land Rehabilitation Network for Environmental Technologies (CLARINET),
    (2002). *Brownfields and Redevelopment of Urban Areas.* Federal Environment Agency
    Ltd, Wien.
Creighton, J. (2005). *The Public Participation Handbook. Making Better Decisions through Citizen
    Involvement.* Wiley, San Francisco.
Custódio, J. (1993). De Alexandre Herculano à Carta de Veneza. In: Coelho, M., *Dar Futuro
    ao Passado.* Instituto Português do Património Arquitectónico e Arqueológico,
    Lisboa.
Davies, A. (2001). Hidden or hiding? Public perceptions of participation in the planning
    system. *Town Planning Review,* 72(2): 193 – 216.
Doick, K., Sellers, G., Hutchings, T. and Moffat, J. (2006). Brownfield sites turned green:
    realizing sustainability in urban revival. *WIT Transactions on Ecology and the
    Environment,* 94: 131-140.
Duffy, K. and Hutchinson J. (1997). Urban policy and the turn to the community. *Town
    Planning Review,* 68(3): 347–362.
Dustin, D. and Schneider, I. (1998). The widening circle: The role of democratic deliberation
    in outdoor recreation conflict management. *Trends,* 35: 27-30.
Ekman, E. (2004). *Strategies for Reclaiming Urban Postindustrial Landscapes.* Master Thesis.
    Institute of Technology, Massachusetts.
Environmental Protection Agency (EPA), (2009a). *International Brownfields Case Study:
    Emscher Park, Germany.* Retrieved December 21, 2009, from
    http://www.epa.gov/brownfields/partners/ emscher.html
Environmental Protection Agency (EPA), (2009b). *International Brownfields Case Study:
    Westergasfabriek, Amsterdam, Netherlands.* Retrieved June 14, 2009, from
    http://www.epa.gov/ swerosps/bf/partners/ westergas.html
European Academy of the Urban Environment, (2001). *Emscher Park: International Building
    Exhibition (IBA).* Retrieved November 09, 2008, from
    http://www.eaue.de/winuwd/137.HTM
European Council for Construction Research, Development and Innovation, (2001). *Building
    the Future.* Office for Official Publications of the European Communities,
    Luxembourg.
Fadigas, L. (2007). *Fundamentos Ambientais do Ordenamento do Território e da Paisagem.* Edições
    Sílabo, Lda, Lisboa.
Faga, B. (2006). *Designing Public Consensus – The Civic Theater of Community Participation for
    Architects, Landscape Architects, Planners, and Urban designers.* John Wiley & Sons,
    Inc, New Jersey.
Ferreira, V. (1998). Património Urbano. A memória da cidade. In: *Urbanidade e Património.*
    Sociedade Industrial Gráfica Telles da Silva, Lda, Lisboa. pp. 53-61.
Fiorino, D. (1996). Environmental Policy and the Participation Gap. In: Lafferty, W. and
    Meadowcroft, J. (Eds.), *Democracy and the Environment: Problems and Prospects.*
    Edward Elgar Publishing, Cheltenham. pp. 194-212.
Forman, R. (2002). The missing catalyst: design and planning with ecology roots. In:
    Johnson, B. and Hill, K. (Eds.), *Ecology and Design: Frameworks for Learning.* Island
    Press, Washington DC. pp. 85-109.

Fuchs, M. (2010). *Industrial Heritage Monuments on the UNESCO World Heritage List.* Retrieved April 12, 2008, from http://www.ihtourism.pl

Gastil, R. and Ryan, Z. (2004). *Open New Designs For Public Space.* Van Alen Institute, New York.

Gaventa, S. (2006). *New Public Spaces.* Octopus Publishing Group, London.

Giddings, B., Hopwood, B., Mellor, M. and O'Brien, G. (2005). Back to the City: A Route to Urban Sustainability. In: Jenks, M. and Dempsey, N., *Future Forms and Design for Sustainable Cities.* Architectural Press, Oxford. pp. 13-30.

Gil, A. (1994). *Metodologia do Ensino Superior.* Atlas, São Paulo.

Green, O. (1985). Our recent past: The black hole in museum collections. *Museum Journal* 85, 1:5-7.

Grumbine, R. (1994). What is ecosystem management? *Conservation Biology,* 8(1): 27–38.

Gunderson A. (1995). *The environmental Promise of Democratic Deliberation.* University of Wisconsin Press, Madison.

Gustafson, K. and Porter, N. (2007). *Practice Profile.* Retrieved March 10, 2007, from http://www.gust afson-porter.com/ showcase6.htm

Hartig, J., Zarull, M., Heidtke, T. and Shah, H. (1998). Implementing Ecosystem-based Management: Lessons from the Great Lakes. *Journal of Environmental Planning and Management,* 41(1): 45-75.

Hayden, D. (2000). Forward: In Search of the American Cultural Landscape. In: Alanen A. and Melnick, R., [eds.]. *Preserving Cultural Landscapes.* Baltimore: Johns Hopkins University Press.

Hester, R. and Blazej, N. (1997). Three Phases of Participatory Landscape Architecture. *Council of Educators.* Landscape Architecture Conference, September 10-13, 1997, Asheville, North Carolina.

Hough, M. (1995). *Naturaleza y Ciudade, Planificación Urbana y Procesos Ecológicos.* Editorial Gustavo Gili. Barcelona.

International Council on Monuments and Sites (ICOMOS), (2008). *Charters adopted by the General Assembly of ICOMOS.* Retrieved March 06, 2007, from http://www.international.icomos.org/home.htm

Kirkwood, N. (2003). Brownfield Passages: From Westergasfabriek to the New Westerpark. In: Koekebakker, O. *Westergasfabriek Culture Park.* NAi Publishers, Rotterdam. pp. 5-7.

Koekebakker, O. (2003). *Westergasfabriek Culture Park.* NAi Publishers, Rotterdam.

Krauel, J. (Ed.), (2008). *Urban Spaces: New City Parks.* Links, Barcelona.

Krinke, R. (2001). Overview: design practice and manufactured sites. In: Kirkwood, N. (Ed.), *Manufactured Sites: Rethinking the Post-Industrial Landscape.* Taylor & Francis, New York. pp. 125-149.

Kuhl, B. (2004). Questões Teóricas Relativas à Preservação da Arquitectura Industrial. *Desígnio,* 1: 101-102.

Landscape Institute, 2007. *Westergasfabriek Park Amsterdam.* Retrieved March 10, 2007, from http:// www.reusebv.com/projecten /Upload/westergas.pdf

Latz + Partner, (2007). *Landscape Park Duisburg Nord: Metamorphosis of the Blast Furnace Plant Thyssen – Meiderich.* Retrieved October 27, 2007, from http://www.latzundpartner.de/projects/ detail/17

Latz, P. (1992). Duisburg North Landscape Park. *Anthos,* 3(3): 27-32.

Latz, P. (2001). Landscape Park Duisburg-Nord: the metamorphosis of an industrial site. In: Kirkwood, N. (Ed.), *Manufactured Sites – Rethinking the Post-Industrial Landscape*. Taylor & Francis, New York. pp. 150-165.

López, F. (2004). *Arquitectura y Naturaleza a Finales del Siglo XX 1980-2000: Una Aproximación Dialógica para el Diseño Sostenible en Arquitectura*. Doctoral Dissertation, Universitat Politècnica de Catalunya, Barcelona.

Loures, L. (2008a). Industrial Heritage: a gear to redevelopment. *Proceedings of the EURAU 08 – Cultural Landscape - 4th European Symposium on Research in Architecture and Urban Design*. January 16-19, 2008, Madrid. pp. 1-7.

Loures, L. (2008b). Post-Industrial Landscapes as renaissance locus - the case study research methods. In: Brebbia, C., Gospodini, A. and Tiezzi, E. (Eds.), *Sustainable City V*. WIT Press, Southampton.

Loures, L. (2008c). Post-Industrial Landscapes: dereliction or heritage? *Proceedings of the 1st WSEAS International Conference on Landscape Architecture*, Universidade do Algarve, June 11-13, 2008, Faro, Portugal. pp. 23-28.

Loures, L. (2011). *Planning and Design in Postindustrial Land Transformation: East Bank Arade River, Lagoa – Case Study*. PhD Dissertation, Universdidade do Algarve, Faculdade de Ciencias e Technologia: Faro, Portugal.

Loures, L. and Panagopoulos, T. (2007a). Recovering Derelict Industrial Landscapes in Portugal: Past Interventions and Future Perspectives. *Proceedings of the International Conference on Energy, Environment, Ecosystems and Sustainable Development*, Agios Nikolaos, July 24-26, 2007, Crete Island, Greece. pp. 116-121.

Loures, L. and Panagopoulos, T. (2007b). Sustainable reclamation of industrial areas in urban landscapes. In: Kungolas, A., Brebbia, C. and Beriatos, E. (Eds), *Sustainable Development and Planning III*. WIT Press, Southampton. pp. 791-800.

Loures, L. and Panagopoulos, T. (2010). Reclamation of derelict industrial land in Portugal - greening is not enough. *International Journal of Sustainable Development & Planning*, Vol. 5(4) 343–350.

Loures, L., Heuer, T., Horta, D., Silva, S. and Santos, R. (2008). Reinventing the Post-industrial Landscape: A Multifunctional Cluster Approach as redevelopment Strategy. *Proceedings of the 1st WSEAS International Conference on Landscape Architecture*, Universidade do Algarve, June 11-13, 2008, Faro, Portugal. pp. 123-129.

Macaulay, R. (1953). *The Pleasure of Ruins*. Weidenfield and Nicolson. London.

Mendes, J. (1995). A arqueologia industrial ao serviço da história local. *Revista de Guimarães*, 105: 203-218.

Meyer, P. (2000). Accounting for Stigma on Contaminated Lands: The Potential Contributions of Environmental Insurance Coverages. *Environmental Claims Journal*, 12(3): 33-55.

Montaner, J. (2001). *A modernidade superada: arquitectura, arte e pensamento do século XX*. Editorial Gustavo Gili, SA, Barcelona.

Nassauer, J. (1997). *Placing Nature: Culture and Landscape Ecology*. Island Press, Washington DC.

Neyret, R. (2004). Du monument isolé au «tout patrimoine». *Géocarrefour*, 3(79): 231-237.

Nickerson, T. (2007). *Landschaftspark Duisburg-Nord*. Retrieved October 27, 2007, from http:// courses. umass.edu/latour/Germany/tnickerson/index.html

Ozguner, H. and Kendle, A. (2006). Public attitudes towards naturalistic versus designed landscapes in the city of Sheffield, (UK). *Landscape and Urban Planning*, 74(2): 139-157.

Pateman, C. (1970). *Participation and Democracy Theory*. Cambridge University Press, Cambridge.

Potts, T. and Harrill, R. (1998). Enhancing communities for sustainability: A travel ecology approach. *Tourism Analysis*, 3: 133-142.

Rae, D. (2005). *City: Urbanism and its End*. Yale University Press, New Haven.

Rahman, N. (1998). *Development of a Riverfront Park Planning Model with Application to Islamic Perspective*. Doctoral Dissertation. Michigan State University, East Lansing.

Ramadier, T. (2004). Transdisciplinarity and its challenges: The case of urban studies. *Futures*, 76: 423-439.

Rea, C. (1991). *Rethinking the Industrial Landscape: The Future of the Ford Rouge Complex*. Master Thesis, Massachusetts Institute of Technology, Cambridge.

Regeneration of European Sites in Cities and Urban Environments (RESCUE), (2004). *Best Practices in Citizen Participation for Brownfield Regeneration*. Retrieved January 10, 2007, from http://www.rescue-europe.com

Secchi, B. (2007). Section 1: Wasted and Reclaimed Landscapes: Rethinking and Redesigning the Urban Landscape. *Places*, 19(1): 6-11.

Selman, P. and Parker, J. (1997). Citizenship, Civicness and Social Capital in Local Agenda 21. *Local Environment*, 2(2): 171-184.

Shaw, R. (2002). The International Building Exhibition (IBA) Emscher Park, Germany: A Model for Sustainable Restructuring? *European Planning Studies*, 10(1): 77-97.

Smith, D. (1974). *Amenity and Urban Planning*. Lockwood Staples, London.

Spens, M. (2007). Deep Explorations Into Site/Non-Site: The Work of Gustafson Porter. *Architectural Design*, 77 (2): 66-75.

Steelman T. (2001). Elite and participatory policymaking: Finding a balance in a case of national forest planning. *Policy Studies Journal*, 29(1): 71-89.

Steiner, F. (2000). *The Living Landscape: An Ecological Approach to Landscape Planning*. McGraw-Hill, New York.

Stilgenbauer, J. (2005). Landschaftspark Duisburg Nord: Duisburg, Germany. *Places*, 17(3): 6-9.

Storelli, C. (2003). The city as heritage. In: *Towns and Sustainable Development – Council of Europe, Naturopa*. Gilly, Bietlot.

Stratton, M. and Trinder, M. (2000). *Twentieth Century Industrial Archaeology*. Spon press, London.

Strauss, A. and Corbin, J. (1990). *Basics of qualitative research: grounded theory procedures and techniques*. Sage, Newbury Park.

Strike, J. (2003). *Architecture in Conservation. Managing development at historic sites*. Digital Printing, New York.

Sustainable Cities, (2008). *Emscher Park: From dereliction to scenic landscapes*. Retrieved October 21, 2008, from
http://sustainablecities.dk/en/city-projects/cases/emscher-park-from-dereliction-to-scenic-landscapes

Taylor, M. (2000). Communities in the Lead: Organizational Capacity and Social Capital. *Urban Studies*, 37(5): 1019-1035.

The International Committee for the Conservation of the Industrial Heritage (TICCIH), (2003). *Nizhny Tagil Charter for the Industrial Heritage*.

Triviños, A. (1995). *Introdução à pesquisa em ciências sociais: A pesquisa qualitativa em educação*. Atlas, São Paulo.

Tymoff, M. (2001). *Reinterpreting the Post-Industrial Landscape Athens' Former Manufactured Gas Plant*. Master Thesis, University of Georgia, Athens.

United Nations Educational, Scientific and Cultural Organization (UNESCO), (2010). *World Heritage List*. Retrieved April 12, 2010, from http://whc.unesco.org/en/list

Vasconcelos, L. (2001). New forums out of sustainability – recent trends at local level. *First World Planning Congress – ACSP-AESOP-APSA-ANZAPS*. Tongji University, Shangai, July 11-15, 2001.

Vollmer, M. and Berke, W. (2006). *Ruhr Picturebook. Industrial Heritage – new life in old buildings*. Klartext Verlag, Essen.

Von Haaren, C. (2002). Landscape Planning Facing the Challenge of the Development of Cultural Landscapes. *Landscape and Urban Planning*, 60: 73-80.

Weilacher, U. (Ed.), (1999). *Between Landscape Architecture and Land Art*. Birkhauser Basel, Berlin and Boston.

World Bank, (1992). *Governance and Development*. The World Bank, Washington DC.

13

City Image – Operational Instrument in Urban Space Management – A Romanian Sample

Marius-Cristian Neacşu and Silviu Neguţ
Bucharest Academy of Economic Studies
Romania

1. Introduction

One of the fundamental realities that have marked mankind's existence on Earth is the *urban reality*. Throughout the ages, in all geographic regions, the fascination of the city has set in motion people, resources, ideas, generating forces of unforeseen intensity that have continuously modelled the planet's surface. Invoked in lyrics, pinned in eternity – "I know: the city will be" (Russian poet Vladimir Maiakovski, 1893-1930) –, interrogated by sciences, object, subject and support of various professions, the city continuously challenged human knowledge. Virtually, there is no science that has not attempted to unravel its mysteries, either sequentially or wholly, without leaving behind concepts which later evolved into professions or academic disciplines. Over time, a strange blend of academic disciplines and professions accompanied the evolution of knowledge regarding the city and as the study's complexity increased these professions and disciplines also prospered.

Theories and concepts have succeeded, gathered, refined, but some issues remained constant throughout generations of residents or various specialists: the urban space is much too complex, too vaguely defined, too hybridised, with a stunning mix of functionalities, polysemantic, with a much too confused image in its own residents' minds, with an many interests that must be mediated and problems to be managed. Dilemmas remain in this complexity of elements: quantitative or qualitative approach? System or phenomenon? Reality or image? Are residents prisoners of the urban habitat or beneficiaries that actively take part in the urban planning and re-planning? How can we fully and sufficiently address such a complex set of issues?

In the end, all these should serve one philosophy: the city must be a good place to live and work in.

But how can we measure the impact the intervention in the city's quantitative dimension (elements, flows, shapes) is going to have on the residents' spatial behaviour and attitude?; how can we use the residents' perception – "the mental city" – in planning the future ways of urban space organisation?

This study aims to: 1. Emphasize *geography's role as a science in the integrated approach of spatial manifestation phenomenon and processes* – geography is, first of all, the science of places (with all that a place's spatial reality means; the city is a place!) and 2. *Identify and test the role that city image can play in the process or urban space organisation* and in a broader scope, in the integrated process of urban planning.

It is the intrinsic need of today: the integrated approach of the concepts that operate in the urban space, concepts which although make the object of study for various social sciences or their branches in areas such as economy (e.g. *urban marketing and branding*), town-planning (*urban design*), sociology (*urban segregation*) or urban psychology (*urban behaviourism*), target, above all else, *processes with a spatial/territorial manifestation.*

In this spirit, geography's participative-constructive role becomes obvious for at least two reasons: 1. most of the processes and phenomenon that are part of the approach to city issues have a spatial character and space is one of the essential variables in socio-human systems (from economic activities to matters of sociology and psychology); 2.this science has specialised, over the course of the last decades, in microscale spatial analyses (neighbourhood, city etc.).

It is thus felt the dire need of identifying an operational instrument to mediate all these urban space approaches and that would provide a link between the residents, the urban planners, the urban subspaces "producers" (economic entities etc.), specialists, theoreticians, in other words an instrument that would connect all interests... Over the course of time, the city "belonged", more or less discretionary, to one category, but never to all categories at the same time. But which would be the instrument that could provide the link between spatial reality and human will, between urban actors and their interests, between the city-system and the city-phenomenon, between function and meaning...? One possible answer: *city image.*

And there is another thing: globalisation. Globalisation catalysed to an unprecedented intensity both the processes and their phenomenology. Today's cities are being restructured according to new rules and forces, overcoming national borders. The city of the present competes both for attracting new residents and for retaining its old ones with cities all across the planet, aiming to increase the standard of life it offers, imprinting a way of life that would distinguish it from the other cities, a way of life that is essentialised and synthesised in an image that would impress in the residents' minds. It is that city's brand, its "signature", a guarantee of quality and added value to the offered conditions of life. It is the image being sold and that can currently determine its place in the global hierarchy. It is a set of symbols in which the residents can identify themselves. We currently live a genuine image myth. Nothing sells better than image and it has some unsuspected resorts in stimulating decision... All our pieces of information are included in images, as our emotions and feelings likewise, we sell and we buy images, we are worshipers of the image cult.

The aim of this study, as stated above, also substantiates the *topicality of the research*, even the *scientific freshness, analysing the city image and integrating it into a specific conceptual context* (its relation with other operational concepts from the urban sphere such as *urban planning* and *urban design*, with *urban marketing* and *branding*) individualising a complex approach from multiple perspectives – urban-sociologic-economic -, while using a geographic (integrating) thought process. From the science of places (the science of [geographic] space) to the science of planning and creating the place (space).

2. Urban space – Elementary operational entity in analysing regional planning

Space, geographic space, urban space. Concepts, semantics, approach

Starting from one of the questions of this research – what role could geography, as a science, play in organising urban space and which could be the practical valences with which a geographer could take part in a mixed team of specialists that would plan a model of organising urban space? –, we immediately have the opportunity to identify some viable answers. And these could be synthesised as such: understanding space (metabolism, phenomenology) and using an integrated approach for it.

Above all else the city represents "an objective form of existence of a human community on Earth" (Neacşu, 2010a), a highly anthropic "piece" of space resulted and modelled in time by the action of all the geospheres (with a clear dominance of the anthropic or socio-sphere component). The city occupies a concrete space, precisely located, visible through its morphology and components – *urban landscape* -, is the result of the corroboration and interaction of several geographic conditions – *urban environment* -, it acts and functions as an optimally open thermodynamic and informational system, with a dissipative structure (Ianoş, 2000) – *urban system* –, it represents such a specific way of life that it influences attitudes, behaviours, ideas and value systems becoming a true phenomenon – *the urban phenomenon* -, generating through its dynamic countless new urban subspaces.

Even though there are some answers, the city still remains sufficiently complex and complicated, with a larger number of unknowns than known (sort of a "grey box"), which trouble urban spaces specialists and managers, still being too difficult to answer questions such as: *but, still, what is the city? How can we control it?* A true methodological and semantic thicket has accompanied, over time, analytic studies of urban space, so defined and yet without definition, so clarified to the smallest of details and yet obscure from an analytic point of view, the city is in every époque, for every generation of specialists, always surprising and seeming to increasingly sediment the idea that a city is more auto-organising itself than it can be organised, managed.

It is not this study's purpose to attempt a more comprehensive, synthetic and essential definition, but in the spirit of the analytic process we could select two definitions of the city, one from Antiquity and the other from the beginning of the third millennium, both synthesising the method of understanding, the semantics and the expectations of an era. Thus, if in the 4th century BC Aristotel (384-322 BC) saw the city as a desideratum – "a city must be built to offer its inhabitants security and happiness" (as cited in Cucu, 2001) –, present times maintain the desideratum in an implied manner, insisting more on accepting the city as an objective reality – "the city represents a superior form of organisation of space with concrete attributes, quantifiable, more or less delimited from an administrative-judicial point of view (…) imprinting new qualitative characteristics to life (…) [of type] *urbs"*(Cucu, 2001).

Many meanings can be extracted from the previous two definitions, some becoming advanced subjects of analysis for sciences that operate today with the urban conceptual arsenal– for example urban sociology and the residents' perception approach of the state of happiness or public security (could the present studies referring to these aspects or recent events in European cities such as Paris, London, etc. have disappointed Aristotel?!) –, but we will only stop to analyse two of them: 1. The city must be built (planned!) and 2. The idea of space.

If we take a look at the two phrases – *planning* and *space* – and we track them in several schools of thought we will notice very important nuances (Table 1).

	The philosophic category – frequently used terminology		
School of thought	Space	Urban space	Urban planning
German school	*Raum*	*Städtischen Raum*	*Stadtplanung*
French school	*Espace*	*Espace urbain*	*Aménagement du territoire urbain,* *Aménagement urbain*
Anglo-Saxon school	*Space*	*Urban space*	*Urban planning*
Italian school	*Spazio*	*Spazio urbano*	*Pianificazione urbana*
Spanish school	*Espacio*	*Espacio urbano*	*Planificación urbana*
Romanian school	*Spaţiu*	*Spaţiu urban*	*Organizarea spaţiului urban*

Table 1. Notions of *space, urban space and urban planning* in various schools

Thus, if the phrase "space" or "urban space" is similar, even keeping the same root, even in the case of different linguistic families (the case of Roman and Germanic families, just Anglo-Saxon!, probably showing the same linguistic source of the expression; no longer the case of German language), in terms of the phrase "urban planning" things change significantly, the terms' semantics inducing a net differentiation. As such, if in the Anglo-Saxon literature the most used term for the concept of urban space organisation is *urban planning* (similarly in the German school, but also the Spanish and Italian ones), in the French one appears the phrase *aménagement*, while Romanian authors frequently use the term *organisation*. This semantic translation from the original phrases of "space" and "urban space" can only confirm both the complexity and the prudence in approaching the urban space and space in general, when it comes to planning and management.

What are the identified nuances of the three schools of urbanism and urban geography in the specialised literature? The three phrases – *planning, organisation* and *improvement* – have different meanings, which express three intermediate phases of the same process: *voluntary intervention in the relocation or resizing urban elements/components, shapes and flows*, synthesising the definition of urban planning. Thus:

-   *planning* represents the immediate action following the intention to intervene; it is the chain between thought, intention and deed, the actual action; it represents the theoretical and mental laboratory in which the action plan that will be used in all phases of the intervention is being elaborated, sketched, prepared; as an actual product we can have the millimetre paper on the architect's draw board, or the computer software that foresees and includes all the details, structures them on phases and well delimited stages, with well drawn and predicted effects and expectations resulted from mathematic models;

- *organisation*, the frequent phrase in Romanian literature, is a follow-up of planning (but it can include it as well, representing the activity between the crystallisation as a thought of the intention to intervene and obtaining effects) indicating the set of coordinated actions in virtue of the elaborated plan, in order for the objectives to be achieved; from among the intrinsic associated phrases to this concept the following cannot be missing: making it work organically, as a whole, methodically coordinate after a well thought plan, synchronise, systematise, rank, prioritise, sort actions and activities;
- *improvement*, the phrase used by French authors leaves the passive spectrum bringing an increased active participation; already designates the preparation of a terrain (space, territory) for a certain use.

All this terminological and semantic interrogation illustrates one aspect: not only the more or less developed predisposition for either the theoretical spectrum or the practical, applicative aspects of the mental type that dominates one school or another (this would have meant that the French and Anglo-Saxon mentality would have inversed their roles, taking into consideration the highly theoretical-analytical inclination of the French and the especially pragmatic and practical approach of the Anglo-Saxons), but the differentiated approach and reference to the concept of *space*, in general, respectively *urban space*, in particular. If we introduce into the equation adjacent phrases such as *geographic space* or *territory* (the city occupies a portion of this on the surface of the Earth, in a specific place, but also interacts with the other geo-spheres), at least from a theoretical point of view, with the afferent practical implications, the things seem to get hugely complicated.

In this context, the geographer could play a very important part in decrypting these phrases and operating with them in some complex trans and inter-disciplinary analyses, such as the ones related to the interventions on the urban space. In the end, the geographic space "seems to be geography's privileged project (...) the king concept of [geographic] science" (Dauphiné, 2004, as cited in Neguț, 2011), more often than not both French and Anglo-Saxon representatives of the schools of human geography slowly "melting" the word "geographic" from "geographic space" for a better cohesion and coherence on the spatial analyses shrine, a fact well revealed also by a representative author of the Romanian school of urban geography who entitled an important part of a paper "Space – central variable in economy and systems of settlements", emphasising the idea that geography is a science of space, the geographer being a mandatory part in any spatial analyses, from micro to macro-scale (Ianoș & Heller, 2006).

We consider this fine delimitation of terminological phrases, the understanding of the metabolism of these urban space entities significant because planning different models of urban planning depended on the way space/geographic space was perceived and understood and ultimately the quality of human life and of the urban habitat depended on the results of those models. On one hand space, through its characteristics, influences human attitude, behaviour and perception, the human's image of space and on the other hand human behaviour falls, as a feedback, on the method of planning, the method in which space is conceived, on the way it is filled with meanings. Again, a new model of planning will influence man's perception and image. It's a causal chain.

As for the notion of *space*, in a general, philosophical approach, this defines *an objective form of existence of matter*. It is an intrinsic law of existence of matter (concrete, visual shape of energy), to which one can add: permanent movement and time (dynamic).

Space is defined by some metric properties (distances, surfaces, volumes), it implies an ordering of objects, processes and phenomenon based on the logic of mutual relations and it has a series of specific qualities, among which: continuity, coherence, multi-dimensionality, **self-organisation** and others (Ianoş & Heller, 2006). From here results the idea that when we approach a series of subspaces (urban, rural, agricultural, industrial, transport, commercial, touristic etc.) in the general context of "regional planning" we're in fact referring to short voluntary human interventions at certain levels of scale (neighbourhood, city ... locally, regionally, nationally etc.), with effects in certain timeframes in which prevail, for a certain moment, the continuous character of self-organisation of space, of macro scale self-organisation.

For a geographer space represents the essence of his science, all the analytical approaches of a phenomenon, process or element, referring to space, distribution in space, conditioning this model with the spatial characteristics and meanings etc. Individualising *geographic space*, as a particular space came by itself. This is the space on the surface of the planet (terrestrial space) where the geo-spheres meet and where the complexity and intensity of their interactions is at maximum. To facilitate its analysis, geographers resorted to a necessary reductionism substituting the terminological notion of geographic space with *territory*, which became a fundamental operational entity and at the same time a "total barometer" of the condition human-space-time – "the localisation of the economic space, the materialisation of the psycho-social space and the temporisation of the historic space" (Beaujeu-Garnier, 1971, as cited in Cândea & Bran, 2006) –, as French geographer, plastically and in a reductionist manner, stipulated, in defining geographic space.

In this epistemological order, *urban space* is only a subspace of the geographic space, a taxonomic level for which it takes and keeps all the rules of the first, but with its own capacity to develop certain dominant properties.

Thus, there are at least four perceptive dimensions of space, as they were crystallised over time, four different ways of reporting man and society to space, or, citing another French geographer (Dauphiné, 2004), terrestrial [geographic] space is *given* (the "first space" at Thrift, 2003 and the "absolute space" at Cocean, 2002), *produced* ("relative space" at Cocean, 2002 or "space of connections" at Thrift, 2003), *perceived* (reflecting a "second reality" – the subjective reality –, which derives from the "first reality" – the objective one – being filtered by human sensors and receptors) and *lived* (at Frémont, 1976; the "third space" at Soja, 1996 or "interpretative space" at Zierhofer, 1999; it is a synthesis of the first two physical spaces – absolute-relative – and imagined, a hybrid space, a space of direct, lived experience; it is a space of simultaneous representations, of simultaneous, transposed spaces).

Space becomes a social product, made up of many subjective spaces, generated by multiple actors. Soja (1990) introduces the notion of "hyperspaces", putting forward the idea that space is not given, is not a stage on which a drama is being played, neither a box that needs filling, but a cultural product, part of the second nature (the subjective one, "the second reality") which transforms both the physical space as well as the psychological one (imaginative).

City paradigms. The mix between urban "systemology" and "phenomenology"

Urban space, as a particular subspace, made no exception from this evolution of the approach philosophy, from this dynamic of understanding and deepening the rapport between man and space in general, all currents of thought of the 20th century marking the researches of urban geography, urbanism and so on. The semantic translation of the rapport man-space was different, being able to be enrolled in a general binomial: physical-mental, modern-postmodern, system-phenomenon, function-image, given-product, determinism-possibilism etc., every epistemological terminal marking the periods and studies that Minca (2001) named "academic fashion" in one of his works.

All these approaches of urban space, from the determinist current ("environmentalism") that marked empirical studies, to matters of the "philosophy of morality" from the beginning of the new millennia, from the modern approach to the postmodern one, no new current of thought escaping profound critics, with numerous returns to epistemological reflections, crystallised several ideas:

- it is not a problem of accepting or rejecting one or another of the approaches because the human society's way of thinking is in a constant process of change and the philosophy of perceiving and accepting urban space has the same dynamic based on the necessities, purposes, values and mentality of every generation. Countless spatial analyses studies have emphasised new research niches and have brought new ways of understanding and managing urban space.
- although the dynamic of these currents of thought has followed a chronological line, one current or another dominating a certain period, based on whether the urban space was in focus, either as an objective geographic reality or as a social construct, or the residents and their spatial behaviour were, reality proved that quantitative analyses must not exclude the qualitative ones, that "positivism" cannot fill the role of "behaviourism", ultimately that a positive statistic – a large number of kilometres of street, trams, water pipes, wireless networks or various public utilities – does not guarantee a better perception of life in the urban environment, a residents' state of happiness and satisfaction (remembering Aristotel's definition from the end of Antiquity: the city must offer its residents security and happiness! This desideratum has remained valid to this day...).

"What makes a good city?... Cities are too complicated, too far beyond our control and affect too many people, who are subject to too many cultural variations, to permit any rational answer. Cities, like continents, are simply huge facts of nature, to which we must adapt. We study their origin and functions because that is interesting to know and handy for making predictions." (Lynch, 1984) – starting from this well-known American urban planner theoretician's assertion we just cannot help but repeat his thoughts, almost obsessively "What makes a good city?...": a larger number of kilometres of urban infrastructure and town network or a stronger place identity?; public security, urban accessibility, efficient transport or a stronger feeling of membership to that place and that urban community?; uphill statistics or meaning, orientation?; a town drawn by specialists, technicians, urban managers or a city of the residents? Are the people just prisoners in the urban habitat or active players in urban planning?...

The city between quantitative and qualitative approaches: both (Fig. 1). It is an equation with many unknowns, but which functions as a two-stroke engine: one piston is the quantitative part, the other is the qualitative one; they strengthen each other.

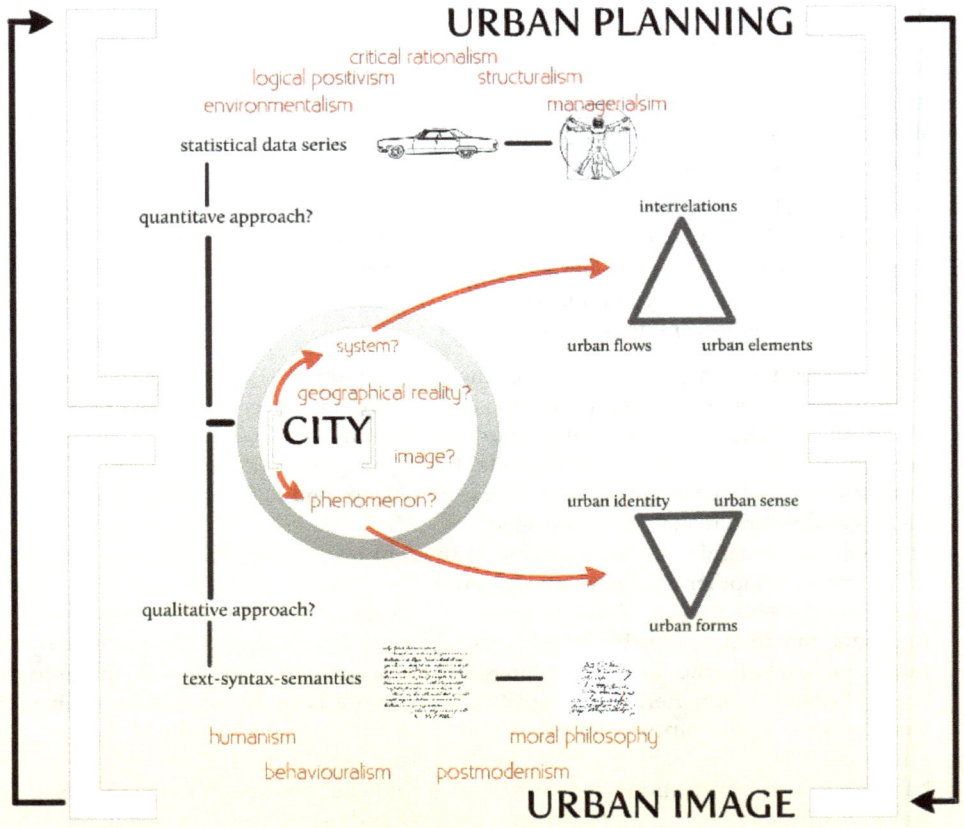

Fig. 1. City paradigms: The mix between urban "systemology" and "phenomenology"

The physical city is doubled by the mental one. The mental city shapes or should indicate the dysfunctions of the physical city. People are no longer the "prisoners" of a predefined, wholly determined urban model, but instead they like or manifest a repulsive attitude towards certain urban subspaces, shaping several types of spatial behaviour: topophilia, topophobia or topoindifference. Inhabitants thus become an active element in reshaping and resizing the urban architecture and shapes.

3. City image: A possible "key" to a good urban planning

3.1 *City image* concept

Short literature review

The American urban-planner Kevin Andrew Lynch (1918-1984), one of the most representative theoreticians in the field of urban design and planning, with a rich publishing activity, coins the term and concept, using it as a title for his 1960 study – *The image of the city* –, although the frequently used expression throughout the book is that of "image of environment" (in reference to the urban environment, having three representative case studies – the cities of Boston, Jersey City and Los Angeles). Also, studying the specialised literature (see Neacşu, 2009, 2010b), reveals at least three main directions in which the concept can be pursued:

1. *proper city image* as it was entitled by its founder, as a mental projection of the city – „(...) every citizen has had long associations with some parts of his city, and his image is soaked in memories and meanings" (Lynch, 1960). Lynch also initiated several theories in the field of urban design, among which the most expressive is the one related to good city forms, in which he attempts to prove that urban landscape design is not a purely physical determinant, but social and psychological factors also weigh in considerably, the essence of his publishing and professional activity consisting of the way the habitual environment is perceived by its inhabitants (opening a research area – besides the work aforesaid he also published *What Time is this Place?* in 1972, *A Theory of Good City Form* in 1981 and several articles – see also Banerjee & Southworth, 1995 – to which other authors adhered, such as: Jacobs, 1961; Sommer, 1969; Downs & Stea, 1973; Alexander, 1987; Hiss, 1990; Nasar, 2001; Ianoş, 2004 etc.).

2. *in the sphere of mental maps and connected concepts.* In this context it experienced a larger publishing activity, starting from the field of psychology and neuropsychology (worth mentioning the *cognitive map* idea's and phrase's debut moment – Tolman, 1948 and the monumental academic „debate" between Kosslyn and Pylyshyn from the 80's regarding the origin, structure and properties of the mental image) and finding itself as an idea in numerous phrases: *topological plans* (Griffin, 1948), *environment images and city image* (Lynch, 1960), *topological representations* (Shemyakin, 1962), *space plans and cognitive plans* (Lee, 1968), *conceptual representations* (Stea, 1969), *cognitive representations* (Downs & Stea, 1973), *mental images* (Pocock, 1973), *mental maps* (Gould şi White, 1974), *orientation plans* (Neisser, 1976), *cognitive configurations* (Golledge, 1977), *cognitive systems* (Canter, 1977), *spatial representations* (Allen et al., 1978), *cognitive images* (Lloyd, 1982), *mental representations* (Gale, 1982), *world's graphics* (Lieblich & Arbib, 1982), *cognitive atlas* (Kuipers, 1982), *cognitive space* (Montello, 1989), *abstract maps* (Hernandez, 1991), *configuration representations* (Kirasic, 1991), *place plans* (Axia et al., 1991), *cognitive collage and mental spatial model* (Tverski, 1993) and so on.

3. *the city image, as a technique of urban marketing – urban branding,* where it was found as an idea in works of authors such as: Ashworth & Voogd (1990), Ashworth (1998, 2005), Kotler et al. (2002), Metaxas (2002), Kavaratzis (2004), Deffner & Liouris (2005) and so on.

Deffinition

The city image makes its appearance on the scene of scientific research when the humanist currents of thought (*behaviourism, humanism*) bring a different light over the accepted urban space, as a reply to the rationalist, machinist, determinist approach (based on *environmentalism* and even more on the *positivist* current).

The city is not only a given space (absolute), but also a socio-cultural construct imprinting in its residents' minds a certain image that generates specific spatial attitudes and behaviours. The operational instruments with which a geographer can "map" the *image* or *mental map* and/or the *cognitive map*, thus capturing the territorial differences regarding the individual or group perception and meaning (identity) over a certain geographic space (city, neighbourhood etc.). As such, the *image*, for the above purposes, essentialises the information from the "lived space" („l'éspace vecu" used by Frémont, 1976, also used by Lefebvre or Soja) as a result of daily experience and as a hybrid complex between reality and its perception, experimented simultaneously. Most of the time the daily city life does not assume us directly perceiving it (as we did the first day we moved in), but indirectly, through a lens, by mentally rebuilding past sensory experiences, however continuously enriched by new information or new sensations, feelings imposed by (re)perceiving a certain place currently undergoing a certain change, modification, forced by its natural evolution (either by its internal dynamic, or by factors that control and manage urban space).

Thus, *the city image* represents "an essentialised reality at city level, filtered by a subject and put in circulation as information" as Ianoş (2004) stipulated.

- "reality" expresses life lived in that urban space, in all its complexity, with all daily events and emotions, feelings and sensations generating by those experiences. This reality is essentialised at the lived space level (the interaction between the physical space, of direct experience, and mental perception, of the emotions forced by this experience).
- "the subject" is the individual, but also the human group, and based on their source they can be: residents (that perceive the city directly, daily) and non-residents (that perceive the city directly, through limited experiences – temporary residence for study or work, commuting, touristic transit etc. or indirectly, by means of oral communication, mass-media, advertisements, articles or books etc.). The variables through which the generic subject synthesises the information are very complex and from a sociologic point of view they can be classified into at least two categories: independent (sex, age, health condition and so on) and dependent (education level and cultural "background", income and so on).
- "the information", extremely complex as a construct and content, is also synthesised at the impact generated by the city's visual personality (a result from the balance between the urban space's functionality and its malfunctions, between a planned way of organising and using terrains and the city's evolution), at the general impression people have of a place, which quite frequently expresses the positive or negative characteristics that the place's name invokes mentally and sensory (Cowan, 2005).

Characteristics and specific elements

The perception of a place is complex and is not a simple result of direct observation, visual or otherwise, but a stored construct, constantly updated with new information, that keeps the memory of past experiences and carries emotions, sensations, meaning and identity. But in order for a city to be "good", according to Lynch, it must present certain properties obtained from practices of urban planning and urban design such as:

- *legibility*. It is a notion that expresses the *urban landscape's degree of clarity*, meaning the ease with which any part of the city can be identified and its image organised in a coherent model. Thus, the city is capable of generating a lower or higher visual quality for the receiving subject.

To this end, Lynch associates the urban space to a grammatically and literary coherent and correct text, which through its coherence and logic is capable of producing strong impressions to the reader. A city's residents or passers-by must be able to, in a similar manner, easily "read" the city, resulted from the current planning method, which means that understanding and organising information, orientation in the urban landscape are facile. Landmarks, strong visual elements are easily recognised and at a mental level, symbols are assimilated after a coherent cognitive structure.

On the city's degree of legibility depends the city image's quality, registering in a complex spectrum from the negative dimension to the positive one. It is intuitive that a city or neighbourhood with a good, positive image will always attract investors, tourists and new residents at the expense of another with a less favourable image.

Urban legibility also carries an important social role through the strong collective image it can generate, increasing the sense of identity, of affiliation to a certain place – "(...) the city must be readable, legible. If the residents do not see the purpose of urban spaces and these do not generate a certain sense of identity that can help them easily find a certain route, the city loses an important part of its informative capacity, making it hard to reach" (Bohigas, 1999, as cited in Cowan, 2005) –, thus guaranteeing a sort of *emotional security*. Expressing a city's visual quality reception, capable of generating unified and coherent image (Miles, 2004) legibility becomes an *imperative function of urban design and planning*.

-   *imageability*. It is also a notion introduced in the work „Image of the City", targeting *a city's quality to provoke a strong image perception to an external observer* (Lynch, 1960). To this quality contributes not only the respective urban landscape's visual personality, but also its coherent structure and the sense of identity that the city is able to generate. Because a "good place" is the one that can be mentally mapped by individuals, with a spatial organisation easy to remember, the American urban-planner fixes several elements of the urban space that can become strong landmarks of mental maps, among which: *routes, ways, directions of movement, limits* or *discontinuities, neighbourhoods, nodes* and *landmarks* (for more details see Lynch, 1960).
-   *liveability*, invariably associated with a city's ability to promote a good quality of life, a high standard of living, through qualitative urban services and infrastructure (for everything that means living conditions in the urban environment – the quality of the neighbourhood, of the street, of the utilities network, accessibility, public security, public spaces for recreation and so on). The relation with *city image* is also based on mutual conditioning: if a city is a good place to live in, in terms of living conditions and standard of living, it will generate a good image and an attractive attitude; in reverse, a low quality of life will promote a negative mental image and a repulsive attitude towards that city.
-   *linkability*, introduced by Nasar (1997, as cited in Nasar 2001) refers to a place's ability to generate a positive feedback (similar to Lynch's "imageability", a place's ability to generate a strong image), this indicating several essential characteristics of the urban space, with a prominent role in an observer's perception and evaluation of a certain place: degree of naturalness (the more an environment keeps a higher accuracy of the natural shapes and elements, the more its attractiveness level increases, individual perception feedback is stronger and the role in qualitatively evaluating that place is higher); the elements' degree of order (and the coherent distribution of urban elements and shapes in space); complexity (a certain space degree of complexity can generate a certain degree of interest from the observer and can become an emotional stimulus); degree of spatiality (openness); historical meaning (spaces with historical

significance have a strong impact on human perception generating a strong mental feedback).

As any cognitive process, *the construction of the city image* implies the existence of two entities: 1. The city (the observable reality) – urban elements and shapes, processes and phenomenon (that are combined in a unique manner conferring distinctiveness) in a constant dynamic, perceived at different spatial-temporal observation scales, a dynamic that maintains a permanent change of information; 2. The receiving subject (resident or non-resident) - perceives the city through his own variables, selecting, organising and loading the "urban information" with meaning and importance (Fig. 2). Based on the image's genesis, the predominant element, the type of observer, the generated type of spatial attitude and behaviour etc., the city image can be classified as such: natural or built image ("brand"), economic image, cultural or mixed, endogenous or exogenous, positive, neutral or negative etc.

3.2 City image and its role in urban planning. Relations with other concepts

First of all, urban planning targets the city's functionality, as a system: either redesigns the disposition of urban components, or resizes the inter-conditionings between them and the territorial flows. Thus, the target is always the maximum optimisation of the functionality of urban spaces and the spatial configuration of the new model of urban planning is projected at the residents' mental level, influencing their spatial behaviour by developing certain attitudes towards certain places in the city, attitudes that can be negative, repulsive, or diametrically opposed, positive, attractive.

These mental meanings of the residents transform a city's population into a pressure component over urban spaces, generating territorial imbalances, respectively balances, in a successful space management. As such, the city image keenly receives all the malfunctions in the actual way of urban planning.

Operationalising the image is done by means of *mental maps*, which individualise the attractive areas (topophilias), the neutral and the repulsive (topophobias) ones. But, a place's image, by itself, has no strong practical operational value if it's not put in a larger context and looked at as a *philosophy of space management* with all that implies, respectively the interface between the physical, real city and the human perception of it; from among the space "managers" and "producers" (government, local authorities, private entities, economic entities etc.) and "consumers" (residents and non-residents); a current and past management; a current model of space organisation and a future one, emphasising the malfunctions perceived by the individuals, lastly an interest mediation platform which is individualised at the urban space level.

Geography's contribution becomes so much more important because of the *city image* concept's spatial character, mental maps being able to find various territorial imbalances regarding the perception of aspects such as: the quality of life, the quality of public services, meanings, feelings induced by the manifestation of certain phenomenon – the perception of the state of fear, insecurity, affiliation to a place, attractive areas (topophilias) and repulsive areas (topophobias) and so on, matters of fact without which no correct philosophy of the place's management can be perceived. As such, mental maps can become an operational "key" in *the decision making process*, a difficult and complex process today if we take into consideration the number of variables involved in what is called control and benefit regarding the model of organisation of a territory (authorities, law, owners, beneficiaries, functionality, design, experts, conflicts of opinion and interest, technical methods etc. etc. etc.).

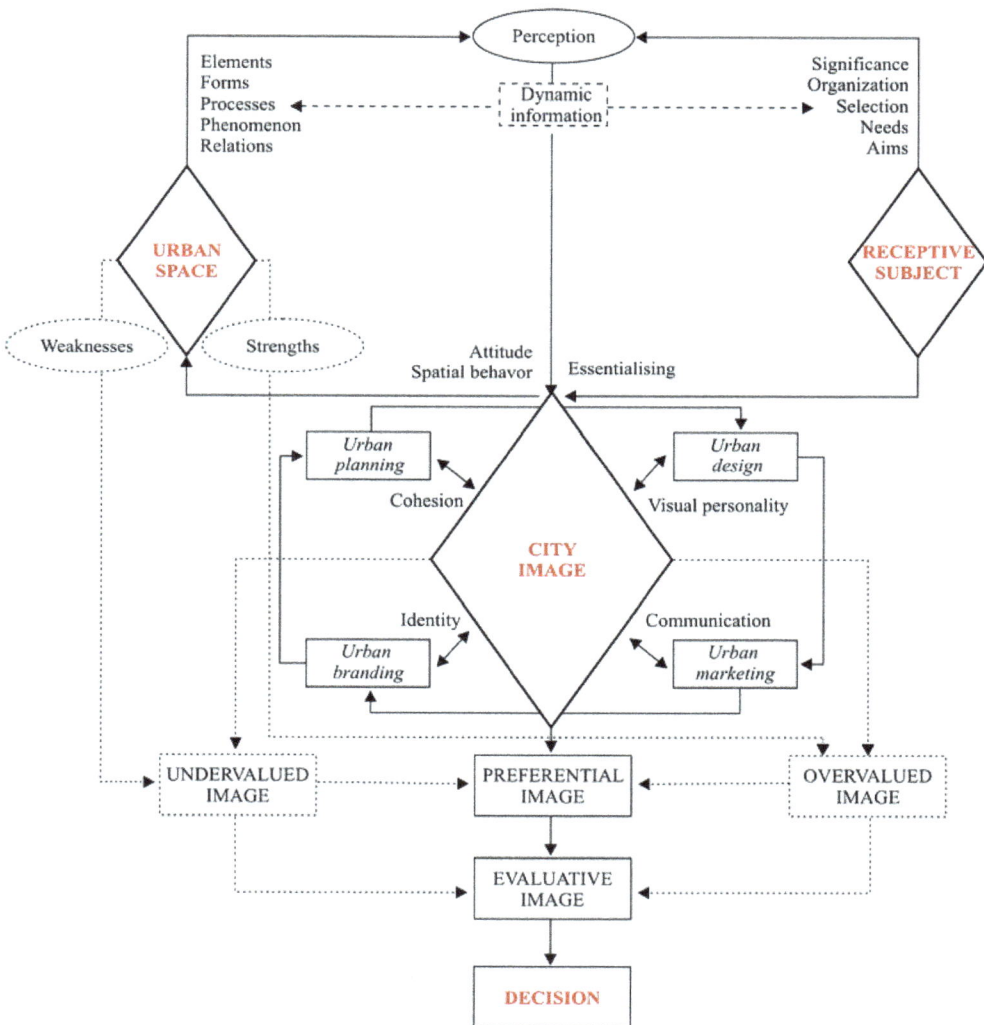

Fig. 2. City image, an instrument in the decision making process. Relations with other concepts (see also Ianoş, 2004; Neacşu, 2009, 2010a)

Thus, the city image can become an "extremely useful instrument in identifying strengths or weaknesses" of urban subspaces, the first generating, by exaggeration, an "idyllic" image at a perceptual level, while the latter an "inhibiting" one, a mediation between the two leading to a "constructive image", as operational as possible in the decision making process (Ianoş, 2004).

Another strength of the *city image*, in the spirit of the territorial reality approach philosophy mentioned above, is the fact that it can amplify other operational concepts and instruments in the urban space approach (Fig. 2) and it can have an integrator character, the urban planning model, the city's visual personality generated by urban design practices have repercussions in the city image, while the urban marketing techniques can influence how the city is perceived. In turn, city image can function as a pressure element in the decision making process to intervene in the current way of urban planning.

4. Case study: Ploieşti city (Romania). Integrating the *city image* into *urban planning*

4.1 Methodological aspects

Theoretical substantiation implied, as a method of research, using *interpretative analysis* in scanning specialty literature. This route was individualised by underlining the main markers in developing "the *city image* theory", basically, the development of notions such as: *urban space*, urban space perception – *mental image, mental map*.

The city of Ploieşti has been chosen as a case study, considering the mapping of the city image at a micro-scale level (city and neighbourhood) and the forecasting of some future scripts of evolution through the integration of the city image obtained in their elaboration.

Mapping the city image elements and, in general, the perception of residents (endogenous) and non-residents (exogenous) upon the living conditions of Ploieşti, but also the general impression towards the city has been done by using the *questionnaire method*.

Thus, a survey was done, through a questionnaire, in two temporal sequences and two different ways: a) in 2001, through direct approach, on the street, stochastic (350 interviewees, 36 have refused the interview, the rate of representativeness being 89,7 %) and b) in 2004-2006, by using a web questionnaire unfolded over the Internet (200 interviewees).

The results of the two researches are almost similar, but we can still enumerate a few limitations of the obtained results such as: the comparison between the two moments is relative taking into account the different techniques of sampling; samples are experimental, one being interested by the tendency and not by their generalisation.

The application of this questionnaire aimed to, in essence, shape the image evoked at a mental level by the city of Ploieşti, also resulting a few other secondary objectives: the people's perception of the urban habitat (living conditions), the correlations that can be established between the perception of the city and different independent variables selected in the questionnaire's header – sex, age, studies, civil state, for how long has the subject been living in the neighbourhood etc., identifying new relations of causality between the perception of different living and dwelling conditions in the city (urban habitat) and the general image of the city, shaping new territorial disparities and mapping a new mental map.

Regarding urban planning, the diagnostic analysis of the current model was followed by the use of the SWOT analysis through which, thanks to the main factors that condition the city's evolution, identified and oriented into the four categories (favourability, vulnerability, opportunity and risk), have been identified four possible (theoretical) evolution scenarios – 1. Sustained development, 2. High development potential, but in a risky environment, 3. Favourable environment for development, but with a low potential and 4. The regress situation, of impossible development. The real evolution of the city will be within these limits. This method has been chosen for forecasting a future evolution model of the urban system and a future organisation model of the urban space.

4.2 Ploieşti city – Illustrative sample of the Romanian urban environment. Urban planning model

Rank: municipality, county seat; **Localisation:** Southern Romania – 60 km North of the capital, 100 km North of the Danube, 300 km West of the Black Sea; **Population:** > 230 000 inhabitants; **Area:** 58 km^2; **Resources:** oil, gas, salt (the Northern Subcarpathian area), agricultural resources (Southern plains) and others. **Urban dynamic:** before 1600 – village; 1600-1800 – borough (commercial and trade activities); 1800-1850 – trade center (commercial, trade and transports); 1850-2nd WW – industrial city (oil boom); 1945-1989 – municipality (socialist economy, county capital, complex industrial city); 1990-2007 – transition to capitalism (strong industrial restructuring); 2007-present – mix functional city (EU regulation alignment). **Urban functions:** complex (economical, social, cultural, administrative, residential etc.) **Other characteristics:** polarising centre for the cities in the Southern part of the county (a ratio of 6/1 of the hypertrophy index of county urban network), development centre of regional significance. Website: **www.ploiesti.ro**	Ploieşti – urban profile

The Romanian urban space presents a "fertile ground", in general, for studies of urbanism, urban geography, social geography, behavioural geography, being a true laboratory, if we take into consideration the effects of transitioning from "communist systematisation" towards western "urban design" practices, that combine the community's economic and social needs with those of cultural and historical preservation.

The city of Ploieşti is illustrative for the Romanian urban environment, synthesising it through its dominant characteristics: large city (over 230 000 inhabitants), complex functional structure, polarising role in the superior administrative unit it is part of (Prahova county), having passed through all the urban evolution phases specific to the Romanian urban environment, ever since its documentary attestation in 1503. Also, being a part of the generation of medieval cities and, among those, one of the demographically larger, Ploieşti, from an urban dynamic point of view, can be considered a "miniature" evolution of

Romania's capital, given its proximity to Bucharest (60 km), but also its similar organisation and architectural personality.

The analysis of the functional-territorial structures of Ploieşti, as well as the morphology of its street network, lead to the individualisation of a specific spatial organisation model that could be synthesised to a *radial concentric model* (tolerated by the lack of natural constraints imposed by the configuration of the permissive plain relief) with certain distortions imposed by its evolutionary and contextual dynamic in the regional and national urban system frame (the historical periods which had different impacts and urban policies, the communication axes that penetrate the city in a radial manner from all cardinal points and that have "altered" the ring morphology, territorial decentralisation etc.).

By comparing it with the fundamental theoretical models of urban planning, Ploieşti's urban planning model presents many similarities with E.W. Burgess' „concentric zones model" elaborated in 1923 (Fig. 3).

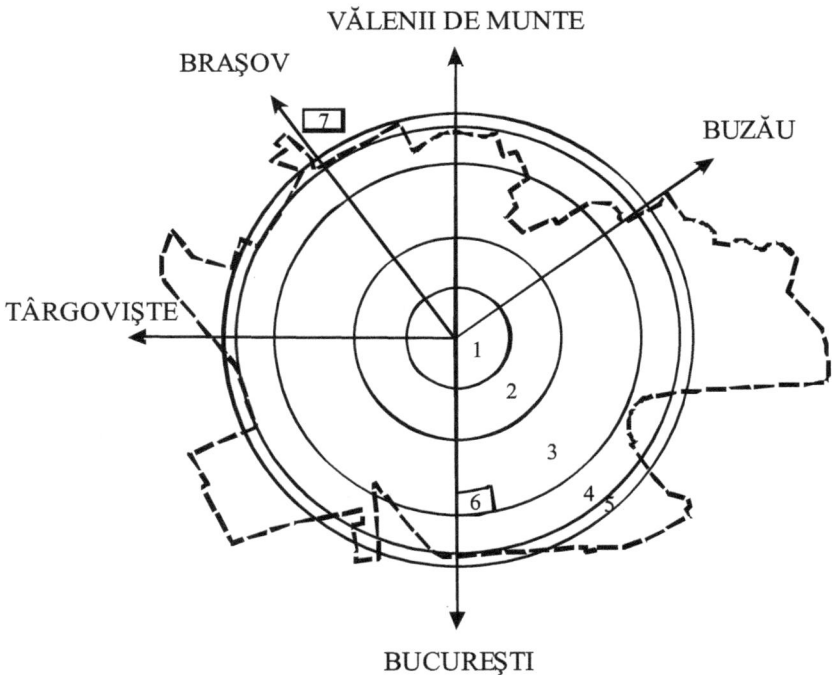

1. Mixed space: business, administrative, tertiary, cultural, residential (a "local" and modified version of CBD). 2. "Traditional" residential space (old neighbourhoods). 3. New residential space (large communist urban habitats – so called "worker districts"). 4. Specialised industrial space – petro-chemistry (discontinuous) and machine construction, chemical and food industry etc. 5. Transport space (railway ring and the highway system ring road) and residential (fragmented). 6. University space 7. Commercial space (Metro)

(a)

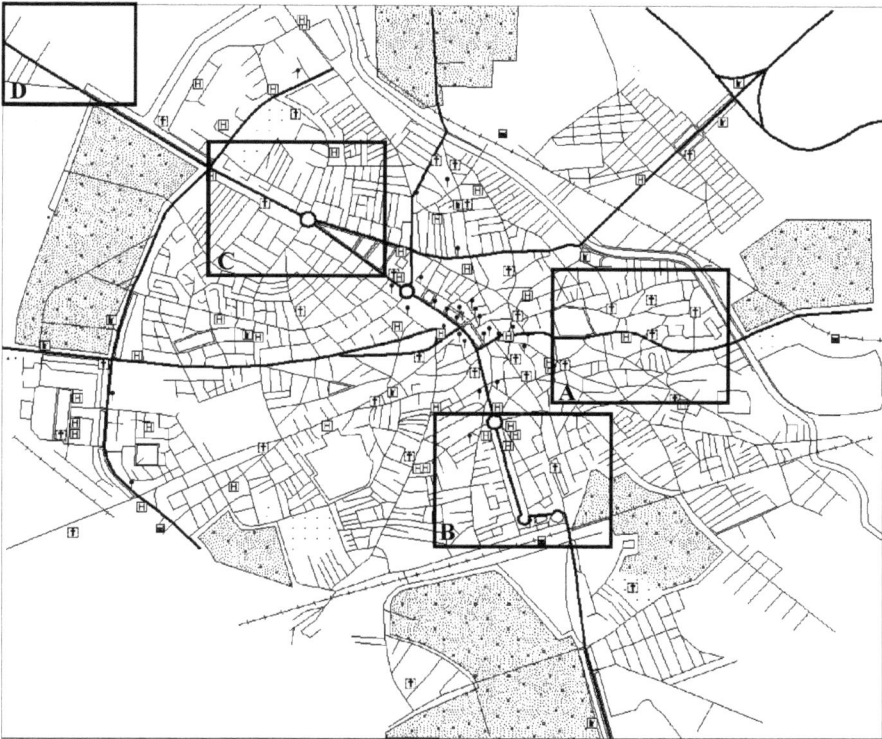

A. The oldest part of the city, the central settlement in existence for over five centuries. The irregular street network with short and tangled roads, the traditional inspired toponymy (tanners, fellmongers etc.) express the old age of space organisation. **B.** Interwar organisation sequence, but "perfected" by post interventions. The street network has a somewhat linear geometry imposed by the development of this Southern part of the city along the axis that connects the Centre to the South Railway Station and continues with E60 towards Bucharest. **C.** Geometrised street network – parallel streets. It represents the communist organisation sequence (1945-1989). Unjustified toponymy from the national history registry prevails (names of heroes, rulers or historical places that have no connection with the city space). **D.** New type of residential area (villa type houses on the city outskirts or in its immediate proximity) resulted from decentralisation, after 1990.

(b)

Fig. 3. Radial-concentric model of urban planning in Ploieşti (a) and street network (b) Therefore, the city's central part with a complex and mixed functionality (from administrative to services and from cultural to residential) is surrounded by the first residential ring (old neighbourhoods, in which individual houses are predominant, "retouched" reminiscences from previous centuries), doubled by a second residential ring, mostly built in the communist period, with large urban habitats that had to host the influx of new residents attracted by the opportunity of having a job on the industrial platforms. Towards the exterior there are the industrial space, organised in "compact platforms" in all the four cardinal points, transport space (the railway and road ring), university space (at the city's Southern periphery) and commercial space.

4.3 Integrating the *city image* in *urban planning* in Ploieşti

Ploieşti's visual personality was imprinted by its spatial-temporal dynamic, its effects being found in the urban landscape configuration, but also in the community's mental projection. The centralisation and interpretation of the two analyses mentioned above, for the analysis and evaluation of the city's image, showed the following:

-   individualising an **attractive space**, relatively continuous, overlapping the Central Civic Area, with extensions on the city's main functional axis, North-South (Republicii Boulevard –Independeţei Boulevard), towards which the residents and other categories show a topophilic behaviour, the flows of people converging towards this territorial "corridor";
-   identifying and demarcating three clear cores that represent **repulsive areas**, located on the city's outskirts, beyond the industrial and railway rings, spaces heavily marked by the residents' topophobic attitude and behaviour;
-   shaping a **neutral space**, concentrated between the previous two, larger and relatively continuous, marked by topoindifferent spatial attitudes and behaviour (Fig. 4).

The Central Area represents "the heart of the city" towards which all the roads converge, being best known for its functionality – a space for shopping, financial-banking services, administration (the City Hall and Local and County Councils can be found here), culture (The Palace of Culture which holds the largest public library in Ploieşti – Nicolae Iorga, the Toma Caragiu theatre, the Philharmonic, main high schools from the entire county - I.L. Caragiale, Mihai Viteazul, the Clock Museum, the National Museum of Oil etc.) and is also, in a smaller degree, a residential area. The second and third places in the *attractive areas* hierarchy converge in the residents' perception because of the presence of green space, Independenţei Boulevard surnamed the "Chestnut Boulevard" (in weekends the road traffic is forbidden) and the Mihai Viteazul Park (Republicii Boulevard), with recreational valences, to which we can add the semi-commercial function of Republicii Boulevard. Important to add is also the fact that the two boulevards represent the city's vital axis, oriented from North to South, on which an intense traffic is done both towards Braşov (North), but also towards Bucharest (South).

As for the neighbourhoods that are mentally identified as negative, generating repulsive attitudes and topophobic urban behaviours (Bereasca, Mimiu and Vest I), the three areas are located on the Eastern, Southern and Western outskirts of the city, beyond the "limits" that separate them from the city's body – Dâmbu stream and the railroad (East), the railroad and industrial platform (South) and the West road (West), outskirts situated in proximity of the industrial platforms, true "colonies" in which the problems of infrastructure and urban services (street quality, cleanliness state, public lighting and security etc.) are numerous. Also, it was noticed a distortion of perception regarding these areas, the main reason being ethnical segregation (the presence of "Roma" communities in the area of these neighbourhoods).

Urban elements that evoke strong mental images fall under either one of the following categories of urban spaces: impressive through their functional characteristics (e.g. "Central Area", "Chestnut Boulevard", "green spaces"); identity sphere ("the city's history", "oil", "old buildings", "cultural personalities" associated with the city's name); cultural sphere ("theatre", "festivals"); utilities facilities; social – personally motivational ("it's a tranquil city", "I was born here" etc.). In the same context it is worth mentioning as another attractive element the city's proximity both to the capital and to touristic areas (Prahova's valley and others). Urban elements that provoke a topophobic attitude are also numerous, but can be classified in several categories. Predominant are the ones of a social nature (ethnic segregation, not integrating the Roma communities, the more or less acute feeling of public insecurity) and the ones regarding utilities equipment (lack of parking spaces, poor road condition, insufficient sanitation, insufficient green spaces, pollution, poor sewage network etc.). Another voiced issue is the lack of interest in rehabilitation of patrimony buildings.

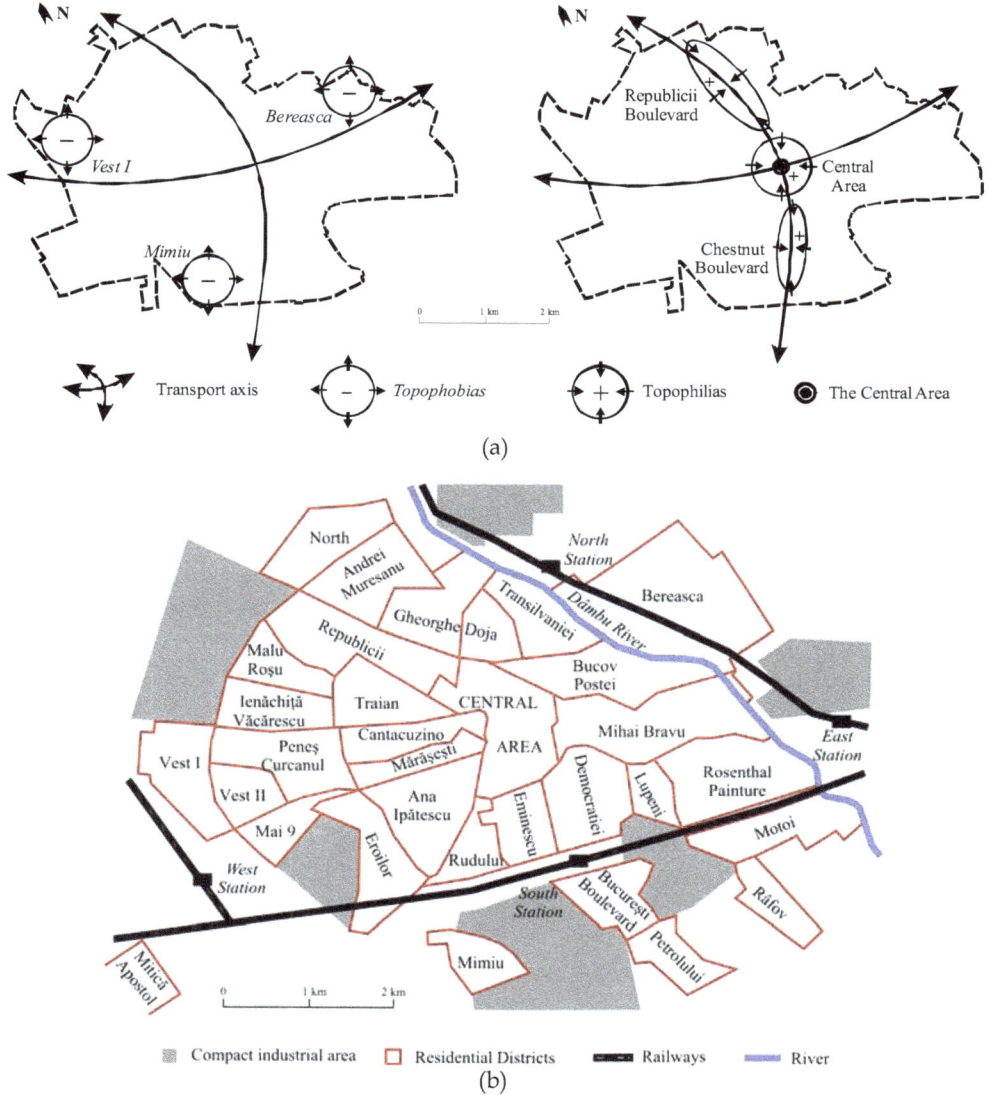

Fig. 4. Topophilias and topophobias (a) identified in Ploieşti (chorematic representation) and main residential districts (b)

City image and urban planning in Ploieşti city. SWOT analysis

With over 230 000 inhabitants and with an economic profile dominated by services and industry, the city of Ploieşti has, in its territorial relations, the role of a convergence centre for human and material flows and as a centre for diffusing information flows.

These processes of concentration and dispersion of territorial flows define Ploieşti's urban space configuration, its organisation being a result of the constraints imposed by a series of components, classified in four categories: 1. Core components – geographic constraints (geographic position, permissive topography, resources complementarity), human (over five centuries of tradition), political and administrative (obtaining preferred ranks in the regional urban network at different moments in history – fair, leader city, county seat, municipality, communist systematisation, community laws, the appearance, from a functional point of view, of some new territorial unities such as the "metropolitan areas" or the "development regions") and economical (oil "boom"); 2. Restrictive components – the high degree of urbanisation of the Prahova county and the competition over the same local resources, proximity to the capital (until 1989 Bucharest concentrated most of the territorial flows in Ploieşti's detriment, but the transition towards market economy transformed this element from a restrictive one towards a favourable one, Ploieşti benefiting from the moving of some activities from the capital – the presence of some production units of strong multinationals, more of them choosing to bring their "decision centres" here as well); 3. Incentive components – being located on one of the main urban and economical axes in the country Bucharest – Braşov, an axis found in all historical sequences as part of the intra-extra-Carpathian transit space, from Central Europe towards the Danube and the Black Sea, local resources, oil manufacturing "tradition", 150 years old in this area, "alive" in its residents minds (a fact highlighted by the questionnaires), existent infrastructure and the city's technological calling, its proximity to the capital; 4. Pressure components – the demographic element and massive industrialisation (representative for the communist period), the real-estate "boom", the political element (laws, community regulations), foreign investments, being the components that generated "pressure" on present urban spaces.

Taking into account these reasons and the above mentioned premises, we can "visualise" a future model of planning and management of Ploieşti, taking advantage of the city's radial-concentric structure. To this end, using the SWOT analysis in order to identify several possible evolution scenarios and forecasting several future models of urban planning and management in Ploieşti has revealed the following (see also Neacşu, 2009):

Scenario 1. It is the ideal future development situation, resulting from the perfect convergence of the city's strengths and the capitalisation of all the opportunities, together with the full minimisation of the risks and the elimination or change of the vulnerabilities' tendency to slow the settlement's economic and human growth.

As strengths, the city of Ploieşti mainly "counts on" its geographical position. It is a location that enables intra and extra-Carpathian transit (between Transylvania and Muntenia and from here on towards the Danube or the Black Sea), transit on Prahova's Valley, with profound historical roots, a corridor that is a part of the urban-industrial axis Bucharest – Ploieşti – Braşov, but also a segment of the pan-European communication corridors TEN IV (Berlin – Budapest – Bucharest – Sofia – Istanbul), heading from West to East and IX (Helsinki – Sankt Petersburg – Chişinău – Bucharest – Plovdiv), heading from North to South and in proximity of corridor VII (transcontinental waterway Danube – Main – Rhin). This location and reconfirmation of Ploieşti as one of the most important road and railroad nodes, allowing access from the Capital towards Transylvania and Moldova, is doubled by its proximity to Bucharest, respectively to the "Henri Coandă" International Airport (approximately 35 km).

To the favourable geographic position we can add: the permissive topographic surface of the natural units in proximity which ensures a complementarity of resources, the intra-city available space and the polarising role of the city, being the residence of the most populated and urbanised county in Romania, which assures a pretty consistent human resource pool, thus the disadvantage of demographic aging and natural population deficit can be minimised by attracting young students from the University in Ploieşti and "settling" them through programs that would guarantee a profitable relation between studies – research and the market of work – production. On the other hand, as far as the workforce quality goes, the city has at its disposal a specialised workforce, a great deal of it working in the manufacturing industry (one third), while most of it works in the tertiary system.

Although it has a manufacturing tradition, seconded by an industrial one, the city has a poly-functional structure presenting economical alternatives, for instance the boom of the tertiary sector (in which case, at least hypothetically, the workforce specialisation could have worked as a "break", through the stiffness manifested towards professional reconversion). Also, even if it concentrates only a third of the county's population, Ploieşti produces more than 80% of the overall county commercial value.

Another strength is the good equipping of the territory, the modernising and extension process following an ascending curve, to which we can add the pronounced dynamic of the constructions sector, as proof of a high development potential. Also as a plus, we can add the municipality's capacity to apply for various European projects and attract investment funds, for instance the high number of projects unrolled during the last few years, such as: infrastructure projects (for example „The Municipal Project – Ploieşti Begins From the Neighbourhoods Towards the Centre", which had as a subject the infrastructure rehabilitation and redefining the city image at a neighbourhood level (were also included in the project the neighbourhoods identified in this study as spaces that generate repulsive attitudes), projects within partnerships between public institutions, projects within public-private partnerships, European projects („The Local Agenda 21", „CiVitas-SUCCESS", „SpiCycles", „Practise") and so on.

Scenario 2 represents the situation of a high development potential, but in a risky environment. The main risk categories which might affect the city of Ploieşti are either natural (earthquake, floods, global warming etc.) or technological (the proximity of the industrial platforms to the residential areas, secondary activity, plus traffic jams ensure a high atmospheric pollution), legislative (the dependence of the economic area on the political one and the clear inseparability of the two, the frequent substitution of policies, instability or legislative ambiguity etc.), social (social segregation, economical poverty, stiffness to professional reconversion) or economical (Ploieşti and other cities of the county, compete on the same niche of resources and activities).

Scenario 3 is the obvious reverse of the previous one, the environment being very favourable, but the development potential rather low due to the identified vulnerabilities. The favourable environment is maintained by the general historical and socio-economic context in which our country currently is (member of the EU, situated at the Black Sea on the route of Caspian energy from the Asian space towards the large Western European consumers), this offering the possibility to obtain structural funds for developing and modernising the infrastructure to the Union's standards. On the other hand, the "appetite" for foreign direct investments is rather high, taking into account that Romania is a rising market, unsaturated, therefore a "fertile field" in terms of costs, corroborated with certain facilities offered by the authorities to attract investors.

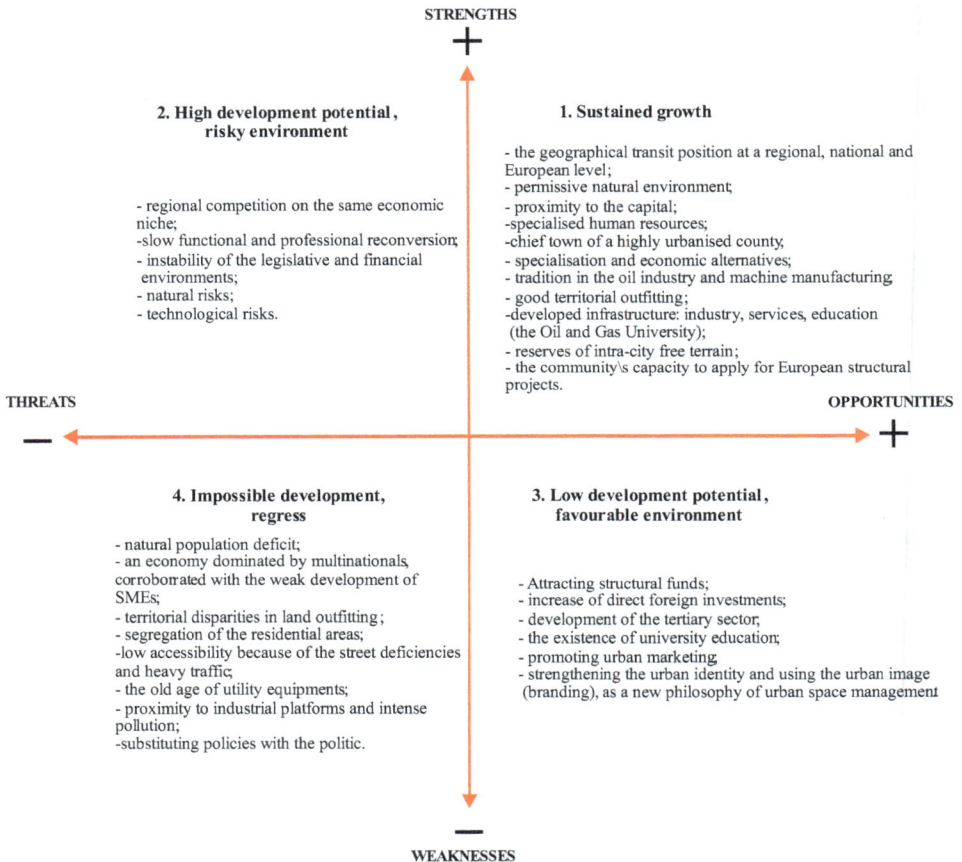

STRENGTHS
+

2. High development potential, risky environment

- regional competition on the same economic niche;
-slow functional and professional reconversion;
- instability of the legislative and financial environments;
- natural risks;
- technological risks.

1. Sustained growth

- the geographical transit position at a regional, national and European level;
- permissive natural environment;
- proximity to the capital;
-specialised human resources;
-chief town of a highly urbanised county,
- specialisation and economic alternatives;
- tradition in the oil industry and machine manufacturing;
- good territorial outfitting;
-developed infrastructure: industry, services, education (the Oil and Gas University);
- reserves of intra-city free terrain;
- the community\s capacity to apply for European structural projects.

THREATS   **OPPORTUNITIES**
—   ◄───►   **+**

4. Impossible development, regress

- natural population deficit;
- an economy dominated by multinationals, corroborated with the weak development of SMEs;
- territorial disparities in land outfitting;
- segregation of the residential areas;
-low accessibility because of the street deficiencies and heavy traffic;
- the old age of utility equipments;
- proximity to industrial platforms and intense pollution;
-substituting policies with the politic.

3. Low development potential, favourable environment

- Attracting structural funds;
- increase of direct foreign investments;
- development of the tertiary sector;
- the existence of university education;
- promoting urban marketing;
- strengthening the urban identity and using the urban image (branding), as a new philosophy of urban space management

—
WEAKNESSES

Fig. 5. Theoretical diagram of the SWOT analysis in the Ploieşti city

Also, the industrial restructuring has allowed a fast paced development of the tertiary sector, services rising considerably in volume and value over the last few years. At the same time, in the urban space, different marketing techniques have broken through, urban marketing becoming a main component of space management in the city. To all of the above we add the attempt to revitalise the city from an identity point of view, creating images, brand-images to revive the consciousness and civil spirit of its residents. Worth mentioning are: *UNESCO Year „I.L. Caragiale"* (cultural program conducted with the occasion of 150 years since the birth of the great dramatist), *The International Festival of Poetry „Nichita Stănescu", The National Contest of Classical Music „Paul Constantinescu", The International Festival of Caricature with the theme „Home at Caragiale", The "Chestnut" Festival* (with contests and folk music), international festivals of classical music etc.

Vulnerabilities have been identified in several fields, such as: *demographic* – accelerated population aging and natural deficit both in Romania and in Europe. These have repercussions on the workforce, the problem of developing a coherent policy of attracting immigrants from the Eastern part of the continent or Asia becoming more serious every year; the effects are: emergence of ethnical segregation, problems with cultural integration in the adoptive community etc. In Ploieşti the rate of the natural growth was of approximately -1 $^0/_{00}$ over the last few years; *economical* – local economy is dominated by monopolistically multinational companies, the SMEs field being weakly represented; *social* – social segregation, manifestation of the poverty phenomenon etc.; *city-planning* – low accessibility in the intra-city space and in the city outskirts, the road network presents discontinuities in its radial and ring-like structure, exponential growth of motor traffic surpassing by a lot the existent infrastructure leading to traffic jams, the old age of utility equipments and so on; *ecological* – intense atmospheric pollution, the risk of polluting the waters or soil with oil, phonic pollution etc. and *political* – excessive politicisation of local structures and interference in the economical sector.

Scenario 4 is totally situated in the negative area and can appear as a result of the convergence between risks and vulnerabilities, economical development of the city being impossible under the conditions of an unfavourable, risky environment, with a low profitable capacity and having a low development potential.

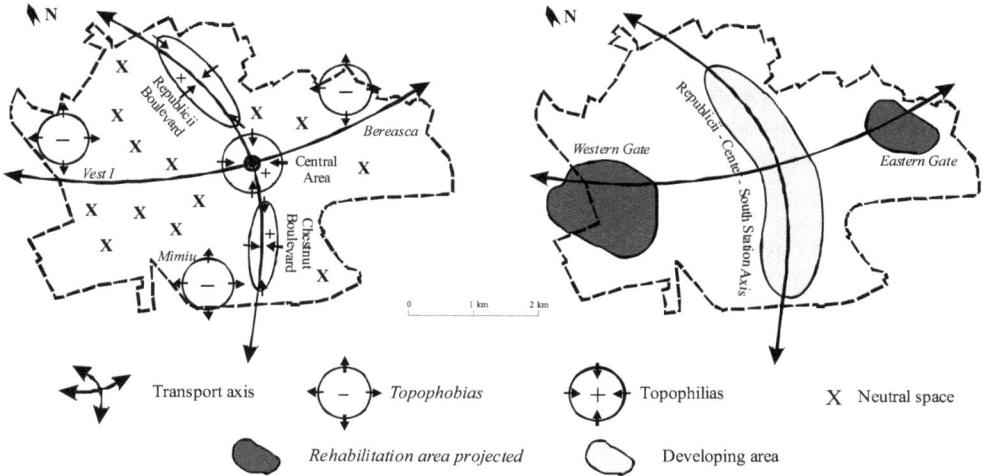

Fig. 6. City image (left) and urban space management (right) in Ploieşti city

The above analysis of the possible evolution scenarios, but also of the main categories of components that could influence future development, show several priority directions regarding urban planning and management. Captured in "The development strategy of Ploieşti 2007-2025" (Institute for Housing and Urban Development Studies Romania [IHS], 2007), the above mentioned priority directions fully illustrate the integration of the city image, of the city's mental maps individualised in this study:

- *spatial-territorial integration through a connection to the European transport network.* Thus, four major categories of infrastructure investments have been identified: 1. the „Mărăşeşti" road passage: continues the homonym street and connects the city, over the Bucharest – Braşov railroad, with the National Road 1 (E 60), on which the TEN IV pan-European corridor is overlapped. Thus, approximately 12% of the intra-city space re-enters economic circulation and also, the Ploieşti Vest Neighbourhood (Mitică Apostol), at the moment "isolated" from the city's "body", is territorially integrated; 2. the ring road South Railway Station – West Railway Station – North Railway Station – Bereasca – Râfov – South Railway Station which will enable the cleaning and regulation of the Dâmbu stream, increasing mobility and accessibility, plus eliminating an important part of the traffic from the city centre, with effects such as reducing pollution, preventing traffic jams etc.; 3. completing the tram ring (or surface subway) which will connect the South Railway Station and the West Railway Station, with positive long term effects: a viable, high capacity, fluent and ecological public transport system. 4. rehabilitation and introduction in real estate circulation of fields from the South part of the city, today enclosed between the industrial and residential areas, situated beyond the Southern railroad.
- *completion of integrated projects – „The City's Gates".* Western Gate. Integrating in the urban texture the Ploieşti Vest Neighbourhood (Mitică Apostol) by increasing accessibility (Mărăşeşti passage) and by raising the residents' living standard (introducing basic utilities, specific to the city, water, sewage, gas or modernising the existent ones; the neighbourhood is today more of a suburban village, with a rural aspect). Given its proximity to the Ploieşti Industrial Park, developing complementary economic activities, high end industries or research institutes integrated into a research – IT – production area, is compulsory. *Southern Gate and Central Axis.* Developing the urban axis South Railway Station – Central Area – Republicii, based on a future relation (centred on the Oil and Gas University) – CBD (Central Business District), with a mix of activities, taking into account the attitude of attractiveness towards this area revealed by the results of the research. *Eastern Gate.* Rehabilitation of Bereasca and Râfov neighbourhoods, identified in the residents' perception as repulsive areas, development of the ecologic axis of the Dâmbu stream by cleaning and regulating it, modernising the infrastructure and raising the quality of life.

As far as the East and the West of the city are concerned, the research field materialised through the mapping of the city image has emphasised the strongly repulsive attitude of the residents towards these areas, their rehabilitation becoming a priority.

- *Imposing a new vision towards urban marketing* by implementing new marketing techniques, such as urban branding.

5. Conclusions

The analytical in-depth analysis of the *city image*, both as a theoretical concept in itself and also in its immediate relation with *urban planning* and its role in future planning models, has led to several conclusive ideas that crystallise the nature of the conceptual relations between the analysed notions. Thus, *city image*:

- **expresses a spatial analysis necessity, synthesising and essentialising a territorial and phenomenological reality in manifestation.** As most concepts that interact in the city's theoretical field, due to the decline of some "traditional" concepts that no longer managed to capture reality, city image expresses a spatial analysis necessity, managing to mediate two states of the city – system (function) vs. phenomenon (meaning) -, in the context of mutations in the urban dynamic and morphology stimulated and catalysed by globalisation (thus a change in conditions). Even more so with today's approach philosophy oriented towards sustainability – environmental balance: man – territory – and micro-spatial analyses.

What synthesises the *city image*? The total number of perceptible malfunctions in the urban habitat, which create certain spatial attitudes and behaviours: of attractiveness, indifference or repulsiveness.

What essentialises *city image*? The global image's integrator character of a place (neighbourhood, city etc.), this being a result of the interaction between individual images. A *positive image* points to an (mentally) *attractive space* which gives birth to a *topophillic behaviour*, a *neutral image* defines an *indifferent space*, in which case the individual develops a *topoindifferent attitude or behaviour*, while a *negative image* shapes a strongly *repulsive space*, towards which the individual or a certain community show a hostile attitude, a *topophobic behaviour*, of rejection (which affects territorial mobility within the city, investments, real-estate prices etc.).

- **leads to generating mental maps.** Zoning the perceptions of urban spaces or subspaces, expressed through different emotions, states, attitudes and behaviours, leads to the generation of *mental maps*. Mental maps help an individual organise all the information regarding a place, despite its complex structure, to "navigate" in an external environment and it influences his spatial behaviour or can modify his attitude towards that place. Also, these products of mental mapping complete the images that form the whole picture of a certain territory. The city or neighbourhood are perceived by the individual or community as generally being "good" or "bad". Local authorities using mental maps in follow-up settlement development policies (or sectors of it) can become the measure of success. Neglecting or avoiding them can lead to major dysfunctions in "understanding" that space, dysfunctions that can generate repulsive attitudes or topophobic behaviours, in which case the respective city or urban area begins to decline.
- **quantifies the mistakes in understanding urban space.** In this regard, the city image has the ability to express the way residents, both individually and collectively as an urban community, perceive the city/neighbourhood as a whole, mediating between objective necessities (community's interests) and subjective ones (individual's interests). Any mistake in "reading" the urban landscape has repercussions, at a mental level, in the form of either an overvalued image, or, on the contrary, an undervalued one. In this

case, the positive spectrum generates topophilic spatial attitudes and behaviours of attractiveness, while the negative one generates topophobic and repulsive ones.

- **operationalises the relation between individual/urban community and space.** Zoning the residents' perceptions of urban spaces and mapping them shows urban attitudes and behaviours (topophobias, topoindifferences or topophilias) that individualise associated mental spaces: repulsive, neutral, attractive. Thus, the city image concept's applicative valences result from the fact that it can become an instrument in planning the future model of space organisation, an instrument that will have to accompany a city's development policies in the spirit of good management.
- **it is an urban habitat's barometer.** At micro-scale level the city image individualises the main malfunctions, imbalances between activities, functions and afferent urban spaces. In fact, it only crystallises glitches of the current model of urban planning. Thus, the "strengths" or "weaknesses" of an urban space can be detected, even though they do not result exclusively and univalent from quantitative analyses.
- **it is an expression of urban identity.** Identity is what gives an urban community's life coherence and continuity, what makes the city appear like a structured and stable system. Translated into image, the city's image, perceived by the local community, identity represents the system's internal coherence. Identity cannot be separated from its structure and coherence, from the visual configuration it gives to urban shapes and forms, with a strong impact on individual's and community's perception. City image coherence mostly reflects the structure's coherence.
- **a stimulator decision factor to intervene in urban planning.** Any urban space or component is characterised by certain strengths and weaknesses that, at an analytical-theoretical level, cannot be the privilege of statistical data or individual or group perceptions, but of a balanced combination between the quasi-objective reality of spreadsheets and the perception in the residents' minds, who, through experience and meanings, feelings and sense associated to places, complete the urban reality picture. By overlapping the two images – the objective one, resulted from mathematical modelling of statistical strings of data and the subjective one (at the urban community level), obtained by mapping perceptions of urban elements and shapes – we can emphasise the following distortions: either the trend of *overestimating the city's image* (generated by an ultra-positive image based on personal emotional considerations – a place of birth or a place loaded with certain meanings), or the *under valuating* one (given by an extremely negative image). Correcting such distortions leads to a *preferred* or *evaluative image* as close as possible to the objective territorial reality.
- **is a mediation interface.** The city image can become a communication platform between various actors – residents (urban community), local authorities (political decision, urban planning), specialists (theoreticians and experts) –, mediating between territorial reality and all these categories of interests that converge on the urban spaces level. All these aspects iterate the idea that the city image is a component element of both the diagnosis process of the urban planning mode, functioning as an important factor in the decision making process and of the prognosis one, the positive city image being also a goal in the marketing process (city branding or rebranding).
- **create "couple of elements" with other operational concepts in the spirit of community oriented urban planning.** The mix between the physical and mental spaces, between systemic and semantic understanding of the city, between the city as

a machine-system, a product of rational organisation and scientific logic, and the city as a "text", its meaning interpretable at a perceptive level, between the city as a territorial support of men and their activities and the city as a result of man's perception and imagination, loaded with meanings, enhance the spectacular approach of the concept of *city image*. This gains a special value only in relation with the other operational concepts of the urban space, especially when each one's strengths empower the other's.

Taking the functional cohesion given to the urban space by the planning model, the visual personality generated by the shapes' aesthetics of which urban design is responsible, loading itself with identity meanings and the sense given by the plus value of that "place's story" (its brand) and gaining an excellent communication potential, as information, through marketing techniques, *city image* validates once more its exceptional ability to capture the territorial reality and, in this respect, its remarkable applicability potential.

- **can become a success factor in the urban marketing, in a visionary acceptance of urban planning.** If we imagine the city as an "urban product", the municipality as an urban places image "producer" and the residents and other inhabitants as future "consumers", then the *city image*, in a "good" social marketing philosophy (oriented towards the consumer's needs, not towards the product), can represent a *success factor* and the image can become a symbol, a *brand-image*, for several reasons:
- it carries the semantic load of the *city's identity* and the brand represents the synthesising of the place's "story" at a symbol level; if the residents find identity anchor points in this story then the brand-image is durable, consistent, viable and becomes an image of the standard of life promoted by that city, even a warranty of it;
- through its identity and personality, the brand-image gathers the unique set of qualities that confer *distinctiveness* to a city in comparison with others and a certain *rank* in the local, regional or global hierarchy;
- the brand-image, through its unique character and plus value that it brings to urban spaces, insures the *residents' loyalty* (as "consumers" of urban spaces and services);
- the brand-image becomes *a social interaction platform*, inviting people to interact, exchange opinions, ideas, knowledge and, most importantly, to take action. Thus, the brand-image becomes more practical, generating active people and even activists in the respective spirit.

6. References

Banerjee, T & Southworth, M. (Eds.). (1995). *City Sense and City Design. Writings and projects of Kevin Lynch,* MIT Press, ISBN 978-0-262-62095-6, MIT Press, Cambridge, USA

Cândea, M.; Bran, F. & Cimpoieru, I. (2006). *Organizarea, amenajarea şi dezvoltarea durabilă a spaţiului geografic,* Universitară Press, ISBN 973-749-022-3, Bucharest, Romania

Cocean, P. (2002). *Geografie regională,* Cluj University Press, ISBN 953-7619-34-3, Cluj-Napoca, Romania

Cowan, R. (2005). *The Dictionary of Urbanism,* Streetwise Press, ISBN 978-095-44-3300-0, London, UK

Cucu, V. (2001). *Geografia oraşului,* Cultural Foundation „Dimitrie Bolintineanu" Press, ISBN 953-7619-34-3, Bucharest, Romania

Dauphiné, A. (2004). *Espace terrestre et espace geographique*, In: *Les concepts de la géographie humaine*, Bailly, A., pp. 51-62, Armand Colin Press, ISBN 953-7619-34-3, Paris, France

Frémont, A. (1976). *La région, espace vécu*, PUF, Paris, France

Ianoş, I. & Heller, W. (2006). *Spaţiu, economie şi sisteme de aşezări*, Tehnică Press, ISBN 973-31-2298-2, Bucharest, Romania

Ianoş, I. (2004). *Dinamica urbană. Alicaţii la oraşul şi sistemul urban românesc*, Tehnică Press, ISBN 973-31-2139-8, Bucharest, Romania

Ianoş, I. (2000). *Sisteme teritoriale. O abordare geografică*, Tehnică Press, ISBN 973-31-1482-0, Bucharest, Romania

Lynch, K. (1984). *Good city form*, MIT Press, ISBN 978-0-262-62046-8, MIT Press, Cambridge, USA

Lynch, K. (1960). *The image of the city*, MIT Press, ISBN 978-0-262-62001-7, MIT Press, Cambridge, USA

Miles, M. (2004). Legibility and Liveability: A Critique. *Perspectivas Urbanas*, No. 5, pp. 7-19, September 4, 2008, Available from
http://www-etsav.upc.es/urbpersp/num05/art05-1.pdf

Minca, C. (2001). *Postmodern Temptations*, In: *Postmodern Geography. Theory and Praxis*, Minca, C., pp. 196-237, Blackwell Publishing, ISBN 978-0631-22560-9, Oxford, UK

Neacşu, M.C. (2010a). *Oraşul sub lupă. Concepte urbane. Abordare geografică*, ProUniversitaria Press, ISBN 973-129-551-0, Bucharest, Romania

Neacşu, M.C. (2010b). *Imaginea urbană. Element esenţial în organizarea spaţiului*, ProUniversitaria Press, ISBN 973-129-550-3, Bucharest, Romania

Neacşu, M.C. (2009). The City Image and the Local Public Administration: a Working Tool in Urban Planning. *Transylvanian Review of Administrative Sciences*, No. 27E, (October 2009), pp. 172-188, ISSN 1842-2845

Nasar, J.L. (2001). Images of cities, In: *International Encyclopedia of the Social & Behavioral Sciences*, Smelser, N.J. & Baltes. B.P. (Eds.), pp. 1822-1825, Elsevier Science Ltd., ISBN: 978-0-08-043076-8, USA

Neguţ, S. (2011). *Geografia umană*, Romanian Academy Press, ISBN 973-27-2029-5, Bucharest, Romania

Soja, E. (1996). *Thirdspace. Journeys to Los Angeles and other real-and-imagined spaced*, Blackwell Publishers Inc, ISBN 1-55786-675-9, Malden, USA

Soja, E. (1990). *The Postmodern City/Bonaventure Hotel* In: *Open University program (BBC2)*, August 12, 2008, Available from www.youtube.com/watch?v=hhyQ0HES8mM

Thrift, N. (2003). *Space: The Fundamental Stuff of Human Geography*, In: *Key Concepts in Geography*, Holloway, S.L.; Rice, S.P. & Valentine, G., pp. 96-107, Sage Publications, ISBN 0-7619-7389-3, London, UK

Zierhofer, Z. (1999). *Human Geography – A Social Science?*, In: *Alexander von Humboldt lectures in Human Geography*, August 10, 2008, Available from
http://socgeo.ruhosting.nl/colloquium/WoZReferat1.PDF

*** (2007). *Strategia de dezvoltare a Municipiului Ploieşti 2007-2025*, Institute for Housing and Urban Development Studies Romania (IHS), August 23, 2008, Available from
http://www.ploiesti.ro/ploiesti_document.pdf

14

Agglomeration Economies Versus Urban Diseconomies: The Case of the Greater Metropolitan Area (GMA) of Costa Rica

Arlette Pichardo-Muñiz[1] and Marco Otoya Chavarría[2]
[1]The International Centre on Economic Policy (CINPE, by acronym in Spanish)
Universidad Nacional,
The Costarrican Society of Urbanism and Land Use Planning
[2]The School of Economics (ESEUNA, by acronym in Spanish)
Universidad Nacional
Costa Rica

1. Introduction

The process of agglomeration economies proceeds in increasing manner, indirectly affecting productive activities such as education, public services, transportation and all of its linked components. Nevertheless, there is a limit to this accumulative process where the gains from economies of scales are reversed and this is related to the accumulation of decreasing costs in the areas of agglomeration. These costs include price of factors (i.e., commuting) with scarcer factors (land and labor); it also includes costs related to transportation (traffic, stress, crime, etc.) (Polèse, 1998). Generally speaking, a city, just like any other economic resource, begins to enter a phase of decreasing returns of scale and cost of essential urban services like transportation, adopting an U-shaped trend, as explained in microeconomic analysis (Camagni, 2005).

This chapter is structured as follows: The first section begins with a review of the concepts required to define and understand agglomeration economies and urban diseconomies, from the standpoint of economy as well as, from the perspective of the individuals; and includes the main focus which helps build indicators commonly used to measure diseconomies. The second section describes the area of study, including the distribution of the main economic activities and population centered in the territory. The third section explains the methodology used to assess the extent of diseconomies, indicating the conceptual framework, the dataset, variables, indicators, and methodology used. The data comes from various sources, and a statistical model was built to obtain estimates (2005-2009). The fourth section covers the results obtained and the findings of the analysis of agglomeration economies and urban diseconomies in the case of The Greater Metropolitan Area in Costa Rica. An analysis is provided of city life and its ever-growing problems demanding new styles of planning, regulations, and urban management, along with physical intervention based on comprehensive, innovative, technical solutions. The last section presents the most significant conclusions, including their implications for design and regulation of urban planning, in the case studied, and others with similar characteristics.

2. Agglomeration economies versus urban diseconomies

From an economic point of view, Roberto Camagni (2005), reassessing the principles that govern the city, has done a spatial analysis and, in theoretical terms, has reinvented the study of the economics, from an urban perspective. To this effect, and from the premises of Marshall (static efficiency), Schumpeter (dynamic efficiency), and Marx (the conundrum of power), he introduces five basic principles to understand the city and the urban policy design: i. *Agglomeration* (or synergies), corresponding to the basic question of why the city exists; ii. *Accessibility* (or spatial competition) linked to the question of where productive and domestic activities should be located; iii. *Spatial Interaction* (or the need for mobility and contact); iv. *Hierarchy* (or city order); and, v. *Competitiveness* (or export as an economic driver).

The benefits of spatial agglomeration may be maximized, by taking advantage of economies of scale when people commute, and lessen the effects of dependency on fossil fuels; this can also reduce urban diseconomies minimizing the costs linked to public transportation systems which are a result of population growth and urbanization. Every single interrelation between city components must be taken into account; they include the agglomeration of enterprises, services, industries, and urbanization, among others. If there is a positive reaction, positive external economies such as the following can emerge: company clusters, availability and accessibility of productive inputs and workforce, generation of employment, provision of services and infrastructure, logistic of telecommunication networks, rise of virtual circles generated by closer residential zones and concentrated jobs, and anything else linked to time saving associated to people's commuting habits.

Externality is a concept introduced in economy (Smith, 1776) as amended (Marshall, 1890); and the question of who should pay the external costs is still a controversy since the origins of welfare economics (Pigou, 1920) and reexamination (Coase, 1960). By definition, it is generally used to refer to the consequences or economic side effects that the actions of one or more operators produce on other(s) that do not participate in the exchange, namely, it does not quantify the benefits and losses to externally affected individuals, families and businesses and they do not receive compensation or payment. In other words, externalities arise when there is no market for the exchange of goods or services, since there are no defined property rights over a large amount of goods and services when there is presence of public goods in the context of common resources, therefore it does not build a market around them, producing effects outside the market that are not compensated or paid for in commercial exchange. This makes the economic management of external approaches approximate to the presence of public goods and common resources.

From a practical perspective, the strategy that cities propose to tackle their growth, land expansion and public transportation infrastructure may trigger positive and negative outcomes. The first outcome is related to concentrated economic activity and maximizes the positive externalities emerging thereafter (Moomaw, 1981, 1983; Richardson, 1995). The second outcome is a consequence of negative externalities normally increasing private and social costs, therefore considered a diseconomy.

In short, urban diseconomies, or negative externalities, are the result of raises in average total cost over time as production and the use of certain factors increases, without increasing scales. The externalities generate change in the quality of the infrastructure and provide the required services according to demand, without incurring in inefficiencies during the process and without increasing production, coordination and transaction costs due to urban inefficiencies. The cost of urban diseconomies related to transportation and public safety problems affect both external conditions contributing to quality of life (e.g.: level of income, access to services, resources and productivity) and people's subjective perspective of quality of life in various domains (e.g.: stress, time use, leisure time and so on). These deficiencies have led to the reversal of gains from concentrated economic activities and positive externalities achieved by agglomerated dwellings, thus becoming urban diseconomies.

From the perspective of the population, the people of Costa Rica, - in comparison with other countries in the region- declare low levels of dissatisfaction with the standards of living and with the city overall; however, areas associated with urban diseconomies, reveal important levels of dissatisfaction and mistrust, especially in transportation and public safety. In fact, Table 1 shows that, while in standard of living and the city overall, less than 20% of the population expresses feeling unsatisfied; in infrastructure, and public transportation system is 37% and 27% respectively; in confidence in the Judicial System is 47% and in other aspects of public safety it surpasses 50% (Gallup, 2011).

Domain	Indicator	%
General	(Dis) Satisfaction with Standard of Living	19
	Overall City (Dis) Satisfaction	16
Transportation	(Dis) Satisfaction with Roads and Highway	37
	(Dis) Satisfaction with Public Transportation	27
	(Dis) Confidence in Judicial System	47
Public Safety	(Dis) Confidence in Local Police	56
	(Un) Safety when walking alone	59

Table 1. Costa Rica: Selected indicators by domain of (dis) satisfaction with urban diseconomies; elaboration based on Gallup (2010).

With regards to diseconomies product of public insecurity, the National Public Insecurity Survey 2006 (Madrigal, 2006) reveals that one third of the population surveyed considers this issue as the country's main problem, even over issues like the state of the economy (28%) covering topics such as unemployment, poverty and high cost of living, among others. The latest Global Competitiveness Index (2011-2012), developed by the World Economic Forum, considers the insecurity and poor transport infrastructure as the main factors affecting the competitiveness of the country, resulting in Costa Rica losing five positions, from the position 56 to 61 of a total of 142 nations studied.

Nevertheless, Civil Society Organizations have not defined urban diseconomies as one priority area of interest, although some business associations and professional organizations have prepared and presented technical proposals to the national government (ACCCR, 2006; CFIA, 2010 and Costa Rican Chamber of Construction 2011). Professional training is not geared specifically to the issue of urban diseconomies, as national research, with the exception of the International Center on Economic Policy (CINPE, by acronym in Spanish) of the *Universidad Nacional* for studies on urban diseconomies in The Greater Metropolitan Area of Costa Rica (EU/PRU-GAM / CINPE-UNA 2007c).

2.1 Description of area studied

2.1.1 The greater metropolitan area (GMA) of Costa Rica

Costa Rica is a small country (less than 50.000 km[2]), and The Greater Metropolitan Area (GMA) occupies less than 4% of its territory (CCP/INEC, 2002). Subsequently, in the context of giant cities (Brennan and Richardson 1989), which give rise to the megalopolis (Gottmann, 1962), world cities (Friedman, 1985) or global cities (Sassen, 1991), it could be considered a large city (according to ECLAC)[1]. Although, the GMA exhibit the majority of the mega-problems seen in mega-cities with urban diseconomies: road congestion, traffic accidents, air pollution, public unsafety, and others problems associated to urban expansion and the worsening of disorderly growth which lacks adequate planning and regulation.

Costa Rica, ever since colonial times, has been characterized by concentrating its population and activities in the Central Valley Area (where the most fertile land is found). The structure of what is currently known as The Greater Metropolitan Area (GMA) – main center for the attraction of population and national economic space – began during early republican life[2], and was officially established in 1982, by decree of the Executive Power. Its limits are configured according to geographical factors and it's currently comprised of 31 cantons (this is the way; municipalities are called in Costa Rica). A relatively planned growth can be admitted[3], it is based on one particular characteristic: the comparative advantage granted by the proximity of four of the six main localities of Costa Rica (San José, Cartago, Heredia and Alajuela, including the Juan Santamaría International Airport; until recently Costa Rica's only international hub) located within a radius of approximately 40 km, which have gained more ground and have become leading centers, both within the GMA and throughout the entire country[4]. Due to population growth and the resulting process of economic activity agglomeration, and, with the specificity acquired by that process, we have witnessed a rupture in the traditional spatial structure of the capital (San José) as the sole urban center of importance, and the emergence of a spatial organization with three additional population cores.

---

[1] This classification is based on socio-demographic criteria identifying three types of cities: a *metropolis*, with over 4 million people; a *large city*, with 1–4 million inhabitants, and an *intermediate city*, with 50 thousand to one million people (Rodríguez & Villa 1998).

[2] Indeed, in 1848, by legal mandate, the main towns of the central areas of the country were turned into cantons (roughly equivalent to municipalities or counties in other countries), which define the nation's main economic space.

[3] As seen in the grid identifying historical areas, and particularly in the coverage of basic electricity and drinking water networks, as well as other urban services, as in the lattice frame for distribution of traffic in urban centers.

[4] Four metropolitan areas have emerged, each with the name of its main city (San José Metropolitan Area, Cartago Metropolitan Area, Heredia Metropolitan Area, and Alajuela Metropolitan Area); together they compose the Greater Metropolitan Area (GMA).

By 2010, the GMA was estimated to have a population of 2.5 million, 57% of the country's total number of inhabitants. This also includes 86% of factories, stores and head offices (for control or processing) of the 50 most important export companies in Costa Rica, which as a whole generates 92% of the FOB-export value of these companies (EU/PRU–GAM/CINPE-UNA, 2007a).

General consensus shows that the current urban-regional approach is the result of poorly planned organizations; consequently, space-occupation processes, manifested in deep metamorphoses in terms of land use for multiple purposes, are constantly vindicated. The current organization is attributed to several factors, among which we can point out existing low interest in urban planning and regulation. This lack of interest caused decisions that would have contributed to deter unhealthy tendencies that were seeping, either openly or stealthily, into the GMA's urban space, to be postponed.

Simultaneously, it harbors socio-economic and spatial imbalances in its territory, which limit its own metropolitan integration, as well as its integration with the rest of the country. This leads to different levels of development and reveals mechanisms that are insufficient for activating this process. They include the design and execution of programs for sustainable urban planning in different cities, urban occupation oriented to integrate the main cities and other small towns within the metropolitan city system, the consolidation of institutional coordination mechanisms for increasing the management capacity of municipalities, the development of service infrastructures and systems, and the improvement of personal security and asset security, among other aspects to be considered in public, private and mixed policies and programs.

3. Methodology

3.1 Conceptual framework

To estimate the total cost of urban diseconomies, generally, methods of valuation analysis are utilized. In the case of diseconomies of traffic congestion, applied socio-economic valuation models to estimate, on one hand, the time lost in congestion through the use of socio-economic indicators such as labor income and, on the other, the additional consumption fuel. In the case of diseconomies generated by traffic accidents and air pollution from vehicle emissions, as a rule, comes as direct or indirect methods of assessment.

Direct methods (primarily contingent valuation or revealed preference method), focus on the willingness to pay by individuals in order to experience changes in their welfare, through the application of complex questionnaires and direct questions, that provide information on the value or cost of the resulting deadweight loss, including material costs for medical care and estimates of so-called *intangibles* costs such as loss in quality of life, pain and suffering. This has been the preferred method for economic valuation of externalities of the transportation system product in those countries with high-income levels (mainly European) (GRSP Focus 2003). It has applications, among other limitations, which depend on subjective judgment of people, as it is they, who determine the value to avoid or mitigate the direct and indirect effects of externalities, in addition to being influenced by the ability to pay of the people being interviewed. Hence, the difficulty utilizing it, because it is based on considerations of probability and statistical risk commonly known. Due to these limitations its use is ruled out in Costa Rica.

On the other hand, indirect methods of assessment are based primarily on the complementarities and substitutability relationships between good or service and market value (Azqueta, 2007). Unlike direct methods, this approach allows the inclusion of intangible costs such as suffering, pain or loss of quality of life that the person considers to be affected by when the environment varies; it is not based on subjective considerations. Principally, it is possible to assess the costs related to health care and lost labor productivity, according to market prices or observable medical costs. Generally, among these methods are: the human capital method and the method of replacement cost.

The human capital approach is a methodology used in economic evaluations known as *human costs*, measured by the value of a statistical life, that is to say, what society as a whole loses with the early death of one of its members, referring to the loss of productive capacity. Therefore, considering the loss in productivity, taking into account the age of death of the person and the maximum age for employment, according to the definition used in the economically active population (EAP), valuing life years lost as a function of income or Gross Domestic Product (GDP) per capita, thus reflecting the economic losses product of death[5]. This method can be complemented with other approaches to estimate - for example - damage to property and administrative costs and medical care to victims. Toward this end usually the method used is the cost replacement, since what it seeks is to return the asset to its original condition (Azqueta, quoted). Which is used to estimate all necessary costs to be incurred to return the asset to its initial state, the above in the case of physical damage but it also considers the costs necessary to restore health to people.

3.2 The data set

The quantitative information used for the estimation of different urban diseconomies in the present chapter, derives from diverse public institutions, producers of primary character information, mostly for the period of 2004 – 2009. Socioeconomic information comes from statistics from the National Institute of Statistics and Census (INEC, by acronym in Spanish), which is the main entity for statistical matters in Costa Rica. The investment information in urban infrastructure is from Costa Rica's General Comptroller's Office; the highway conditions and measurements were provided by the Public Works and Transportation Ministry (MOPT, by acronym in Spanish). The information regarding the flow of vehicles is from the National Public Transportation Safety Board (COSEVI, by acronym in Spanish), whereas, the vehicle count in congested traffic at specific locations was provided by the Department of Sectorial Planning. Both entities are under the umbrella of the MOPT; COSEVI also provided details on traffic accidents (deaths, material damage, and injuries). The statistics on morbidity and mortality from acute respiratory infections were obtained from the Costa Rican Social Security System (CCSS, by acronym in Spanish). The information on the total number of felonies and related costs is from the Planning Department of the Judicial Branch, and hospital costs associated to murders and suicides were obtained through the CCSS, additional information concerning material damage to

---

[5] It is clear that in no way human life is valued in monetary terms; it is something besides meaningless, contrary to ethical and moral considerations. The attempt with this method is to have a rough idea of how society relinquishes in terms of income and production potential.

private property, and investment to prevent the social insecurities was provided by the insurance companies and local government. The Central Bank of Costa Rica maintains the economic data for each year.

Similarly, the additional information required for the calculation of estimations that are not generated periodically by public entities, has been taken from studies or special publications for a specific year. The population data, utilized in the period of the analysis, was taken from estimates and projections (CCP/INEC, 2002) taking as the base the year 2002. Labor income data correspond to the National Survey of Multiple Purposes (ENPM, by acronym in Spanish), and the data on transportation expenses from the National Survey on Income and Expenses (ENIG, by acronym in Spanish) conducted in 2004: however, for the purpose of the estimations, the values have been indexed by annual inflation. The fuel consumption data is from the Survey of National Energy Consumption in the transportation sector of Costa Rica 2004, conducted by the National Energy Directorate (DNE, by acronym in Spanish), Ministry of Environment, Energy and Telecommunications (MINAET, by acronym in Spanish). The statistics on percentage of emission of pollutants from the vehicle fleet with human health effects were taken from previously conducted studies (Allen, *et al* 2005).

3.3 Variables and indicators

Assessment depends on the quantity and quality of the statistics available. It also depends on reliability, accuracy, and continuity; in many cases, this represents a restraint when making estimates. In Costa Rica, the institutions responsible for record keeping do not update information continuously, and at times the data is not gathered with the purpose of measuring costs. Quantitative methods were used for the calculation of diseconomies and its expression in monetary terms in regards to GDP, considering a trailing period from 2005 to 2009, based on other information that enables, making reasonably reliable and consistent inferences regarding to:

1. *Transportation System: a)* Road Congestion (Time spent in traffic blockage and additional fuel expenses due to that); b) Traffic Accidents (cost of fatal victims with severe and minor injuries and material damage); and, c) Air pollution from motor vehicles (incidence on respiratory diseases and hospital care and Carbone Dioxide Emission – Green House–).
2. *Public Safety: a)* Murders, b) Suicides, c) Injuries, d) Vehicles Thefts and d) Expenses on public and home safety.

The methodological approach used in the quantitative estimation of urban diseconomies in the GMA and its expression in monetary terms are responsibility of the authors. In that sense, the proposed approach pretends, first of all, to discover the cause and effect relationship associated with the transportation system, land traffic and public insecurity, to determine its socioeconomic and environmental impact and its posterior expression in monetary terms; using indirect methods of economic valuation, as in the case of human capital method, avoided costs and replacement costs; which are based on the complementarity and substitutability relationships that exists between markets good and those without price. Given that the interest in the analysis of urban diseconomies in Costa Rica is recent and lack up-to-date studies, by the application of statistical models, must be taken into account the limited information available for periods greater than 10 years, which prevent from using more sophisticated analysis models, such as econometric model, utilizing time series.

3.4 Model and estimations

Traffic Congestion can be defined as the inconvenience and increased costs that travelers impose on each other, because of the intrinsic relationship between density and speed of traffic on the road network capacity (Department of Transportation, 2011)[6]. In broader terms, congestion is the condition that prevails if the introduction of a vehicle in a traffic flow increases the circulation time of others (Thomson & Bull, 2001 as cited in ECLAC, 2002). A more precise technical definition, considers congestion as the condition that prevails if the introduction of a vehicle in a traffic flow increases the delay of others in more than x% (ECLAC, 2003). In other words, congestion can be said to be the situation that occurs when the hourly traffic demand exceeds the maximum sustainable yield of the given time limit and occurs when vehicles are stuck to each other, because of the relationship flow rate under conditions where the use of a transportation system is nearing capacity (Grand-Muller & Laird, 2007)[7].

There are several ways to measure congestion, among the most commonly used; traffic delay is seen as a situation in which an incremental position over the status of "free flow" occurs. The latter is the uninterrupted flow condition occurred when the road has sufficient capacity so that any car that passes does not affect the travel time of another.

The approach to obtain the total cost of time lost in traffic[8] has been developed in two stages: the first refers to the quantification of the number of vehicles circulating in congested conditions, as well as the occupancy level[9]. Subsequently, and once the total occupancy in congestion (number of persons in congestion) is known, proceed to determine the average time lost in congestion and provide an economic value through an indicator of proximity to income. What was described above, allows the inclusion of a monetary valuation of loss in productivity as a result of congestion.

The total cost of time lost in congestion is obtained by applying the following expression:

$$CTTPC = [((TDP *TDPC)*TC)*1,4]*IL$$

Where:

CTTPC =   Total cost of time lost by traffic congestion (US$/year).
TDP =   Average daily traffic measured at different points of congestion.

---

[6]This definition coincides with that used by Transit Authorities in countries like Scotland, and corresponds with the one applied by the Canadian Transportation Authority.
[7]Traffic congestion can be recurrent and non-recurring. The first occurs regularly in a daily cycle, weekly or monthly, allowing it to be manageable. The second is caused by traffic accidents or special events, such as construction or road closures. Traffic congestion is, therefore, a marginal phenomenon and shows how, under certain conditions, the incorporation of new vehicles has an impact on all other vehicles that are traveling at the time.
[8]The value of travel time is defined as the cost borne by society due to the fact that a person travels between two geographical points.
[9] It is considered that on average 1.4 people travel each in personal vehicles.

$TDPC=$ Average daily traffic congestion circulating, calculated not only in terms   of peak hours (6, 7, 8:30 am and 4, 5, 6, 7 pm) but also in terms of   schedules which have a behavior similar or higher than the peak hour traffic.
$TC=$ Average time in congestion.
$IL =$ Average hourly labor income.

Given that Costa Rica is not systematic in its measurements and counts, current and continuing, on average daily traffic flow and in fair condition and congestion, the available data correspond to 90 points of congestion on the GMA taken in 2006, the calculation was performed taking as a base that year and estimating the number of vehicles up to 2009 by the rate of growth of vehicle fleet.

A second effect on congestion refers to the total cost for additional fuel consumption[10], which is obtained by applying the following expression:

$$CTCGC=(CA_{gas} *TDPC_n)*P_{gas}$$

Where:

$CTCGC =$ Total cost of additional fuel consumption by traffic congestion (US\$/year).
$CA_{gas} =$ Additional fuel consumption (in liters).
$TDPC =$ Annual average daily traffic congestion circulating
$P_{gas} =$ Average price of liter of gas

Traffic Accidents are sudden event that cause involuntary harm to people (fatal or nonfatal), property and administrative damage. Various factors influence the amount and severity of them, including traffic congestion, climate, infrastructure, vehicle conditions, and the behavior of those who drive, among others[11].

The total cost of traffic accidents is determined by socio-economic cost of casualties, both fatal or nonfatal and severe and minor (cost for medical care and lost productivity costs), as well as for the cost of damage. This lost productivity is estimated not only for road fatalities, but it is also necessary to consider the lost productivity of those severe and minor injuries, which can vary from one day, to one week, or months (referred to as temporary disability), but it also must consider those permanent disabilities that involve loss of productivity for years.

---

[10] Assumptions used in this calculation are as follows: The average performance of gasoline per vehicle (36.6 miles per gallon) (DSE, 2005). The additional fuel consumption per km increases in congestion by 25%. Likewise, it is assumed, as data marked moderate and conservative, that on average, each vehicle travels 10 km daily in congestion.
[11] Externality valuation of traffic accidents in different countries, examined in each case in particular its adaptability to the availability of information and the situation of the country concerned. Thus, the total cost of a traffic accident varies with the methodology involved and the information available, so the amount of such external costs depends on the institutional arrangement of each country (Rizzi, 2005). For example, the European Commission has standardized a method for its member countries (COST 313), which determines the cost of road traffic accidents by victim considering medical costs, non-medical rehabilitation, and loss of productive capacity, along with other economic and human costs. However, this approach yields very different results depending on the member country in which it applies, by way of illustration, to perish in traffic accident in Norway is 15 times higher in cost than in Spain (Asociación Española de Carreteras 2005).

To calculate the total cost of traffic accidents, first, it is necessary to estimate life years lost as a consequence of a traffic accident, from the human capital approach, which results from the difference between the age of death and the economically active age according to the age range used in the definition of the Economically Active Population (EAP), which usually defines working age between 15 and 64. Since the GMA does not record data on all deaths in traffic accidents, by age range; were taken as references the number of accidents and the respective ranks of death by age in traffic accidents, for the total of the country, in order to determine the probability that a person who dies in a road accident is within a particular age range.

Along with the available information we proceeded to construct a relative frequency distribution by age group for deaths from traffic accidents, then was assigned - based on that distribution - the probability that a randomly selected person dies within a specific age range as a result of a road accident. These probabilities are used to distribute the total of accidents in the GMA by age range, so that it can better estimate the number of life years lost, and not rely solely on the average.

The total years of life lost in traffic accidents is calculated by using the following formula:

$$TAVPP = \lfloor (M_n * \theta) * AVPP_i \rfloor$$

Where:

$TAVPP =$ Total years of life lost due to traffic accidents.
$Mn =$ Number of deaths in n years.
$\theta =$ Probability to die within a specified age range.
$AVPP_i =$ Years of life lost according to age range i.

After obtaining the total years of productive life loss, proceed with the assignment of a monetary value to them. It uses a *proxy* to measure the loss in monetary terms. In this case, years of potential life lost are valued considering the average labor income of the central region, which belongs to the GMA; in addition to considering the rate at which the economy grew in recent years and discounting future income present value. This is so, to count the total product cost of the lost productivity traffic accidents cost yearly. The average total cost of lost productivity for road fatalities is the sum of each future productive year that is lost once the rebate is realized, estimates are as follows:

$$CTPP = \sum_{i=1}^{N} \frac{I\left(\frac{\llbracket (1+\beta)^i \rrbracket - 1}{n}\right)}{(1+r)^i}$$

Where:

$CTPP =$ Total cost of lost productivity.
$I =$ Average Labor Income.
$r =$ Discount rate.
$\beta =$ Average annual income growth.
$i =$ Years of life lost in traffic accident.

Should be clarified, that the social discount rate used is different from the market discount rate that is traditionally used, for what it is, is to determine the present value of goods or services not necessarily tangible, or that its value increases with time or not, experience the normal process of physical capital depreciation, for example, human capital and environmental goods and services; the meaning of such fee differs, since in practice, is used as social discount rates relatively low compared with those used by the economic valuation of traditional project[12]. Such differentiation is justified since high discount rates tend to punish heavily the future, while low discount rates lower the value less accelerated by providing utmost importance to its value in the future than at present, especially when dealing with resources whose expected future value will be greater than the current (Eskeland, 1994; Cifuentes, Rizzi & Vergara et al, 2004; Oliva, 2004; De Rus, Betancor and Campos, 2006).

To calculate the economic and social costs of severe and minor victims, both the cost of medical care and the cost of lost productivity are considered. In the economic assessment of severe and minor victims the procedure was to consider the information on costs incurred by traffic accidents reflected in emergency care, outpatient and hospitalization, with information from the COSEVI and the Social Security Fund. We used the average cost of care for traffic accidents reported, such as hospitalization, acute or emergency care, provided by the CCSS and in particular in the *Calderon Guardia* Hospital. For purposes of the estimate, it is assumed that every person injured by an accident has a cost for emergency care. Adding hospitalization costs by only computing people with severe injuries and costs for outpatients care only for people with minor injuries. In consequence, we proceed to estimate the total cost of care for people injured in traffic accidents using the following formula:

$$CTHAT = \sum_{i=1}^{n}\left[\left[(HAT_n) * Cme_{ri}AE\right] + \left[(HGAT_n) * Cme_{ri}AH\right] + \left[(HLAT_n) * Cme_{ri}AA\right]\right]$$

Where:

$CTHAT =$	Total cost of hospital care for road traffic.
$HAT =$	Number of persons injured in traffic accidents in the n year.
$Cme_iAE=$	Average cost for emergency care in traffic accidents rank i.
$HGAT =$	Number of persons seriously injured in traffic accidents in the n year.
$Cme_{ri}AH=$	Average cost of hospital care for rank i.
$HLAT =$	Number of people slightly injured in traffic accidents in n year.
$Cme_{ri}AA=$	Average cost of outpatient care by rank i.

To medical cost is necessary to add the total costs per absence from work. Whenever an accident occurs, the person involved must stop working during the recovery period, in some cases this period may include days, weeks, months or even years. Given the lack of data on the subject, it is considered, according to the international benchmark, that the average recovery period a person with minor injury has is 3 days, and serious injury is of days.

---

[12] At the present time, there is no consensus on the TSD to use when assessing the loss in labor productivity or natural resources; the rule states that should be less than the discount rate the market usually utilizes. In several countries, the choice of the social discount rate is between 3% and 8%. The variability is relatively high and simultaneously, in Latin American countries with similar characteristics have coexisting rates ranging from 5% to 8% (De Rus, Betancour & Campos *et al* 2006).

$$CPPH = HLAT * DI * I + HGAT * DI * I$$

Where:

CPPH =   Total cost for lost productivity.
HLAT =   Number of people slightly injured in road accidents.
HGAT =   Number of people seriously injured in traffic accidents.
DI=   Average days of disability.
I=   Average labor income.

To estimate damage costs it is considered the total traffic accidents with material damage and their respective average cost as an approximation for their replacement cost. In turn, the replacement cost was estimated at consultations in different workshops in the metropolitan area, the average cost incurred by a person to repair their vehicle following a road event[13].

Although in these cases the insurance companies should be the best reference, using the information held on compensation for these estimates with the recent opening of the insurance market in the country makes dealing with such information confidential and therefore not available to the public.

$$CTDMA = AT * CR_{me}$$

Where:

CTDM =   Total cost for material damage.
AT =   Number of traffic accidents with material damage.
CRme =   Average cost of replacement.

From a public policy standpoint, another relevant diseconomy is air pollution, which surpasses acceptable levels and affects people's health and the atmosphere. In the case of diseconomies of air pollution from vehicle emissions, it is interesting to estimate what society least has obtained in exchange for the work of the person, if that person had not died prematurely by sickness as a result of Acute Respiratory Infectious (ARI, IRA by acronym is Spanish) due to emissions, or if he or she had not been absent to work for the same reason. Apply the same principles used in the calculation of mortality and morbidity from road traffic accidents (Equation 3 y 4) and attribute that 20.5% of morbidity and mortality from ARI due to vehicular emissions from a direct or indirect source (Allen, *quoted*).

When estimating the costs associated with morbidity due to ARI, after establishing the numbers of cases attributable to vehicle emissions, it is necessary to identify the main costs incurred by the health care system for services provided and identify the respective average costs for each type of service. In this sense, we consider the outpatient medical services, emergency consultation and hospitalization and expenses for ARI and their respective costs provided by the Social Security Fund.

The total product cost of morbidity due to ARI, and measured by the costs incurred by the public health system is given by the following equation:

---

[13] The insurance companies should be the best reference, using the information held on compensation for these estimates; the recent opening of the insurance market in the country makes dealing with such information confidential and therefore not available.

$$CTMIRA = (((HIRA * \theta) * CPH) + ((CEIRA * \theta) * CCE) + ((MIRA * \theta) * CPC)$$

Where:

$CTMIRA$ =	Total cost of morbidity due to ARI.
θ =	% ARI attributable to vehicle emissions.
$HIRA$ =	ARI Hospitalization Cases.
$CEIRA$ =	Emergency cases handled due to ARI.
CCE =	Average cost of an emergency consultation.
$MIRA$ =	Morbidity from ARI cases.
CPC =	Average cost of a consultation for ARI.

The productivity costs of morbidity due to ARI are estimated as follows[14]:

$$CPPH = MIRA * DI * I$$

Where:

$CPPH$ =	Total cost for lost productivity.
$MIRA$ =	Morbidity cases due to ARI per year.
DI =	Average days of disability.
I =	Average Labor Income.

While the cost of *greenhouse* gas emissions (CO_2) is estimated considering the costs of compensation, defining the fuel consumption and emissions of greenhouse gases is associated with evaluating the establishment of such emissions at market prices.

To determine the impact of transport on the environment sector, it is necessary to consider the CO_2 emissions, as this is a major greenhouse gas. For this estimate we must determine the annual fuel consumption by type: Gasoline (Super and Regular), Diesel and Liquefied Petroleum Gas (LPG). Due to lack of information available on sale or consumption of fuel for the GMA, we proceeded to estimate the fuel used as a bridging factor to the population. To calculate the CO_2 emissions multiplied fuel consumption by type, by their respective emission factors. The method determines the expenses that would be incurred to compensate.

The total cost for the emissions of greenhouse gases (CO_2) is given by the following equation:

$$CEGEI = (CC_i * FE_i) * P_{ton}CO_2$$

Where:

CEGEI=	Total cost emissions of GEI (CO_2).
CCi =	Total fuel consumption i.
FEi =	Emission factor for fuel i.
$PtonCO_2$ =	Price ton of CO_2.

Public Safety is an issue that in recent decades has attracted attention worldwide and can be defined as the fear people feel to be threaten by aggression and violence from the environment, affecting the full enjoyment of life, psychological implications and socio-economic costs, both repair and prevention, which must be tolerated by families and businesses.

---

[14] In the absence of national data on temporary disability, it is estimated that the average number of days not worked due to this cause can be between 4 to 6 days (PAHO, 2005); we use the average of 5.

The total cost for the public safety is given by the following equation:

$$CTIC = (QCDA*CMA) + (QH*CMAH + QS*CMAS) + CAPHS + CSP + CSPH$$

Where:

$CTIC =$	Total Cost of public insecurity (US$/per year).
$QCDA =$	Number of cases of crime and aggression
$CMA =$	Average cost of care for cases of crime and aggression
$QH =$	Number of homicides
$QS =$	Number of Suicides
$CMAH =$	Average cost of medical care for homicides
$CMAS =$	Average cost of medical care for suicides
$CAPHS =$	Cost in life years lost due to homicides and suicides
$CSP =$	Cost of Public Safety
$CSPH =$	Cost of private security for homes

The estimated economic cost of lost productivity that society experiences with the deaths of its members as an anticipated product of a suicide or homicide are estimated following the methodological approach presented in equation 3 and 4 of this section. The total cost of reported cases as crimes or assaults were regarded as injuries, for which we estimated the respective health care costs, distinguishing between minor and severe injuries, and in these cases, it is also considered the result of lost productivity of temporary disability.

To calculate the cost of public safety, it is considered the country's total expenses incurred by the Ministry of Public Security, Judiciary, Ministry of Interior and Police, Ministry of Justice and Interior. Given that, this information is only available at national level for purposes of charging a cost only for the GMA, the costs were multiplied by the relative weight of crimes reported in the GMA with respect to the whole country, resulting in an average 55%.

For the estimation of expenditure made by households as a preventive measure of protection against insecurity, it is considered the information and assumptions made in the study "Economic aspects related to insecurity" (Sauma & Chacón, 2006), which, based on National Survey of Income and Expenditure (ENIG, by acronym in Spanish), which was conducted by the National Institute of Statistics and Census (INEC, by acronym in Spanish), technical governing body of the National Statistical System, estimates the minimum average cost in different areas linked to insecurity. Given that, there is no disaggregated information at the level of GMA, since it is done at the national level[15], we proceeded to update the data from the cited study, considering the growth of the GMA in terms of population, the number of dwellings and vehicles, also the prices level and the assumptions made in the baseline study, using the transfer of results of 2006 to later years.

Among the items considered to estimate household spending on security are private security, installing fences, alarm and monitoring systems for homes, the installation of razor wire, the alarm system in family vehicles, the installation of locks and electric gates, construction of garages for cars and payment of car theft insurance and housing.

---

[15] For the realization of ENIG, INEC uses a stratified sample to establish as a domain the central region Central (urban and rural), Rest of the urban country and Rest of rural country, so it is not possible to disaggregate the information in a specific way for the GMA.

4. Results and discussion

4.1 Transportation system diseconomies

Transportation system is an important diseconomy, especially in cities, like GMA, that have not given priority to investing in transportation infrastructure and networks. This disregard for investment affects city living and structure cost (socially and economically), which becomes a burden for enterprises, families, and society as a whole.

Road congestion originates from interference and friction among vehicles in comparison to the normally expected vehicle flow. This is interrelated to highway conditions and a reasonable speed required for traveling from one place to another. Additional cars make it more difficult for other vehicles to advance in traffic, and congestion becomes more evident. The stress passes on to people by means of traffic problems and longer transportation times, especially during peak hours; reduces the amount of time that could very well be spent on alternative activities (productive or leisure). This situation also leads to greater fuel consumption, and increases the probability of other costs related to human injuries and material damages due to more traffic accidents. Another apparent cost is air pollution due to toxic vehicle emissions and their effects on the atmosphere, the respiratory system (diseases) and building structures (decay). In other words, traffic congestion caused by a larger number of vehicles and a limited transportation network, unravels a series of factors that could hinder the sustainability of the urban system and affect people's quality of life, thus reducing productivity and the efficiency required in cities, with higher economic and social costs as time goes by.

Table 2 shows the results of the assessment and its expression in monetary terms of the transport system diseconomies in the GMA, for the years 2005-2009, on a cumulative basis, amounts of U.S. $ 3,805,484,646, representing 13.5 % of GDP, being the higher cost attributed to the time lost in congestion (71.5% of total).

Costs	2005	2006	2007	2008	2009	Total
Time Loss in Traffic Congestion	504,780,388	515,082,029	531,011,376	578,212,387	590,012,639	2,719,098,818
Additional Fuel in Traffic Congestion	45,557,171	46,486,909	47,924,556	52,184,516	53,249,507	245,402,658
Traffic Accidents	85,862,252	100,653,299	106,351,163	124,704,692	122,316,355	539,887,760
Air Pollution due to Motor Vehicle Emissions	53,739,866	58,835,214	59,141,786	63,663,457	65,715,086	301,095,409
Total	689,939,676	721,057,450	744,428,880	818,765,052	831,293,587	3,805,484,646
% with respect to GDP	2.64	2.63	2.66	2.79	2.84	13.56

Table 2. Costs of the main transportation diseconomies in the GMA between 2005 and 2009 (in US$, and as a % of GDP), own calculation based on information from COSEVI, EU/PRU-GAM/CINPE-UNA, DSE, CCSS, INEC and BCCR.

4.2 Public safety diseconomies

Public insecurity is a major urban diseconomy, as manifestations of violence are a constant in the daily activities of the population, which hinders the normal development of people in daily tasks, as a consequence becoming a disincentive to private investment and individual, as the damage to people and property generates an increase in operating costs, both preventive (personal safety devices, insurance and others) and repair, involving both an opportunity cost for families and businesses, since those resources could be devoted to improvements in productivity and quality of life.

In recent decades, the GMA has shown a steady increase in all variables considered in the calculation of insecurities diseconomies, both the number of murders, suicides, injured persons and stolen vehicles, and the households and public spending on security.

Table 3 shows the results of the assessment and its expression in monetary terms of the Public Safety diseconomies in the GMA, for the years 2005-2009. On a cumulative basis, it amounts US$ 2,020,151,841, representing 7.89 % of GDP.

Costs	2005	2006	2007	2008	2009	Total
Murders – Suicides	89,836,942	98,063,197	118,543,256	168,118,618	192,501,616	667,063,630
Severe injured	3,811,422	4,347,415	4,915,683	5,150,947	5,130,861	23,356,328
Minor injured	148,664	215,597	252,050	273,362	284,668	1,174,340
Vehicle Theft	3,725,699	4,867,152	5,472,677	6,456,835	5,587,701	26,110,063
Household Expenditure	7,161,695	7,957,439	10,193,905	11,080,951	4,978,771	41,372,760
Public expenditure	179,524,625	211,071,980	246,533,332	322,999,449	300,945.333	1,261,074,719
Total	284,209,047	326,522,780	385,910,902	514,080,162	509,428,950	2,020,151,841
% with respect to GD P	1.51	1.45	1.47	1.72	1.74	7.89

Table 3. Total costs due to public unsafety in the GMA, 2005–2009 (in US$, and as a % of GDP); Own calculation based on UE/PRU–GAM/CINPE 2007b/ SAUMA 2006, CGR, CCSS, PJ, INS, INEC and BCCR.

4.3 Urban diseconomies versus infrastructure investment

The estimation of socio-economic and environmental costs of externalities of transportation system diseconomies and public unsafety diseconomies in the GMA, permit to have a rough idea of how they represent urban inefficiencies for Costa Rican society, though the calculations should be considered conservatives, because both the difficulties of systematic information, current and reliable, as the inability to quantify the suffering and the moral and psychological damage caused, and the distortions that they generate in the income of the families and enterprises. These deficiencies can be reduced with policies aimed at sustainable investment in urban infrastructure.

It is interesting to compare the estimated total cost diseconomies cumulative for the years 2005-2009, with the cost of some investment projects that have been estimated and proposed during the same period, many of them still undeveloped. Subsequently, the investment options discussed, despite their cost, once compared with diseconomies generated by the current system of transit and land transportation are justified in terms of potential cost that would be saved by society and Costa Rican economy.

The assessment by the concept of diseconomies resulting from poor infrastructure and road transportation system, measured in terms of congestion and delay in the period 2005-2009 amounted to US$ 2,964.5 million; if this is added the costs to traffic accidents and their aftermath, in a country that has a mortality rate higher than the average in Latin America, gives a total of US$ 3,504.3 million, if they accumulate and are added to the above costs the direct and indirect effects of vehicle emissions have on human health and the environment, measured in terms of greenhouse effect gases (CO_2), the estimated cost is US$ 3,805.4 million. The estimation for public safety diseconomies in the GMA, for the years 2005-2009, on a cumulative basis, amounts for US$ 2,020.1 millions.

Table 4 presents a list of proposed investments for five years in analysis with a time scope until 2015 (EU/ PRU-GAM /CINPE-UNA 2007c), but the steep cost of investments should be made in the first five years. Projects identified in the present time by the relevant authorities could help reduce the costs of diseconomies concept, involving an initial investment of US$ 1,488.9 million (for the first 5 years), the latter figure are clear benefits that could result from such investment in terms of reducing diseconomies. This is without considering the effects of linkages that such reductions could generate and its impact on the economic performance of the country through increased efficiency of urban performance of the GMA.

Diseconomies US$ (millions) 2005-2009		2006	2007	2008	2009	2010
Time loss and fuel consumption	US$ 2,964.5	-422,69	-216,27	-216,27	-216,27	-0,56
Traffic accidents	US$ 539,9					
Air pollution	US$ 301,1	-62,43	-156,19	-96,58	-100,32	-1,35
Public Safety	US$ 2.020,1	I.N.A.	I.N.A.	I.N.A.	I.N.A.	I.N.A.
TOTAL	US$5.825,4	-485,12	-372,46	-312,85	-316,59	-1,91

I.N.A. = Information No Available.

Table 4. Summary of diseconomies and investment projects in the GMA between 2006 and 2010 (in US$); Own calculations based in diseconomies estimation and EU/PRU–GAM/CINPE-UNA 2007c

As for urban infrastructure investment needs, local authorities exhibit some degree of consciousness, but - not necessarily - it is reflected in the investment priorities at national level. In this regard, it is interesting the point of views of the Mayors of San Jose and Heredia, two of the biggest Municipalities in the GMA.

The Mayor of San Jose (Pichardo, 2010), states *"neither the repopulation of San Jose and the Promotion of Tourism would have any meaning if there is no improvement of public safety; henceforth, we are raising efforts at two levels: first, at the level of the Municipal Police, to strengthen it, both in number of police officers, equipment, etc., and in that I see a question about the law, but basically we have found in the alarm monitoring service, an alternative in financing growth of the municipal police; it is not only a new service that technology helps improve security in the city but also potentially a revenue for the municipality to increase public safety. I would like to have in four years, ten thousand customers of the alarm monitoring service, which would imply one billion colones a year that we'd all be investing in public safety, it would help a lot. But also another level at which we would constitute the Security Council of San Jose which was an initiative in my government's agenda and we united in a single Council all police there is: the Judicial Investigation Organization, the Ministry of Public Security with its three polices (the Migration police, Drug Control and the Police Forces and Fiscal Control) with the Municipal Police to begin working together, to plan joint operations to avoid duplication. This will help improve public safety and the safety of San José, passing through the recovery of the social framework, and the reactivation of the city, a city that is abandoned after six in the evening is not safe, even with more police officers; then, it is precisely a process that is integral. Safety is not only more police, but also more people living and walking around town. "*

Another project is the tram launched by the Municipality of San Jose as part of the efforts to make the city more livable, competitive and functional[16]. History dates back to 1984, the date in which the first feasibility study was conducted. Political reasons linked to the need for investment, because it is a public works project that goes beyond municipal powers, have limited their development. Today the President of the Republic has made it a national priority project and the initiative is in the field of climate change facing the need to reduce CO_2 emissions as part of the country's efforts to meet demands for carbon neutral by 2021. The French government has again offered support to update the feasibility study. The initiative complements the reactivation of the rail system that propels the Instituto Costarricense de Ferrocarriles (INCOFER, by acronym in Spanish) between metropolitan areas, which must be provided of exchange nodes and build a multimodal system integrated with the urban bus routes, so that is a distribution hub and they do not enter the city center. It is expected that the tramway will have a cost of US\$ 10 million/Km for a total of 8 Km and a construction time of 30 months, and, could initiate construction in 2014, just when the concessions to public transport operators in the GMA expire. However, we need the financing scheme and a complete economic study to assess its impact on reducing negative externalities.

Similarly, the Mayor of Heredia (Pichardo, 2010) said *"Public safety is a major problem nationally and in the case of the canton of Heredia affects much more to have as part of its territory one of the largest, more precarious in the country ... This situation raises the need to increase and modernize municipal police teams ... and, most importantly, the cameras project because we know that we must avoid crime. In every school, every church, wherever people congregate cameras are being installed ... In this sense, we are also maintaining an excellent relationship and coordination with the Judicial Investigation... we have to closely coordinate to succeed in this issue of combating crime. We are also operating in and around colleges and schools to prevent drug trafficking. In*

---

[16] The Mayor of San Jose was interviewed for these purposes, April 12th, 2011.

addition to all this a matter of vital importance is the organization of communities. In that sense, we are working with technical advice from an Israeli company in the Eyes and Ears Program, with organized communities... As for the viability of the Project Radial Heredia - San José, that was dismissed because of costs... and what is considered in coordination with the Ministry of Public Works and Transportation to the 2011 and 2012, is a tunnel at the entrance of Heredia, and the expansion to 4 lanes on the road to Alajuela and to restore Heredia's Train. "

5. Conclusion and recommendations

In conclusion, the formation of cities involves consideration of a specific geographical area where a confluence of economic, recreational, cultural and commercial, facilitated by a number of structural conditions such as infrastructure, transportation system, safety and institutional policies, among others, facilitate the development of agglomeration economies among other things, including the establishment of clusters and various production chains back and forth between various actors (companies, industries, service providers, people, etc.) allowing a better use of the available resources and lower production costs and transaction costs.

From a theoretical point of view, under appropriate urban planning cities can generate various economies product of agglomeration. Improvements in the location, whether in business or family can generate gains in terms of improved productivity, reduced transportation costs, and reduced costs of recruitment and training, also in search of suppliers and information. At the industry level, agglomeration economies even allow a divisibility and / or distribution of the costs associated with location, production, marketing and distribution.

Another important element of economies result of agglomeration is the profit of having the production and distribution of public services and infrastructure in an area that is focused on activities, which obviously reduces their cost, allowing areas to diversify the supply as diverse as telecommunications, energy, financial services and outsourcing. The population also benefits, by shortening travel time between their working and living, in addition to a variety of services concentrated and accessible, reducing travel costs and improving their material conditions of life.

However, economies of agglomeration can decrease, disappear and even become diseconomies or negative externalities due to inefficient urban systems that begin to unravel. Whenever the capacity of the city is saturated, either because they reduce or limit investment in road infrastructure and transport by land, traditional means of transportation become inefficient, creating congestion, lack of planning and other factors such as insecurity, accidents and pollution levels exceeding "tolerable" levels.

In the case of GAM of Costa Rica, this area has experienced a series of changes in structure and composition, which have generated positive benefits in terms of attracting investment, job creation, service provision and innovations in general, but have also developed a set of negative factors given the degree of saturation and the complexity of disordered use of the territory as well as insufficient capacity to invest in urban development.

As an example, in 2005, it was invested in transportation 39%, less than the average for 2000-2005. More recently a study by the Comptroller General of the Republic (2008), determined that Costa Rica has not reached to invest more than 0.83% of GDP in transportation infrastructure in the period between 2000 and 2007.

Subsequently, the permanence and increased costs due to diseconomies of urban GAM are limited by deficiencies in proper urban investment, preventing or diminishing the use of profits generated by economies of agglomeration.

Some of the measures that could improve the territorial competitiveness, the quality of life of the people, the environmental sustainability, and the integration and social cohesion of the GMA, at the institutional level is the creation of a Managing Authority for the Great Metropolitan Area, with functions of policy formulation and management, with autonomy and authority to establish guidelines, programs and investment projects for the territory of the GAM; in order to implement a program of investment projects in urban development and enactment of a law for the reorganization of metropolitan areas.

The participation of local actors is crucial, as it will facilitate the establishment of a balanced and competitive territory (regulating the growth and expansion of urban areas) through proper strategic planning with emphasis on road infrastructure and the transportation system, appropriate location of human settlements and economic activities. Plus, prevent and correct the location of infrastructure and activities in areas of risk and vulnerability conditions.

An action to reorganization and urban regulation of GAM in the terms proposed, would allow the country to join the international Right to the City and the emerging rights derived therefrom: the right to the four basic fundamentals of life, right to mobility and free access, right to public space, housing rights, right to cleanliness, order and decoration, right to collective identity within the city and the right to security.

In this context, it is important the formulation, implementation and systematic evaluation of policy guidelines in critical areas to the economic development of the GAM, such as: (i) mobility and displacement, (ii) local economic development, (iii) urban metropolitan management, and (iv) financial sustainability.

A policy for the mobilization and movement must recognize the specialization of the territory, promote its intensive use while it is possible to correct or contain the current flooding processes to move towards a compact city, with greater capacity to exploit agglomeration economies generated by the industrial areas and major corridors and areas or corridors of commercial animation and offices that are currently observed in the GAM.

Such policy of territorial rearrangement for greater effectiveness must necessarily be able to articulate the improvement of the metropolitan road infrastructure supply (maintenance and rehabilitation of the primary road network, extension and / or expansion of the local road network), with an integrated policy for the reorganization of public transportation at the metropolitan scale of both people and goods (implementation of an intelligent traffic signal coordination system, coordination of different nodes of transportation, fare integration, improved comfort and quality of services, incentives for intensive use and information to the user in real time, building centers of charge transfers, definition of zones and roads to transport goods, among other measures).

The territorial system defined by the convergence of these two elements (improvement in the supply of road infrastructure and reorganization of metropolitan public transport) would make for smart growth of the GAM that would decrease travel time and additional expense in fuel, turning it in improved productivity, competitiveness and territorial quality of urban life and a decrease in costs associated with urban diseconomies. On the other hand, it is recommended that this policy also integrate a component of maintenance, repair and modernization of road infrastructure.

Meanwhile, in terms of local economic development, it is advisable to encourage the implementation of support programs rooted in the local government and implemented in the framework of concerted actions with the private sector, chambers of commerce, universities along with vocational training centers and financial entities, acting as promoters and catalysts for development of new business ventures based on the exploitation of resources related to knowledge, experience and information. This would improve the urban living conditions, increasing the competitiveness of urban areas and promoting the cities of the GAM as a destination for business and international and domestic tourism. Similarly, could act as enablers of urban development by providing stimulus to the renewal of urban spaces, some of them depressed, and reinstate the economic dynamics, avoiding further social and cultural segregation.

Additionally, it is desirable that local economic development of the GAM is grounded in the formulation of a Metropolitan Strategic Plan for Tourism that would integrate initiatives of information about attractions and tours in the hands of business chambers and municipalities at the local level and its transformation into tourist circuits. In that sense, we would recommend government action oriented to support the tourist circuits in the GAM, articulating the cultural attractions with the nature ones, for the express purpose of contributing to increase the accommodation and spending segment of international tourism in socio -economic and educational as well as increasing domestic tourism.

Finally, a third area of policy delineation should focus on the metropolitan urban management and financial sustainability, both by the Central and local governments. To maintain an important activity in terms of investment in urban development should be important to consider the goal of increasing investment in urban development equivalent to 1% of GDP, so as to ensure minimum availability of resources to carry out works of urban infrastructure necessary to recover the functionality of the GAM and even more to line up with the demands of globalization and the demands of different social groups living in it. This will require making a long-term financial planning to ensure the commitment of the different institutions involved and new forms of organization for metropolitan management. Whereas the endowment is not sufficient to ensure effective and efficient action for the metropolitan urban management, it is necessary, especially for the beginning.

6. Acknowledgments

This chapter is the result of various research projects by the authors, at the International Center on Economic Policy (CINPE, acronym in Spanish) of the *Universidad Nacional*, Costa Rica. The opinions expressed herein are the author's responsibility and do not necessarily respond to institutional positions. The authors are grateful to Lizzy Solano for translating the article to English and proofreading it, and to Carlos Eduardo Espinoza for his help during the data collection.

7. References

ACCCR. (2006). *Propuestas para un Plan Vial Nacional 2006–2020.* San José: Asociación de Carreteras y Caminos de Costa Rica.

Allen, P.; Vargas, C.; Araya, M.; Navarro, L. & Salas, L (2005). *Costos en salud por la contaminación del aire.* San José: Ministerio de Salud.

Asociación Española de la Carretera (2005). *Morir en un accidente de tránsito en España resulta 15 veces más barato que en Noruega.* Madrid, Available from http://www.aecarretera.com/Costevidahumana%20_2_.pdf

Azqueta, D.O. (2007). *Introducción a la Economía Ambiental.* Madrid: MsGraw-Hill/Interamericana de España.

Brennan E., & Richardson H. W. (1989). *Asian megacities: Characteristics, problems and policies. International Regional Science Review, 12,* 117–129.

Camagni, R. (2005). *Economía urbana.* Barcelona: Antoni Bosch.

CCC (2011). *Infraestructura nacional: "entre lo ideal y lo real". La perspectiva del sector privado.* San José: Cámara Costarricense de la Construcción.

CCP /INEC (2002). *Estimaciones y Proyecciones de población por distrito y otras áreas geográficas. Costa Rica 1970-2015.* San José: Centro Centroamericano de Población e Instituto Nacional de Estadísticas y Censos.

CFIA (2010). *Pensar en Costa Rica 2025. Una propuesta integral de planificación estratégica de la infraestructura nacional.* San José: Colegio Federado de Ingenieros y Arquitectos.

Cifuentes, L.A.; Rizzi, L.; Jorquera, H., and Vergara J. (2004). Valoración económica y ambiental aplicada a casos del manejo de la Calidad del Aire y Control de la Contaminación. Washington: Informe para el Diálogo Regional de Política del Interamerican Development Bank.

Coase, R. (1960). "The Problem of the Social Cost". In *Journal of Law and Economics, Vol. 3,* 1-44. The University of Chicago Press.

Contraloría General de la República (2008). *Informe del estudio efectuado sobre las inversiones de infraestructura de transporte formuladas para el período 2007-2010.* División de fiscalización operativa y evaluativa. Área de fiscalización servicios de obras públicas y transporte.

DSE (2005). *Encuesta de consumo energético nacional en el sector transporte de Costa Rica 2004.* San José: Dirección Sectorial de Energía, Ministerio de Ambiente Energía y Telecomunicaciones (MINAET).

De Rus, G., Betancor, O. and Campos J. (2006). *Manual de Evaluación Económica de Proyectos de Transportes.* Washington: Development Interamerican Bank.

ECLAC (2003). *Congestión de Tránsito, El problema y cómo enfrentarlo.* Bull, A. compilador. Santiago: Economic Commission for the Latin American and the Caribbean.

Eskeland, G. (1994). The Net Benefits of an Air Pollution Control Scenario for Santiago. In *Chile: Management Environmental Problems Report Nº 13061.* Washington: World Bank.

EU/PRU–GAM/CINPE-UNA (2007a). Estructura económica. In *Estudio económico de la Gran Área Metropolitana (GAM) de Costa Rica.* San José: Proyecto de Planificación Regional y Urbana de la Gran Área Metropolitana del Valle Central de Costa Rica.

EU/PRU–GAM/CINPE-UNA (2007b). Deseconomías urbanas. In *Estudio económico de la Gran Área Metropolitana (GAM) de Costa Rica.* San José: Proyecto de Planificación Regional y Urbana de la Gran Área Metropolitana del Valle Central de Costa Rica.

EU/PRU–GAM/CINPE-UNA (2007b). Evaluación de la factibilidad financiera del modelo de desarrollo urbano propuesto para la GAM. In *Estudio económico de la Gran Área Metropolitana (GAM) de Costa Rica*. San José: Proyecto de Planificación Regional y Urbana de la Gran Área Metropolitana del Valle Central de Costa Rica.

Friedman, J. (1985). The world city hypothesis. *Development and Change, 17*, 69–83.

Gallup (2007-2010). *Gallup World Pol*, Available from http://www.gallup.com.

Gottman, J. (1962). *Megalopolis*. Cambridge, MA: MIT Press.

Grand-Muller, S. & Laird, J. (2007). *Costs of Congestion: Literature Based Review of Methodologies and Analytical Approaches*. The Scottish Government, Available from http://www.scotland.gov.uk/Publications/2006/11/01103351/0

Madrigal P. J. (2006). *Resultados de la Encuesta Nacional de Seguridad Ciudadana. Programa de Naciones Unidas Para el Desarrollo (PNUD)*. Ministerio de Justicia. Ministerio de Seguridad Pública. San José, Costa Rica.

Marshall, A. (1890). *Principles of Economics*. London: Mc Millan and Co., Ltd..

Moomaw, R. L. (1981). Productivity and city size: A critique of the evidence. *Quarterly Journal of Economics, 96*, 675–688.

Moomaw, R. L. (1983). Is population scale a worthless surrogate for business agglomeration economies? *Regional Sciences and Urban Economics, 13*, 525–545.

Oliva, J. (2004). Informe sobre las pérdidas para la economía catalana debidas a la enfermedad, Pérdidas de producción laboral ocasionadas por las enfermedades en Cataluña en el año 2004. Cataluña: Universidad de Castilla La Mancha.

PAHO (2005). *Evaluación de los efectos de la contaminación del aire en la Salud de América Latina y el Caribe*. Washington: Pan-American Health Organization.

Pichardo, A. (2010). *Conversando con Alcaldes y Alcaldesas de Costa Rica. Una Mirada a la Gestión Municipal del Desarrollo Urbano*. International Centro on Economic Policy (CINPE). Heredia: Universidad Nacional.

Pigou, A.C. (1920). *The Economics of Welfare*. London: Mc Millan and Co., Ltd.

Polèse, M. (1998). *Economía urbana y regional. Introducción a la relación entre territorio y desarrollo*. San José (Costa Rica): Libro Universitario Regional.

Richardson, H. W. (1995). Economies and diseconomies of agglomeration. In H. Giersh (Ed.). *Urban agglomeration and economic growth*. Berlin: Springer.

Rizzi. L. I. (2005). Diseño de instrumentos económicos para la internalización de accidentes de tránsito. In *Cuadernos de Economía*, Vol. 42, 283-305. Santiago: Universidad Católica de Chile.

Rodríguez, Jorge & Villa, Miguel (1998). "Distribución espacial de la población, urbanización y ciudades intermedias: Hechos en su contexto". In *Ciudades Intermedias en América Latina y el Caribe: Propuesta para la Gestión Urbana*. Santiago; ECLAC.

Sassen, S. (1991). *The global city: New York, London, and Tokyo*. Princeton, NJ: Princeton University Press.

Sauma F. P. & Chacón Á. I. (2006). *Aspectos económicos relacionados con la inseguridad ciudadana*. Programa de las Naciones Unidas para el Desarrollo 1a. ed. San José, Costa Rica.

Smith, A. (1776). *An Inquiry into the Nature and Causes of the Wealth of Nations*. London: Kindle Editor.

Van Houten, G. & Cropper, M. (1996). When is a life too costly to save? The evidence form U.S. environmental regulations. In *Journal of Environmental Economics and Management 30*, 348-368.

Wilk, D., Pineda, C., and Moyer, D. (2006). *Lineamientos estratégicos para la gestión ambiental urbana en Centroamérica*. Washington D. C.: Inter-American Development Bank. Serie de Estudios Económicos y Sectoriales.

15

Healthy Places, Healthy People: Living Environment Factors Associated with Physical Activity in Urban Areas

Helena Nogueira, Cristina Padez and Maria Miguel Ferrão
University of Coimbra
Portugal

1. Introduction

The final years of the 20th century were characterised by the emergence of new socio-demographic profiles in which behaviour-related diseases were increasingly prominent. Societies have now recognised the fundamental role that health-related behaviours – e.g., diet, tobacco consumption and physical activity – play in human health and chronic disease risk. Everywhere, modern societies are becoming more and more sedentary, a trend contributing to the rise in rates of chronic and degenerative diseases such as type 2 diabetes, CVD, hypertension, some types of cancer, musculoskeletal diseases and high blood pressure and cholesterol. Moreover, it appears that sedentary lifestyles also cause a decline in psychological well-being, thereby increasing the risk of mental disorders (Nusselder et al. 2008).

According to the World Health Organization, 1.9 million deaths globally are caused by physical inactivity (WHO, 2002), including 600,000 in the WHO European Region. In Canada, more than two-thirds of deaths share physical inactivity as a behavioural risk (ACPHHS, 2005). In addition to being one of the main underlying causes of death in developed countries, sedentary lifestyles are a powerful contributor to the new epidemic of the 21st century: obesity.

Obesity has increased globally since 1980 (Finucane et al. 2011); its prevalence has tripled in the last 25 years, making it a major public health concern. Among European adults, it is estimated that excess weight accounts for 50% of hypertension cases, 33% of strokes, and 25% of osteoarthritis (Mladovsky et al., 2009). According to WHO, the current trend, if it continues, will decrease life expectancy by five years by 2050 (WHO Regional Office for Europe, 2007). Portugal has one of the most worrisome situations in the EU, with excessive weight and obesity affecting 50% of adults as of 2006 (ACS, 2008) and 31,5% of children (Padez et al., 2004). How do we explain this trend? Researchers are in agreement that weight gain is of complex multifactorial causation. Although there may be a significant genetic component, it does not explain the recent escalation in cases (Cohen et al., 2006). Environmental factors are increasingly considered to be obesogenic, as they facilitate the excessive intake of calories and/or discourage the expenditure of energy in daily life (Giles-Corti and Donovan, 2003; Portinga, 2006).

When analysing physical activity levels across the EU, a pattern of southern disadvantage clearly emerges. The prevalence of individuals who report no physical activity within any monthly period ranges from 4% in Finland to 66% in Portugal, the highest value in the EU, followed by Italy (58%) and Greece (57%). Men tend to have higher physical activity levels than women, and time spent exercising shows a negative correlation with age. The percentage of individuals reporting that they exercise once a week is 60% for individuals age 15 to 25 but only 28% for individuals over 55 (Mladovsky et al., 2009). Furthermore, socioeconomic differences in physical activity consistently have been reported, with the more educated being, in general, more active (Gidlow et al., 2006). Some studies have even pointed to the emergence of more differentiated and nuanced patterns that vary by gender and types of physical activity (Demarest et al., 2007).

1.1 Linking sedentary lifestyles to the local environment (*via* social structure): A socio-ecological approach

General variations in physical activity between countries occur along with specific and differential patterns within countries, emerging from a complex web of causation. A great deal of social science research finds that health-related behaviours are partly a function of an individual's characteristics and local context (Jen et al., 2009). For example, poor people are more likely to be sedentary, and poor people living in disadvantaged areas are more likely to be sedentary than their counterparts living in more advantaged areas. Researchers have hypothesised that disadvantaged people tend to be less active because they have less leisure time, due to their occupations. Further, disadvantaged people tend to be less active in their leisure time because they are less able to afford and access exercise programmes and facilities. Finally, disadvantaged people are more likely to live in areas that discourage physical activity, e.g., homogeneous, non-mixed land-use neighbourhoods, with crime and road safety problems.

The literature has burgeoned in recent years with theoretical models and frameworks to explain socioeconomic inequalities in health-related behaviours (McNeill et al., 2006; Kamphius et al., 2007; Stafford et al., 2007; Maddison et al., 2009). Throughout these studies, most authors highlight similar pathways – e.g., socioeconomic, psychosocial and cultural – that link socioeconomic status (SES), the environment and health-related behaviours.

The present study presents a framework that attempts to clarify plausible pathways to sedentary lifestyles, highlighting the contextual determinants of physical activity. The premise is that, beyond individual characteristics, contextual factors can facilitate or impede the individual's opportunity to lead a healthy, active life. The framework encompasses five main categories of determinants of physical activity that are structured and operate at different contextual levels: (A) a micro (neighbourhood) level and (B) a meso/macro level.

1.1.1 Micro-level (neighbourhood) environmental conditions

a. Geographical/physical circumstances: these factors stress the availability and accessibility of products, facilities or opportunities that promote or impede active lifestyles. Empirical evidence suggests that active living can be affected by population density, street connectivity, land use (access to residence, work, school, shops, food, leisure, sports and other amenities), pedestrian and cycling conditions (pavements, sidewalks and bicycle paths) and, in general, pleasant surroundings (Frank et al., 2004; Cerin et al., 2007; Frank et al., 2007; Leslie et al., 2007).

b. Social circumstances, including safety: social deprivation, lack of social support, lack of social capital, weak social control, violence and generally unsafe conditions can place constraints on individual choices (McNeill, 2006) and are key determinants of an individual's level of physical activity (Catlin et al., 2003; Cohen et al., 2006; Kim et al., 2006; Burdette & Hill, 2008).

c. Psychosocial and cultural circumstances: behaviours and decisions can be influenced not only by physical and social environmental conditions but also by perceptions of those conditions, by culture-specific lifestyles patterns, by the person's general values, his or her sense of belonging and by neighbourhood connectedness. These factors may modify the individual's interaction with his or her environment. It is also likely that psychological disorders (e.g., depression, anxiety, stress or social isolation) have a detrimental effect on one's willingness to engage in formal physical activity (Wilkinson, 1999) and that psychological conditions decrease one's levels of informal physical activity (i.e., activity related to social interactions and social contacts).

d. Socioeconomic circumstances: these are circumstances that can affect levels of physical activity through any of the conditions mentioned above. People of lower SES have limited resources with which to participate in indoor activities at gymnasiums and health clubs (Wilkinson, 1999; Wilkinson & Marmot, 2003). Beyond these constraints, people with low SES are confronted daily with social and work environments characterised by deprivation (absolute and relative), strain or lack of control over social and work life and lack of leisure time - all of which have psychological consequences, including insecurity, anxiety, stress, social isolation and depression (Wilkinson & Marmot, 2003).

1.1.2 Meso/macro-level environmental conditions

e. Political circumstances: levels of physical activity may be influenced by changes in transportation planning and policies (e.g., policies that emphasise collective and active transport), zoning and urban planning (land use policies and urban design) (Maddison et al., 2009). Conditions at this level also encompass campaigns and interventions to support behaviour change.

Assuming that one's level of physical activity is determined contextually (and thereby dependent on local environments), the purpose of this paper is to analyse and understand the "obesogenic environment" in the Lisbon Metropolitan Area (LMA) by identifying micro-level neighbourhood conditions that are associated with increased individual physical activity.

2. Data and methods

2.1 Study area

This study focused upon the Lisbon Metropolitan Area (LMA), Portugal, located in the central-southern part of the country, on either side of the River Tagus (figure 1).

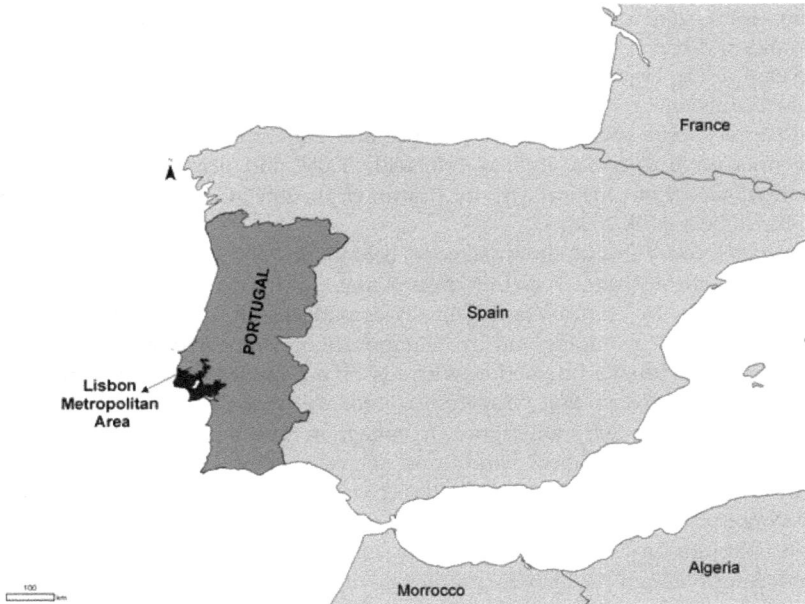

Fig. 1. Lisbon Metropolitan Area, Portugal

The 3,133-km² area, comprising 19 municipalities and 216 neighbourhoods, includes some of Portugal's largest cities: Lisbon, Amadora, Cascais, Almada, Seixal and Setúbal. The LMA is wholly urban in nature, although a few of its parishes are still classified as rural. In 2011, 2,837,626 people, 28.3% of Portugal's population, lived in the LMA, an increase of more than 3% from the time of the 2001 census. The LMA has some features that distinguish it from the rest of the nation: above-average population growth, mostly concentrated in the peripheral parishes; greater population mobility between peripheral and central areas (higher functional integration); higher levels of education and occupational status, mainly in the central areas; and greater purchasing power – the highest value nationally and far above the Portuguese average (in 2004, the national average was 100, while the LMA average was 116, and Lisbon's was 278) (Nogueira, 2007). Of Portugal's 21 healthy cities, 12 are located in LMA. Nonetheless, the region is plagued by social problems: high levels of poverty, deprivation and inadequate housing, mainly in inner-city areas; lack of resources, facilities and amenities, both in peripheral and central areas; low availability and use of public transport in the most peripheral, sprawling areas; and lower levels of social interaction and weak social networks, caused by the region's high level of residential mobility and by the persistent process of urbanisation (Nogueira, 2009).

2.2 Individual variables

All individual data were assessed from the National Health Survey (NHS) for the years 1998/1999. NHS has been carried out by governmental institutions since 1987 (1987, 1995/1996, 1998/99 and 2005/2006, the last with data of small areas not available) and is assessed by the National Center for Health Statistics of USA. NHS drew a random sample of

21,808 households within small areas of the whole country, and 48,606 individuals were interviewed face-to-face by well-trained interviewers during 52 weeks. To the LMA, a representative sample composed by 9,846 individuals living in 143 neighbourhoods (of 216) was collected. From this sample, we selected only individuals aged 15 or over, because in these cases the responses to the inquiry are necessarily given by the individual him/herself (to individuals aged 14 years or younger, responses may be provided by relatives living in the household). Participants with missing data for variables used in the analysis were also excluded; the resulting study population comprised 5,004 individuals from 143 neighbourhoods.

The dependent variable was self-reported physical activity, measured as a binary response: whether the person engaged in or did not engage in physical activity. Individual predictors were age, sex, occupational class (manual, non-manual workers), economic activity (employed, unemployed, and other) education level (less than 4 years, between 5 and 12 years and more than 12 years), tobacco consumption and self-rated health (good/very good and less than good). There were 4,577 individuals in the final sample, of whom 60% were women, 55% were 35 to 64 years old, 67.3% reported sedentary behaviours, 60.5% rated their health as less than good, 44% had less than 4 years of education, 53.7% were manual workers and 21.8% were current smokers. All individual variables were included as dummy variables, except age, which was included as continuous.

2.3 Local environment variables

The generation of ecological data began with a consideration of what humans need in their local environments in order to lead healthy and active lives (Cummins et al., 2005). We listed 14 domains or constructs that described these needs: 1. general aesthetics of the local area, 2. housing environment (indoor), 3. health services, 4. leisure and recreation facilities, 5. sports facilities, 6. family support services, 7. street facilities, 8. house-keeping, 9. work and employment, 10. educational services, 11. crime and policing, 12. road safety, 13. public transport and 14. social capital and cohesion. At this stage, more than 240 contextual variables were assigned to the specified dimensions.

Contextual variables were obtained from a large range of routine and non-routine sources. These include the National Institute of Statistics (INE), local authorities, voluntary and public sector agencies (ministry of health, local and municipal police, the Portuguese social security), commercial organizations and others (the national institute of car insurance and yellow pages). Data refer to 2001-2005 and were collected at the neighbourhood level. We tried to generate accurate and specific contextual measures, beyond aggregate measures, since the latter generally do not reflect the underlying processes that link local environment to health (Cummins et al., 2005). The large amount of data created and assigned to the previously conceptualised domains, reflecting the local environment, had to be processed effectively. This was done using statistical methods following described.

2.4 Statistics

2.4.1 Creating new environmental variables

We collected a large amount of contextual data (246 variables) included in the 14 specified dimensions. Principal component analyses (PCAs) were performed to explore and reduce these data without losing information. All components (or factors) were rotated using varimax rotation to maximise factor loadings. Throughout the procedure, variables with low loadings onto components were discarded from the construct. Our aim was to create a single, strong component in each construct; components were rejected when considered irrelevant using Kaiser's criterion (dropping all those with eigenvalues less than 1.0) (Cummins et al., 2005; Nogueira, 2009). However, some domains revealed a bi-dimensional structure, which resulted in 19 extracted factors composed by 82 variables. These 19 factors, or components, were taken as local environment indicators.

The next step was to evaluate the internal consistency of the extracted factors and their ability to measure the latent contextual domains. Reliability was measured by standardised Alpha Scores, which ranged from 0.51 to 0.98. These high values show that variables within each factor are strongly related, giving us confidence about the factors consistency. Moreover, correlations between the extracted factors were generally low, suggesting single and unambiguous factors, with the ability to reliably capture something unique about the local environment. Statistical analysis was performed using SPSS 17.0.

2.4.2 Linking the neighbourhood environment to physical activity

To examine the relationship between physical activity and individual and contextual attributes, we used the binary logistic regression model. We started with a simple model, considering individual factors as independent variables. In this model, we selected the variables with a conventional significance of p ≤0.05. In a second stage, we extended the model by entering ecological variables when they were significant (p ≤0.05). The final model was adjusted for all of the individual variables, and adjusted odds ratios (ORs) for environmental predictors were calculated. The level of statistical significance was set at 0.05, and statistical analysis was performed using SPSS 17.0.

3. Results

The prevalence of sedentary behaviour in our sample was 67.3%. This percentage was 89% among women and 31.5% among men. Table 1 shows the results from the binary logistic model predicting the likelihood of engaging in physical activity.

In our model, gender showed the biggest influence on physical activity, with women showing less physical activity than men (51% less likely). Age was also found to have a detrimental effect; the odds of individuals reporting engagement in physical activity decreased by 15% with each additional 10 years in age. Individuals with lower levels of education were 24% less likely to be active, while having 13 or more years of education showed a beneficial effect, increasing by 40% the odds of active behaviour. Being in an occupation with manual labour also played a part, decreasing the odds of active behaviour by 31%. Being employed had a negative effect, decreasing the odds of being physically active by 26%. With regard to other health-related behaviours, smoking had a detrimental effect, with the odds of smokers being active decreasing by 24%. Health status, assessed through self-rated health, also had an influence on likelihood of physical activity, and individuals who rated their health as good or very good were 49% more likely to report active behaviour.

Variables	Coefficient	Adjusted Odd Ratios
Sex (female *vs.* male)	-0.71	0.49**
Age (in 10-year increments)	-0.016	0.85**
Occupation (manual worker *vs.* non-manual)	-0.374	0.69**
Education (< 4 yrs *vs.* 5-12 yrs)	-0.277	0.76*
Education (13 + yrs *vs.* 5-12 yrs)	0.335	1.40**
Health status (good/very good *vs.* less than good)	0.402	1.49**
Economic activity (employed *vs.* unemployed)	-0.299	0.69**
Tobacco (smoker *vs.* non-smoker)	-0.273	0.76**
Public transport (less accessibility)	-0.167	0.85**
Safety (more crime)	-0.124	0.88*
Health services (more availability)	0.371	1.45**
General aesthetics (worse quality)	-0.155	0.86*
Social relations (few interactions)	-0.23	0.79**

*p <0.05; ** p<0.001

Table. 1. Coefficients, odd ratios and significance levels from a multiple binary logistic regression model predicting active behaviours in the LMA.

As for neighbourhood conditions, after controlling for all of the individual variables presented in table 1, a significant association was found between physical activity and public transport accessibility, neighbourhood safety, availability of health services, general aesthetics of living local area and level of social interactions. Considering variations of one standard deviation, we concluded that individuals living in areas with lower levels of public transport accessibility were 15% less likely to report an active behaviour. Those living in unsafe areas, with high levels of crime, were 12% less likely to practice physical activity. Living in areas with greater availability of health services increase the odds of being active by 45%. A positive association between general aesthetics of the residential area and physical activity was also confirmed, with the odds of being active decreasing by 14% in areas that are considered unpleasant. Social relations showed a similar influence. Individuals living in areas in which social interactions are scare were 21% less likely to report active behaviour.

4. Discussion

This research presents several important findings that contribute to our understanding of physical activity behaviour. Variations in levels of physical activity arise from individual characteristics such as gender, age, socioeconomic, health-related behaviours (such as smoking) and health status. However, location also plays a role in active living, and our results show that built and social environments are key focal points. The LMA is a rapidly growing region with a diverse population, in which the creation of environments conducive to physical activity depend upon adequate public transport systems, mixed land-use patterns, convenient social environments (safe places with social interactions that provide opportunities to be active) and neighbourhood aesthetics.

In terms of the physical environment, our results from Lisbon, which highlight the availability of health services, general aesthetics and accessibility of public transport as determinants of physical activity, are consistent with previous research. Mixed land-use – e.g., provision of and access to local public services and facilities – and attractive scenery have been found to be associated with higher physical activity levels (Frank et al., 2006; Stafford et al., 2007). Areas with higher diversity and population density, often involving designs that incorporate connected streets, sidewalks and bicycle paths, have been found to be more walkable (Doyle et al., 2006). Authors have also suggested the importance of commuting by public transportation, and of walking or cycling to and from public transport, in helping people to attain recommended levels of daily physical activity (Besser & Dannenberg, 2005; Zheng, 2008). Results in the LMA showed that making public transport accessible and providing more opportunities for active travel may promote and help people maintain active lifestyles. Physical neighbourhood features that can promote active behaviour typically have been summarised by the expression "walkable neighbourhood", which is defined as an area that combines several features: density, land use, street connectivity, transportation, attractiveness and general aesthetics (Calthorpe & Fulton 2001; Jochelson, 2004; Doyle et al., 2006; Frank et al., 2006; Stafford et al., 2007).

Our model also highlighted social and psychosocial factors, corroborating earlier results. Violence, social disorder and the feeling of fear inhibit social interactions and lead people to limit their outdoor physical activity (Parkes and Kearns, 2006). Poortinga (2006) found positive associations between social support, social capital and higher levels of physical activity, and Cohen et al. (2006) suggested a positive effect of social cohesion among neighbours in various positive health behaviours, including physical activity. Van Lenthe et al. (2005) stressed the role of neighbourhood attractiveness and safety in encouraging physical activity. In the LMA, people living in disadvantageous social environments in which violence is prevalent likely decrease their level of social interaction, consequently decreasing their level of social capital and social cohesion. As a result, social isolation, insecurity, anxiety, stress, depression and hopelessness increase, which may decrease individuals' willing to engage in physical activity (Wilkinson, 1999).

4.1 Study weakness

A major limitation of this study is the cross-sectional nature of the data which prevents an exploration into causal relationships. Just because an association was observed, it does not mean causation; we can only argue that these results show a jointed variation of physical activity and some, specific, neighbourhood features. Another limitation is that other important covariates, not considered in our model, may explain the associations. For example, body mass index may be an important determinant of physical activity. It is also important to note that physical activity was self-reported, thus subject to bias.

5. Conclusion

Individuals living in neighbourhoods that are socially and physically disadvantageous are at increased risk of engaging in sedentary behaviour. Local environmental features contribute to the growing rates of physical inactivity and obesity among people living in the LMA. Creating a high-quality developed environment that is walkable, safe and supportive of social interaction would increase opportunities for physical activity and improve the health of all inhabitants. But how do we accomplish this goal?

As empirical evidence on active living environments accumulates, there should be more efforts to translate research into practice. Considering the growing levels of inactivity and obesity of modern societies, and in light of the multiple, diverse features required for an environment that can promote active living, we propose the creation of an interdisciplinary team that can work toward better places and better health: a Task Force for Active Living Environments. The work of the task force would be rooted in empirical evidence. Policy-makers, urban planners, transportations engineers and public health workers would cooperate in order to improve the quality of the local area – its safety, urban design, land use, public transport, social organisation and general aesthetics – and ensure that it provides opportunities for physical activity. Planning that promotes and supports behavioural change is only possible through cooperation from all sectors involved in helping active-living environments to emerge.

Finally, we can suggest a variation on our initial approach. If improvements to the local environment can increase physical activity, increased physical activity can also improve the local environment. Active people change their neighbourhoods by creating more cohesive communities, strengthening social bonds and increasing social control - in short, shaping safer and more pleasant areas. In this sense, *walkability* is more than just a form of community. It is a way of life, one that is healthier, more ecologically sound and more neighbourhood-friendly.

6. References

ACPHHS (2005). *The Integrated Pan-Canadian Healthy Living Strategy*, Advisory Committee on Population Health and Health Security, ISBN 0-662-69384-1, Canada.

ACS (2008). Plano Nacional de Saúde: Indicadores e Metas do PNS. In: *Ministério da Saúde - Alto Comissariado da Saúde*, 3.01.2008, Available from http://www.acs.min-saude.pt/pns/.

Besser, L. & Dannenberg, A. (2005). Walking to Public Transit Steps to Help Meet Physical Activity Recommendations. *American Journal of Preventive Medicine*, Vol. 29, No 4, pp. 273-280.

Burdette A. & Hill, T. (2008). An examination of processes linking perceived neighborhood disorder and obesity. *Social Science & Medicine*, Vol. 67, pp. 38-46.

Calthorpe, P. & Fulton, W. (2001). *The regional City. Planning for the End of Sprawl*, Island Press, Washington.

Catlin, T.; Simoes, E., & Brownson, R. (2003). Environmental and policy factors associated with overweight among adults in Missouri. *American Journal of Health Promotion*, Vol. 17, No. 4, pp. 249–258.

Cerin, E.; Leslie, E.; du Toit, L.; Owen, N. & Frank LD. (2007). Destinations that matter: associations with walking for transport. *Health & Place*, Vol. 13, No. 3, pp. 713-724.

Cohen, D.; Finch, B.; Bower, A., & Sastry, N. (2006). Collective efficacy and obesity: The potential influence of social factors on health. *Social Science & Medicine*, Vol. 62, No. 3, pp. 769–778.

Cummins, S.; Macintyre, S.; Davidson, S. & Ellaway, A. (2005). Measuring neighbourhood social and material context: generation and interpretation of ecological data from routine and non-routine sources. *Health & Place*, Vol. 11, No. 3, pp. 249-260.

Demarest S.; Cox B., Van Oyen H. (2007), *Socio economic inequalities in leisure time physical activity: results of the EUROTHINE project*. Scientific Institute of Public Health, Belgium, Brussels.

Doyle, S.; Kelly-Schawartz, A.; Scholossberg, M. & Stockard, J. (2006). Active community environments and health: the relationship of walkable and safe communities to individual health. Journal of the American Planning Association, Vol. I, pp. 19-31.

Finucane, M; Stevens, G.; Cowan, M.; Danaei, F.; Lin, J.; Paciorek, C.; Singh, G.; Gutierrez, H.; Lu, Y.; Bahalim, A.; Farzadfar, F.; Riley, L. & Ezzati, M. (2011). National, regional, and global trends in body-mass index since 1980: systematic analysis of health examination surveys and epidemiological studies with 960 country-years and 9.1 million participants. *Lancet*, Vol. 377, pp. 557-567.

Frank, L.; Andresen, M.; & Schmid, T. (2004). Obesity relationships with community design, physical activity, and time spend in cars. *American Journal of Preventive Medicine*, Vol. 27, No. 2, pp. 87-96.

Frank, L.; Kerr, J., Chapman, J. & Sallis, J.(2007). Urban form relationships with walk trip frequency and distance among youth. *American Journal of Health Promotion*, Vol. 21, No. 4, pp. 305–311.

Frank, L.; Sallis, J.; Conway, T.; Chapman, J.; Saelens, B. & Bachman, W. (2006). Many pathways from land use of health: associations between neighbourhood walkability and active transportation, body mass index, and air quality. *Journal of the American Planning Association, Vol. I*, pp. 75-87.

Giles-Corti, B., & Donovan, R. (2003). Relative influences of individual, social environmental, and physical environmental correlates of walking. *American Journal of Public Health*, Vol. 93, pp. 1583-1589.

Jen, M.; Jones, K. & Johnston, R. (2009). Compositional and contextual approaches to the study of health behaviour and outcomes: Using multi-level modeling to evaluate Wilkinson's income inequalities hypothesis. *Health & Place, Vol.* 15, pp. 198-203.

Jochelson, K. (2004). The Public Health Impact of Cities & Urban Planning. Report. London Development Agency: King's Fund, London.

Kamphius, A.; Van Lenthe, F.; Giskes, K.; Brug, J. & Mackenbach, J. (2007). Perceived environmental determinants of physical activity and fruit and vegetable consumption among high and low socioeconomic groups in the Netherlands. *Health & Place*, Vol. 13, pp. 493-503.

Kim, D.; Subramanian, S.; Gortmaker, S. & Kawachi, I. (2006). US State-and county- level social capital in relation to obesity and physical inactivity: a multilevel, multivariable analysis. *Social Science & Medicine*, Vol. 63, pp. 1045-1059.

Leslie, E.; Coffee, N.; Frank, L.; Owen, N.; Bauman, A. & Hugo, G. (2007). Walkability of local communities: using geographical information systems to objectively assess relevant environmental attributes. *Health & Place*, Vol. 13, pp. 111-122.

Maddison, R.; Hoorn, S.; Jiang, Y.; Mhurchu, C.; Exeter, D.; Dorey, E.; Bullen, C.; Utter, J.; Scaaf, D. & Turley, M. (2009). The environment and physical activity: The influence of psychosocial, perceived and built environment factors. *International Journal of Behavioral Nutrition and Physical Activity*, Vol. 6, No. 19, doi: 1186/1479-5868-6-19.

McNeill, L.; Kreuter, M. & Subramanian, S. (2006). Social environment and physical activity: a review of concepts and evidence. *Social Science & Medicine*, Vol. 63, pp. 1011-1022.

Mladovsky, P.; Allin, S.; Masseria, C.; Hernández-Quevedo, C.; McDaid, D.; Mossialos, E. (2009), Health in the European Union. Trends and analysis. European Observatory on Health Systems and Policies, WHO Regional Office for Europe, Copenhagen.

Nogueira, H. (2007). Os Lugares e a saúde. Uma abordagem da Geografia às variações em saúde na Área Metropolinada de Lisboa. Ph.D. Unpublished Thesis, University of Coimbra, Coimbra.

Nogueira, H. (2009). Healthy communities: the challenge of social capital in the Lisbon Metropolitan Area. *Health & Place*, Vol. 15, pp. 133–139.

Nusselder, W.; Looman, C.; Franco, O.; Peeters, A.; Slingerlan, A. & Mackenbach, J. (2008). The relation between non-occupational physical activity and years lived with and without disability. *Journal of Epidemiology and Community Health*, Vol. 62, No. 9, pp. 823-828.

Padez, C.; Fernandes, T.; Mourão, I.; Moreira, P. & Rosado, V. (2004). Prevalence of overweight and obesity in 7-9-year-old Portuguese children: Trends in body mass index from 1970-2002. *American Journal of Human Biology*, Vol. 16, pp. 670-678.

Parkes, A. & Kearns, A. (2006). The multi-dimensional neighbourhood and health: a cross-sectional analysis of the Scottish Housing Survey, 2001. *Health & Place*, Vol. 12, No. 1, pp. 1-18.

Poortinga, W. (2006). Perceptions of the Environment, Physical Activity, and Obesity. *Social Science & Medicine*, Vol. 63, pp. 2835-2846.

Stafford, M.; Cummins, S.; Ellaway, A.; Sacker, A.; Wiggins, R. D. & Macintyre, S. (2007). Pathways to obesity: identifying local, modifiable determinants of physical activity and diet. *Social Science & Medicine*, Vol. 65, pp. 1882-1897.

Van Lenthe, F.; Brug, J. & Mackenbach, J. (2005). Neighborhood inequalities in physical inactivity: the role of neighborhood attractiveness, proximity to local facilities and safety in the Netherlands. *Social Science & Medicine*, Vol. 60, pp. 763-775.

WHO (2002). *World Health Report, Reducing risks, promoting healthy life*. World Health Organization, Geneva.

WHO Regional Office for Europe (2007). *The challenge of obesity in the WHO European Region and the strategies for response*, WHO Regional Office for Europe, ISBN 978-92-890-1388-8, Copenhagen.

Wilkinson R (1999). Putting the picture together: Prosperity, redistribution, health and welfare, In: *Social determinants of health*, M. Marmot & Wilkinson RG (eds), pp. 256-274, Oxford University, Oxford.

Wilkinson R, Marmot M (2003). *Social determinants of health: The solid facts. Second edition.* WHO Regional Office for Europe , Copenhagen.

Zheng, Y. (2008). The benefit of public transportation: Physical activity to reduce obesityand ecological footprint. Preventive Medicine, Vol. 46, pp. 4-5.

16

Urban Micro-Spatiality in Tokyo: Case Study on Six *Yokochō* Bar Districts

Jorge Almazán and Nakajima Yoshinori

Center for Space and Environmental Design Engineering, Keio University, Tokyo
Japan

1. Introduction

1.1 Research purpose and background

Inheritors of a disciplinary tradition of grand plans, monumental avenues and boulevards, urban planners and designers have all too often ignored the qualities of small urban spaces. This study aims to contribute to the understanding of those qualities by examining an extreme case of urban micro-spatiality in Tokyo: the (in)famous small bar districts known as *yokochō*. The purpose is not only to document these neglected spaces but also to understand the relationship between the physical setting and the activities taking place within it. This study argues that urban smallness, often considered in negative terms as a *lack of space*, can be seen in a positive light as a support for specific social informal interactions, diversity and creativity.

The goal behind this double purpose is to describe the yokochō space as a "behavior setting" (Barker, 1968). The yokochō space and its unit, the bar, are a stable combination of a recurrent activity and a particular layout of the environment, and can therefore be described as a behavior setting. The behavior in the yokochō space will be in part revealed through the interviews conducted in this study, but these will not be the main object of the investigation. For this research, it will suffice to mention that yokochō spaces are what Oldenburg defines as "core settings of informal public life" or "third places": "a great variety of public places that host the regular, voluntary, informal, and happily anticipated gatherings of individuals beyond the realms of home and work" (Oldenburg, 1999, p.16). The places here, bars and entertainment districts, might be considered as too ordinary or irrelevant for serious architectural and urban research. However, the sociological importance of third places (bars, cafés, taverns, etc.) in sustaining informal public life as a supplement to home and work routines has been widely accepted since the publication of Oldenburg's seminal work.

Casual observation of yokochō districts reveals at least two prominent characteristics of their physical setting: smallness and low profile. The appearance of plainness, low profile, or lack of elegance is an attribute that Oldenburg also mentions in his description of third places. The aging and worn structures of yokochō spaces can be said, similar to third places, to have the function of "discouraging pretension among those who gather there." (Oldenburg, 1999, p.37). But what about smallness? How does the small scale and its associated architectural configuration support informal public behavior in the yokochō setting?

This research aims to answer two questions: How small is the actual scale of the yokochō space and what is the meaning of that smallness for the persons managing them? By providing possible answers to these questions, this study aims to contribute to architectural design theory by documenting a case that can be inspirational for architects and planners. The more general goal, for which this study is but a small first step, is to extend to architecture the theory of smallness, proposed by Schumacher in the field of economics.

1.2 Theoretical framework

An early manifesto for smallness can be found in economist E. F. Schumacher's "Small is beautiful" (Schumacher, 1973). This collection of essays advocates the economic and social benefits of organizing human activities into small units. Several passages dealing with the abstract problem of size can be interpreted from the point of view of disciplines concerned with space. Although Schumacher does not defend smallness *per se* ("For every activity there is a certain appropriate scale"), he considers it necessary, insisting on the small scale as a value for contemporary times: "Today, we suffer from an almost universal idolatry of giantism. It is therefore necessary to insist on the virtues of smallness—where this applies." (Schumacher, 1973, p.70).

In architecture theory we can find one manifesto in the opposite direction, Koolhaas' essay on "bigness" (Koolhaas, 1994), in which he advocates sheer size as a new field for architectural potential. In spite of its ambiguity and the lack of examples to clarify its claims, the text is often cited and seems to have influenced newer generations of architects, such as MVRDV or BIG (Bjarke Ingels Group). This fascination with the mega-scale in architecture has a long history, and almost every culture has developed concepts of monumentality by using sheer size. Japan can be considered as an exception, especially if we follow Lee's observations (Lee, 1992) that small size in itself is a greatly appreciated value in Japanese aesthetics, from the *bonsai* to the *bentō* lunch box.

Additionally, in Japanese architecture we can find an interest in the small scale, which can be best exemplified by the 4.5 mat tea house or the garden pavilions of the Katsura Rikyu palace. In contemporary architecture, nonetheless, an important exemption is the metabolist movement, in which architects tried to solve the emerging problems of rapidly growing Japanese cities by creating architectural mega-structures. Today, the mega-structure has long disappeared from the architectural debate, and the idea of smallness is gaining much attention. One of the practices that clearly articulates their works around this idea is Atelier Bow-Wow. Their directors (Tsukamoto and Kaijima, 2006, p.85) list "smallness" among the leading concepts in their work. This concern with the mini-scale is undeniably cultural, but might be also related to the type of commissions that independent small offices tend to receive in Tokyo; that is, single-family houses in semi-suburban areas on very small lots that have undergone a process of gradual subdivision over the last few decades.

Either as a consequence of high density and small lots or of a cultural inclination, Tokyo offers a rich source of examples of successful urban micro-spaces. It could be possibly said that the Japanese have become accustomed to small spaces and therefore that conclusions from Japanese examples cannot contribute to the more general theory of design. However, the same could be said about any urban space or phenomenon, from the American skyscraper to the Italian piazza. Even if a case is embedded in a specific cultural context, conclusions can always be inspirational and useful as objects of comparison, or as a foundation for understanding spatial manifestations of cultural difference. This paper is thus expected to contribute as a balance to the architectural debate and professional practice which, from the point of view of the authors, seems to forget the small scale too often and overlook its potential.

1.3 Method

The case study methodology is a way to examine complex contemporary phenomena within their real-life contexts. Here, the cases examined are six prominent bar precincts in Tokyo called *yokochō*, famous for their small scale and their informal public life. The research question revolves around dimensions and how those dimensions are subjectively experienced. We combine a design survey (on-site measurements conducted during November and December 2011) with interviews conducted in January 2011. By observing both the physical reality and glimpses of the informal public life through the interviews, we expect to explain the role of smallness in the yokochō space as a behavior setting.

The overall structure of the six yokochō was documented and a total of 12 bars were investigated in detail. The areas documented are *Nonbei-yokochō* in Shibuya (2 bars), *Goldengai-Hanazono* (5 bars) and *Omoide-Yokochō* (2 bars) in Shinjuku, *Sakaechō-dōri* (1 bar) and *Mikuni-kōji* (2 bars) in Ikebukuro. The selected yokochō are those located in the three most remarkable commercial districts near major train stations. The yokochō in these three areas are constituent parts of larger urban centers encompassing major transit hubs or "transit-oriented centers" (Almazán & Tsukamoto, 2009).

On-site measurements were both applied to yokochō streets and to each bar sample. As for the yokochō streets, measurements were realized on the actual street to check their widths. The obtained widths were compared with the dimensions represented in a digital map obtained through GIS software (ArcGIS version 9.1). The planimetric drawings of the sample bars (realized using the CAD software called Vector Works) were entirely based on-site measurements, from the dimensions of exterior and interior architectural elements to the sizes of furniture and fixtures.

As for interviews, the authors conducted a series of in-depth interviews to the owners of the 12 sample bars, who were informed that the survey was developed for purely academic purposes. Interviews contained the same five questions, shown in Table 4, so that replies could compared and key themes could be identify. Interviews were conducted after several visits realized to each bar for the on-site measurements. This allowed the authors to establish a certain personal contact with the interviewees. The authors believe that this previous contact contributed to create a relaxed atmosphere in which interviewees expressed sincerely their points of view.

Combining the two sources of information gives a response to the questions posed in the introduction of this research. How small is small (objectively)? And what are the advantages of smallness (subjectively)? The second question was explored through interviews with bar owners, considered to be the key informants. Owners maintain these spaces, keeping them vibrant with their lifestyles and their resistance against redevelopment. In the interviews, our main concern was to know what makes owners choose these establishments and what their opinion is on smallness.

2. Case study

2.1 Definition: About the yokochō

Yokochō [横町 or 横丁] refers to one, or a group of, small street(s) that intersect a main street, and form a semi-hidden location filled with bars and restaurants (*yoko* means "side" and *chō*, "street, block or town"). Yokochō areas, a usual presence in the entertainment districts around main transit hubs in urban Japan, are often considered as old-fashioned, unsafe, and crowded with people of dubious reputation.

Most yokochō emerged as temporary black markets during the post-war rationing period. Tokyo contained an estimated sixty thousand black market stalls (Seidensticker, 1991, p.153). Every station along the commuter lines had a small market of illicitly acquired foreign goods. Although wholesale rationing finished in 1950 and retail rationing in 1968, the black markets had long become meaningless. Seidensticker offers the example of rice, a measure for other prices, to illustrate this, and says that in 1959 the black-market price of rice fell below the rationed price (Seidensticker, 1991, p.159). Hence, illegal market stalls gradually transformed into bars for snacks and drinks. Some of them changed their location from the station approach to other areas still within walking distance from the station. With fast economic growth starting in late 1950s and continuing up to the 1970s, plots of land lots around stations were rapidly developed into large office and commercial buildings. The multiple-ownership land of the former black markets was difficult to redevelop, since all proprietors had to be convinced to sell their lots in order to form a unified piece of land. This may be one of the reasons why a number of centrally located yokochō have survived up to the present.

Current yokochō have become an urban relic, and some are facing problems. Their wooden structures built in the 1950s are deteriorating, in spite of structural reinforcements and renovations. There is a lack of successors to continue the family business. Regular customers, most of them of the so-called baby boom generation, are close to retirement, which also often means retirement from late night drinking. Narrowness, and an image of dirtiness and danger add to the problems (Ogawa, 2008).

2.2 The feasibility of smallness: Yokochō revival

Yokochō spaces face diverse problems. However, there are also signs of a recovered interest in this type of area and its potential as an activator for urban regeneration (Maeda, 2009). The year 2008 saw the demolition of the famous *Jinsei Yokochō* [人世横丁], active in the Ikebukuro Station area since 1950 (Fig. 1), but also the opening of a *yokochō*-like multi-restaurant in the fashionable district of Ebisu. This *Ebisu Yokochō* (Fig. 2) intends to reproduce not only the visual atmosphere of the old post-war *yokochō* but also its actual management, since the Ebisu Yokochō is run by multiple young chefs in charge of their mini-bars. The small size of each stall allows them to pay the rental fees and affords them the opportunity to develop their own independent business.

Fig. 1. *Jinsei Yokochō* (Ikebukuro) was demolished in 2008

These yokochō-like commercial spaces were planned with the aims of convenience, humanity, and manageability— the type of attributes that Schumacher found in smallness. The business planner Maeda lists four cases of recent commercial initiatives directly inspired by the yokochō model (Maeda, 2009). The first case, mentioned above, is Ebisu Yokochō, which is also extensively covered in *Gekkan rejaa sangyô shiryô* (2009), and the second case is a yokochō project in Akasaka (Tokyo). Maeda also mentions two more cases in provincial cities in Hiroshima prefecture, and cases of renovations of large-scale *izakaya* (a type of Japanese tavern) into yokochō-like subdivided spaces. Advantages of the yokochō model, according to Maeda, can be found in economy, management, and creativity. In the catering trade, the cost of the initial investment is large, and there is an extremely high risk of failure. The new yokochō spaces that she describes are based on the sharing of certain infrastructure or equipment, and therefore the investment for each individual is smaller. Additionally, floor areas are small, and consequently so are the rental fees. For instance, in the Ebisu Yokochō shops have an area of 3-5 *tsubo* (9.93-16.55 m²) (Maeda, 2009). Since shops are small and run independently, they are an attractor or incubator of young entrepreneurs in the catering business, becoming a ground for creativity and innovation in the sector. As Maeda proposes, the yokochō projects offer an alternative to franchised izakaya. In the yokochō projects, the diversity of atmospheres, the communication among customers, and the "retro" atmosphere of the small shops seems to offer opportunities for budding entrepreneurs within the current economic situation of recession.

Fig. 2. Ebisu Yokochō (opened in 2009), a reinterpretation of the yokochō space

3. Results from fieldwork observation and measurements

Through on-site measurements, the overall plans of the yokochō and 12 sample shops have been drawn. Table 1 summarizes the main results from this study, and the plan and location of each yokochō are shown in Fig. 3. Several data are highlighted from Table 1 to understand the scale and character of the yokochō spaces.

- *Density.* All 6 cases show a great concentration of shops. In order to make the quantitative data more meaningful, density has been expressed in number of shops per are (are=100 m²). The *are* is not a common unit of area, but it suits the scale of the yokochō and it is easy to visualize since it corresponds to a square of 10x10 m. Densities expressed in Table 1 are gross densities, they include paths, open spaces, and vacant lots. Results in Table 1 show densities reaching 11.73 shops/are in Nonbei Yokochô and 8.8 shops/are in Hanazono. As an illustration, in the 3,265 m² of Goldengai-Hanazono (equivalent to about half of an international football soccer field) there are 253 establishments.
- *Time.* All 6 examples are clearly catering for dinner or late-night drinking. Opening times can be said to be related to the habit of Japanese office workers, who go drinking after work and before commuting back home. The three stations (Shinjuku, Shibuya, and Ikebukuro) are main hubs to transfer from inner-city office areas to suburban residential areas.
- *Dimensions.* Streets (width between 1.3 and 2.8 m) are pedestrian. In the narrowest sections, walls can be touched by extending both arms. Shops dimensions follow Japanese traditional modules (much contemporary wooden construction still does). Building footprints are from 4.83-66.1 m² (an enlarged planimetric study is shown in next section). Except for the two yokochō in Ikebukuro, all others share toilets, which saves considerable space within the limited dimensions of the shops.
- *Character.* The type of catering is simple food and drinks, and does not require an ample kitchen for cooking. The character of the bar can be read through its degree of permeability. While Omoide Yokochô does not have facades and is completely open, most shops in Golden gai are closed and interiors are barely visible. The degree of visibility acts to filter customers. High permeability facilitates a random visitor, whereas low permeability makes a newcomer hesitant to enter.

Density and time		Dimensions in plan		Character	
a. Total Area					
b. No. shops	f.	Street width (m)	h.	Location and character	
c. Gross density	g.	Shop footprint dimensions:	i.	Type of catering	
d. Opening time		front by depth (F m x D m)	j.	Visibility from exterior	
e. Shared toilet?					
Nonbei yokochō (Shibuya)				h.	Located in an area popular among young people. Used by all ages, including young people.
a. 324 m² (3.24 ares)	f.	1.5			
b. 38 shops	g.	Two types			
c. 11.73 shops/ are	•	2.3 x 2.1 (4.83 m²)→[SHOP 1]			
d. 5 pm to 12 pm	•	2.1x2.6 (5.46 m²) →[SHOP 2]	i.	Yakitori [1], oden [2]	
e. Yes				j.	Different levels of permeability

Density and time	Dimensions in plan	Character
a. Total Area b. No. shops c. Gross density d. Opening time e. Shared toilet?	f. Street width (m) g. Shop footprint dimensions: front by depth (F m x D m)	h. Location and character i. Type of catering j. Visibility from exterior
Goldengai (Shinjuku) a. 1271 m² (12.71 ares) b. 76 shops c. 5.98 shops/are d. 9 pm to dawn e. Yes	f. 2.0-2.7 g. Diverse sizes • 2.8-5.6 x 5.3 (14.84-29.68 m²) →[SHOP 6] • Exception→[SHOP 7]	h. Most stores only offer drinks and appetizers. Rather than the food, customers come to meet the store owners who have developed a distinctive character for each bar. Bar hopping is common. Included in many guides, many tourists visit including foreigners. Mainly alcoholic drinks]
Hanazono (Shinjuku) a. 1994 m² (19.94 ares) b. 177 shops c. 8.88 shops/are d. 9 pm to dawn e. Yes	f. 1.7-2.5 g. Most common types • 2.7 x 4.2 (11.34 m2)→[SHOPS 3 & 4] • 1.8 x 3.8 (6.84 m²)→[SHOP 5]	i. Mainly alcoholic drinks] j. Not visible. Some small windows in some cases allow glimpses of the inside.
a. Omoide Yokochō (Shinjuku) b. 828 m² (8.28 ares) c. 43 shops d. 5.19 shops/are e. 11 am to 12 pm f. Yes	g. 1.3-1.6 h. Diverse sizes • 2.2-4.0 x 5.3 (11.6-21.2 m²)→[SHOPS 8 & 9]	i. Shops have counters on the first floor and tables on the second. Sitting in large group is also possible. j. Yakitori[1], motsu[3] k. Open type, completely visible. Storefronts do not have walls or doors, completely open.
Sakaechō-dōri (Ikebukuro) a. 318 m² (3.18 ares) b. 14 shops c. 4.40 shops/are d. 7 pm to 11 pm e. No	f. 2.4 g. Diverse sizes • 4.9- 8.9 x 3.5 (17.2-31.2 m²)→[SHOP 10]	h. Some bars oriented to young customers and some well known yakitori shops. A certain atmosphere of decay was observed with a number of shops apparently closed. i. *Yakitori*[1], alcoholic drinks j. Not visible
Mikuni-kōji (Ikebukuro) a. 869 m² (8.69 ares) b. 20 shops c. 2.30 shops/are d. 5 pm to 11 pm e. No	f. 2.8 (4 in parts) g. Diverse sizes • 2.9 x 3.8 (11.0 m²)→[SHOP 11] • 2.9-5.8 x 5.0 (14.5-29 m²)→[SHOP 12] • 2.9-8.7 x 7.6 (22.0-66.1 m²)	h. Regular customers rather than bar hopping. Intimate atmosphere, difficult to enter for new-comers. i. Agemono[4] j. Not visible

1. Chicken pieces grilled on skewer. 2. Japanese hodgepodge. 3. Giblets. 4. Deep-fried food

Table 1. Dimensions and characteristics of the yokochō space

From each yokochō area, two or three representative shops were chosen. The selected samples fulfill two criteria: they can be considered representative of each type of shop identified in Table 1, and their owners have shown willingness to collaborate with this survey by allowing the researcher to measure their establishments and by participating in the interviews. Once the shop owners willing to collaborate were identified, sample shops (see Table 2) were selected using the following process:

- *Nonbei Yokochō.* In this yokochō there are two types of buildings in terms of dimensions. According to the author's observations, Shops 1 and 2 can be considered representative of these two types in terms of size and character.
- *Hanazono and Goldengai* are in fact one area having the same character. For each building dimension, one building has been selected. However, shops 3 and 4 have the same dimensions, but are located in different areas and, therefore, both have been investigated. Shop 7 was chosen as an exceptional case in terms of its size and location.
- *Sakaechō-dōri* and *Mikuni-kōji* are two adjacent yokochō that can be treated as a single area. For each building size, one sample shop has been chosen. The largest type (in Mikuni-koji) has dimensions that can be considered usual in Tokyo and thus has not been included.

As a result of the selection process, there is a sample of 12 shops, each representing a specific building size and location. For each sample, floor plans and section views have been drawn, including furniture and fixtures (see Fig. 4 to Fig. 8). By itself, this collection of planimetric drawings represents a visual tool to understand a scale that would be considered too small by normal standards. The most characteristic data and dimensions from the sampled 12 shops are summarized in Table 2.

From the survey summarized in Table 2 we observe the following commonalities:

- Shops are primarily managed by one person, with areas open to the public ranging between 3.8 and 10.5 m². This manager serves 5 to 13 customers.
- Counters are on the first floor and the second floor has diverse uses; on occasion to accommodate bigger groups of customers, and other times for private use.
- Catering is simple, and does not require a large kitchen or working space.

Location	Shop no.	Name	Type of catering	No. staff	1st floor		2nd floor		Total area m²
					Seats	Area m²	Seats	Area m²	
Nombei Yokochō (Shibuya)	01	Yoshinoya よしのや	*Oden*	1	8	3.8	8	3.8	7.6
	02	Tanukō たぬ公	Japanese casual dishes	1	8	4.5	Not public		4.5
Hanazono (Shinjuku)	03	Okutei 奥亭	Drinks & appetizers	1	11	8.3	Not public		8.3
	04	Nakachan 仲ちゃん	Japanese casual dishes	1	7	9.0	Not public		9.0
	05	Kikuko 紀久子	Drinks & appetizers	1	6	4.8	Not public		4.8
Golden-gai (Shinjuku)	06	Kojima 小島	Fish dishes	1	6	12.4	Not public		12.4
	07	Bihorogawa 美幌川	Drinks & appetizers	1	5	5.9	None		5.9
Omoide-Yokochō (Shinjuku)	08	Hinatori ひな鳥	*Yakitori, motsu*	1 or 2	10	9.9	Not public		9.9
	09	Saitamaya 埼玉屋	*Yakitori, motsu*	2 or 3	11	8.6	20	8.6	17.2
Sakaechō-dōri (Ikebukuro)	10	Edoichi 江戸一	*Sushi*	1	10	10.9	Not public		10.9
Mikuni-kōji (Ikebukuro)	11	Hagure はぐれ	Drinks & appetizers	1	6	7.0	Not public		7.0
	12	Ochahana 茶花	Okinawan food	1	13	10.5	Not public		10.5

Table 2. Dimensions of sampled shops

Photographs in Nombei Yokochō

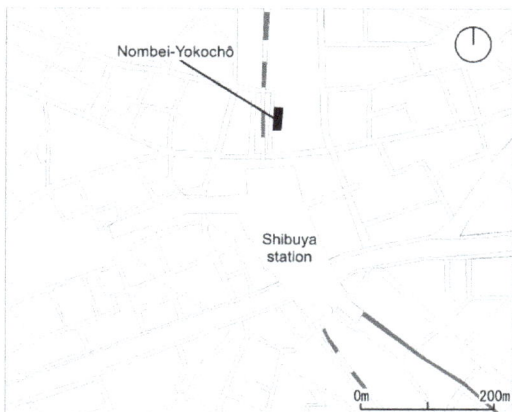

Photographs in Goldengai

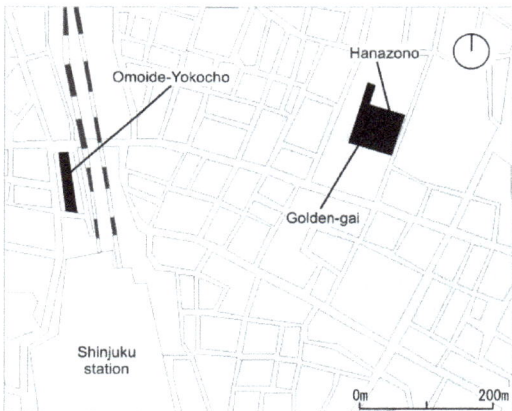

Photographs in Sakaechō-dōri

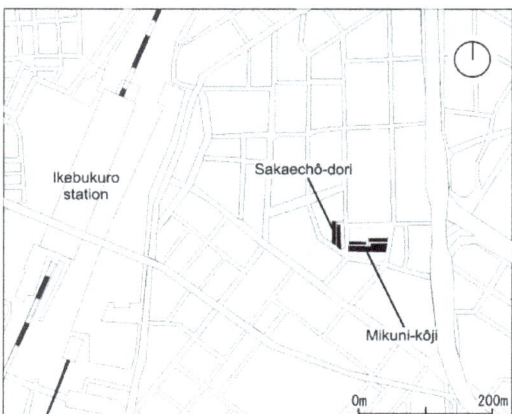

Fig. 3. Site plans of the 6 yokochō, located near train stations

Bar name : Tanukou
(Nombei-Yokochô / Shibuya)

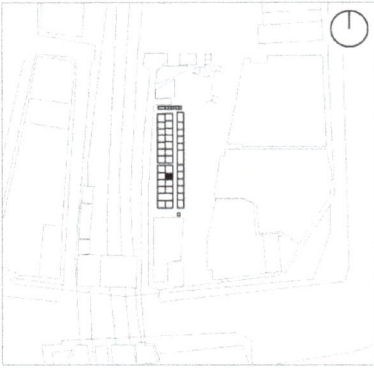

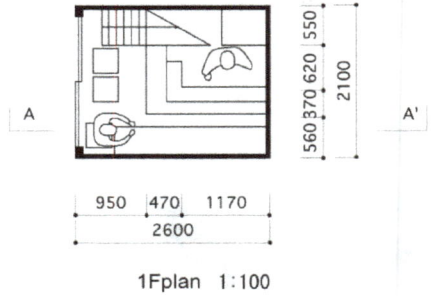

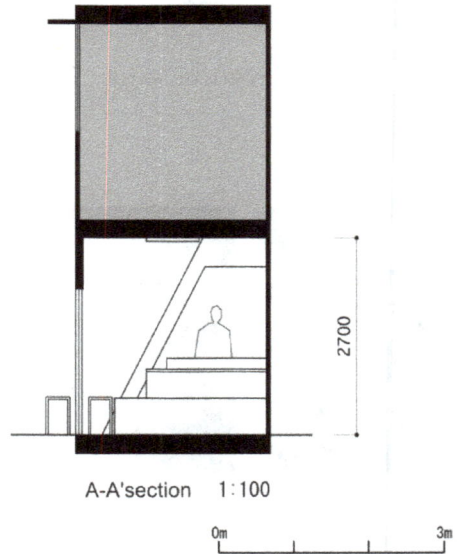

560 370 620 550
2100

A   A'

950 | 470 | 1170
2600

1Fplan 1:100

2700

A-A'section   1:100

0m   3m

Fig. 4. Shop 01 *Tanuko*, in Nombei Yokochō (Shibuya). Plans, sections and photographs on January 18, 2011 (dimensions in mm)

Bar name : Kikuko
(Hanazono / Shinjuku)

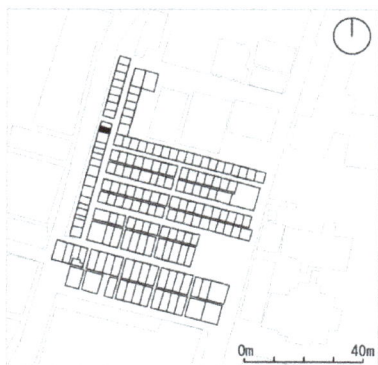

siteplan 1 : 2500

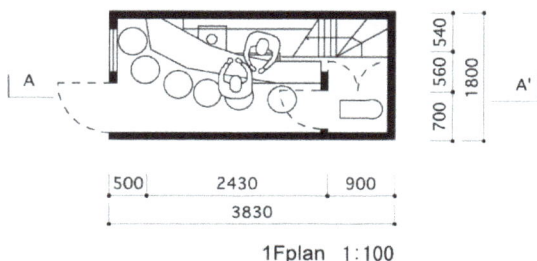

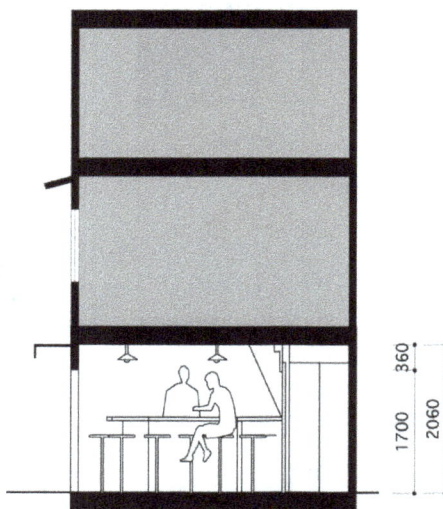

1Fplan 1 : 100

A-A'section 1 : 100

Fig. 5. Shop 05 *Kikuko*, in Hanazono (Shinjuku). Plans, sections and interior photograph on January. 17, 2011. Exterior photograph on February 25, 2011 (dimensions in mm)

Bar name : Bihorogawa
(Golden-gai / Shinjuku)

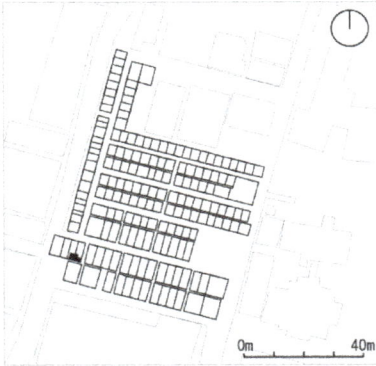

siteplan 1 : 2500

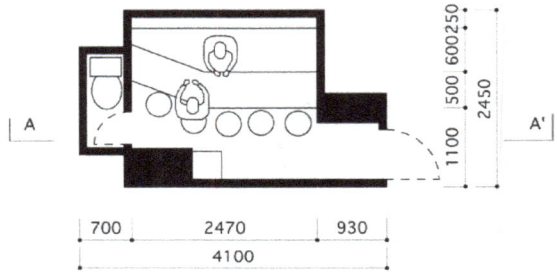

1Fplan 1 : 100

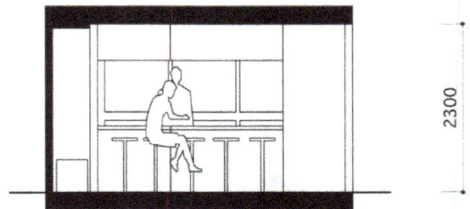

A-A'section 1 : 100

Fig. 6. Shop 07 *Bihorogawa*, Goldengai (Shinjuku). Plans, sections and interior photographs taken on January. 18, 2011 (dimensions in mm)

Bar name : Saitamaya
(Omoide-Yokochō / Shinjuku)

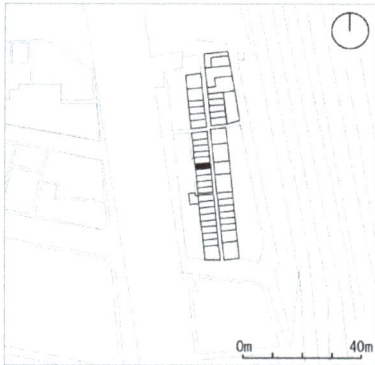

siteplan 1:2500

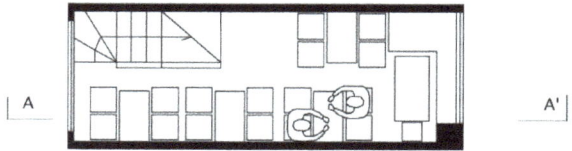

2Fplan 1:100

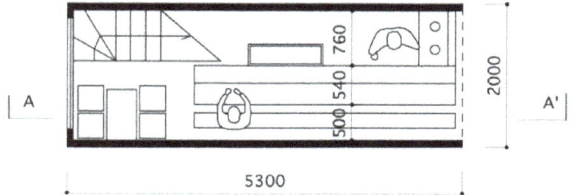

1Fplan 1:100

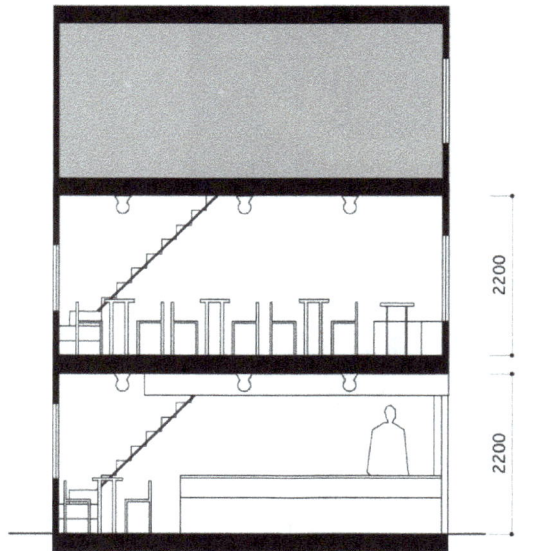

A-A'section   1:100

Fig. 7. Shop 09 *Saitamaya*, Omoide-Yokochō (Shinjuku). Plans, sections on November 29, 2010. Interior photographs on January 19, 2011 (dimensions in mm)

Bar name : Hagure
(Mikuni-kôji / Ikebukuro)

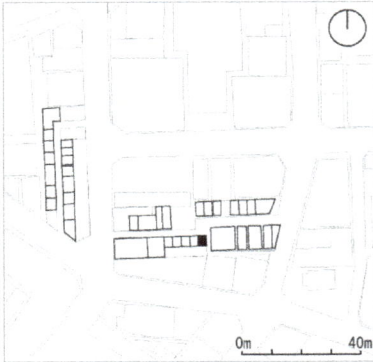

siteplan 1:2500

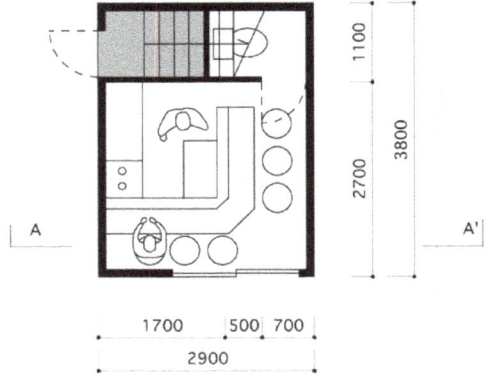

1Fplan 1:100

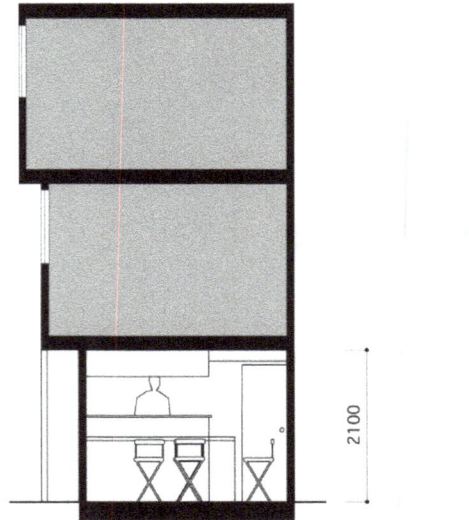

A-A'section 1:100

Fig. 8. Shop 11 *Hagure*, Mikuni-kōji (Ikebukuro). Plans, sections, and photographs taken on November 26, 2010 (dimensions in mm)

4. Results from interviews

In the previous section the character of each yokochō, and the dimensions of bar samples were investigated. Even when compared with other small shops in Japan and, specifically Tokyo, these yokochō bars seem exceptionally small. The results of the interviews conducted with the owners of the 12 establishments in Table 2 are summarized in Table 3, and extracts of their statements are reflected in Table 4. Replies on the perceived advantages of smallness have been organized into seven categories. According to the number of times that those categories were mentioned by the owners, they are organized into the order of importance in Table 3.

		Perceived advantage	No. of responses by different owners
1	(C)	Communication	11
2	(M)	Manageability	9
3	(E)	Economy	3
4	(AS)	Atmosphere of the shop	3
5	(AC)	Atmosphere of the context	2
6	(I)	Incubation of entrepreneurs	1
7	(U)	Urban location	1

Table 3. Summary of attributes perceived by owners

The results showed that communication is the most appreciated attribute. Communication is then followed by manageability, in the sense that a single person can look after an independent space, without hiring extra staff. To a lesser extent, these two attributes were followed by others such as the atmosphere of the shop. Smallness itself, as a spatial condition, was described as "warm", being in a "cave" or feeling "relaxed". This responses suggest that smallness is not only a practical way to make spaces more economic and manageable, but also has positive psicological effects (protection, relax, and warmth) and a proxemic quality that generates conversation. The yokochō district itself, and its power to attract tourists, were also mentioned. One respondent mentioned that he wishes to move to another location once his business improves, which coincides with Maeda's idea of the yokochō as an incubator of entrepreneurs.

Shop no.	What is the typical age of your customers?	Why did you decided to open a store in a such a narrow space?	Do you want to increase or decrease the area of your shop?	What are the advantages of the small sized spaces in yokochō?	What are the disadvantages?
01	30s or older, mostly men	• Because of the warmth of small shops (AS)	• I don't want to increase the area	• I can meet and have contact with customers (C)	
02	40s or older	• Because it is convenient since I can manage it alone (M)	• I don't want to increase the area	• Customers can get acquainted with each other (C)	• Bumping into things • I need more space to store the plates
03	From 20s to 60s	• Because I like the atmosphere of being in like a cave [穴倉] (AS)	• I don't have any particular intention of increasing it.	• I can attend the customers well, even if all seats are occupied (M). • Customers coming alone can get acquainted with others (C)	• The space behind the counter is narrow
04	From 50s to 60s, mostly men	• Since it is so small, I can attend to every nook and corner (M). • I can become friends with customers easily (C)	• To properly attend to the orders of drinks from the customers, 8 seats are the limit. In fact 6 would be best	• Cleaning is easy (M)	
05	Diverse ages	• It is a small place where I can work in a relaxed way (AS) • Cheap rent	• I don' have any intention of extending the shop		• The kitchen has only one burner. I can only work by shuffling sideways

Shop no.	What is the typical age of your customers?	Why did you decided to open a store in a such a narrow space?	Do you want to increase or decrease the area of your shop?	What are the advantages of the small sized spaces in yokochō?	What are the disadvantages?
06	From 30s to 50s, mostly men	• Because it is a size that I can manage alone (M). • Because the initial investment is small (E)	• Since it is my first shop, if it works well, maybe I want to increase the size or go to a bigger place (I)	• It easy to talk with customers, even the ones who enter the store for their first time (C)	
07	From 40s to 50s, mostly men	• Because I was doing a part-time job in Goldengai and I was told about an available shop	• Only 5 people can enter, but if I don't let 7 in, it will be hard for me economically. Personally, I prefer to have a bigger space.		
08	From 50s to 60s, mostly men	• This yokochō has become a tourist spot; many customers come (AC)	• I have 3 more shops of the same size in this yokochō catering the same food	• I can talk a lot with the customers, and customers can also talk to each other (C)	• Bumping into things
09	From 40s to 60s, mostly men. Also tourists	• Many people come because the area is a tourist spot now (AC)	• I have two more shops in this yokochō	• Good communication between customers. Even customers who come alone, end up bar hopping together (C)	
10	From 50s to 60s	• Because this is just the size to manage the show alone, without hiring employees (M, E)	No	• The atmosphere makes it easy to come alone (AS). Customers can become friends (C)	

Shop no.	What is the typical age of your customers?	Why did you decided to open a store in a such a narrow space?	Do you want to increase or decrease the area of your shop?	What are the advantages of the small sized spaces in yokochō?	What are the disadvantages?
11	From 20s to 70s, mostly men	• Since I opened the shop after my retirement, this is the right size to manage without working too hard (M)	No	• I am in close proximity to customers (C)	
12	Mid-30s to mid-50s	• I moved from Jinsei-yokochō, which was demolished. • This is the right size to work alone (M)	• Currently, I am managing 12 seats alone. Someday, I hope to have 40 seats, served by 2 people. • Economically it is ideal (E)	• It is convenient because everything is within the reach of my hand (M). • It is good to be only a short distance from customers (C)	• I have only two burners in the kitchen, I would like two more.

Table 4. Overview of responses by shop owners

5. Conclusions and considerations

This study documents yokochō spaces in Tokyo, famous for their small scale and the informal public life they accommodate. On-site measurements of the streets were conducted and planimetric drawings were developed for 12 sample shops. This documentation reveals the actual scale of "smallness" in numbers (density of shops, areas, interior dimensions) and its organization in plan and section. The owners of those same selected samples were interviewed to understand their opinion on the small scale. In order of the importance, the communication both with and among the customers and the manageability of the spaces were clearly the central factors. To a much small degree, these two principal factors were followed by economic advantages, the atmosphere, the possibility of business incubation, and the location.

Although the shortcomings of such small sizes were also mentioned by the owners, their responses pin-pointed advantages that can be fully attributed only to smallness. Spontaneous conversations afforded by proximity or the intimacy of a small space can be found in the architectural precedent of the 4.5 tatami mats tea house, where smallness plays a fundamental role. Smallness also has the attribute of manageability, which in the case of yokochō permits owners to run their bars alone.

At least three further aspects can be highlighted, which were discovered through the development of this study. Although systematic research was not conducted on these aspects, they may be relevant as suggestions in current urban and architectural debate and for future research.

- *Coopetition.* Accumulation of multiple shops of the same type seems to facilitates *coopetition,* that is, simultaneous cooperation and competition, and this appears to be enhanced by the small size of each unit. Responses did not directly refer to the aspect of coopetition, but they did mention the practice of "bar hopping" −where customers go to several establishments in quick succession to enjoy a diversity of menus and ambiences. This practice suggests the existence of a networked spatial practice, and the working of the district as an emerging entity that has a bigger effect than the sum of its units.
- Sharing: some facilities, such as toilets, are used in common. Smallness seems to be feasible through a balance between individual practice and shared spaces and rules. New ways of sharing can be learn from the yokochō districts in order to preserve urban smallness in the city centres, all too often redeveloped into large sterile complexes or kept as gentrified historic centres.
- *Behavioral architectural values.* The success of the yokochō seems difficult to understand from the point of view of traditional architectural values, which are based on visual qualities (such as views, proportions, treatment of light, detail, etc.). In order to describe the unsightly yokochō spaces as successful, architecture theory needs to develop a new vocabulary not based on visual but on behavioral qualities.

The studied yokochō are in a state of physical decay and in danger of being lost to redevelopment, but they can be considered a model of exploiting smallness while under the tremendous pressure resulting from their central location. These types of establishments may have a bad reputation as drinking places for middle-aged salaried men, but they have managed to create a micro-culture of third places. Furthermore, they might be held up as a good illustration of how to fulfill, in the field of architecture and urban design, the three conditions that Schumacher puts on technology (Schumacher 2010, p.35): "cheap enough so that they are accessible to virtually anyone, suitable for small-scale application, and compatible with man's need for creativity".

6. References

Almazán, J. & Tsukamoto, Y. (2009) Tokyo Public Space Networks at the Intersection of the Commercial and the Domestic Realms (Part III) Study on Urban Content Space. *Journal of Asian Architecture and Building Engineering,* Vol. 8, No. 2 (November 2009), pp. 461-468, ISSN 1347-2852

Barker, R. (1968). *Ecological Psychology: Concepts and Methods for Studying Human Behavior,* Stanford University Press, Stanford, California

Maeda, K. (2009) Yokocho kara no kasseika wo hakaru [Planning district vitalization from the Yokocho] *Fudosan Forum 21,* 2009, Vol.232, 2009 August, pp. 15-17

Ogawa, M. & Kawaguchi, Y. (2008). *Shinjuku no Goldengai-Hanazono Annai* [Guide to Goldengai and Hanazono]

Gekkan Rejaa Sangyō Shiryō (2009) Jimoto saihakken-Chiiki seikatsusha ni kōdō henka no kikai wo unagasu [Title in Japanese characters: 月刊レジャー産業資料, 地元再発見 －地域生活者に行動変化の機会を促す。ケーススタディ恵比寿横丁] [Translation by the authors: Rediscovering the home town - Stimulating opportunities for behavioral change in community life. Case study Ebisu Yokocho] *Gekkan rejaa sangyô shiryô*, March 2009

Schumacher, E. F. (2010, originally published in 1973) *Small is Beautiful: Economics as if People Mattered*, Harper Perennial.

Koolhaas, R. & Mau, B. (1994) Bigness or the problem of Large, In: *S,M,L,XL*, The Monicelli Press, New York

Lee, O. Y. (1992). *The Compact Culture: The Japanese Tradition of „Smaller Is Better"*, Kodansha International

Oldenburg, R. (1999), *The Great Good Place: Cafés, Coffee Shops, Bookstores, Bars, Hair Salons and other Hangouts and the Heart of the Community*, Marlowe & Company

Seidensticker, E. (1991). *Tokyo Rising: The City Since the Great Earthquake*, Harvard University Press

17

Integration of Hydrological and Regional and Urban Planning in Spain

Alejandro Luis Grindlay Moreno

Department of Urban and Regional Planning, University of Granada
Spain

1. Introduction

Planning is related to developing and managing a plan. It defines actions oriented towards specific objectives for a better future. Generally the political-administrative order of the different countries and regions establishes a legal framework to regulate this public activity. One of the first experiences in planning was sectoral, concerned with the development of plans for a particular department of public administration and which would eventually be realized through appropriate programs and/or projects. Usually the sectoral departments and their plans are segmented into different areas: services (healthcare, education, administration, commerce, security, etc.); environment (water and hydrology, protected areas, landscape, waste, etc.); and public works and transport (transport infrastructures - roads, rails, ports, airports, public transport facilities, transport services, buildings, etc.). Since the nineteenth century the developmental needs of a country have led to intense and continuous infrastructural planning and development, for transport, hydraulic works, and all utility and infrastructural systems.

One of the first planners to define the planning process from an urban and broad territorial perspective was Patrick Geddes (1915). He recognized it as a process of survey, analysis, and plan, and since then it has been developing in a similar way. In Spain there have been numerous experiences of urban planning from the success of Ildefons Cerdá's Expansion Plan ("Plan de Ensanche") in Barcelona in the middle of the nineteenth century, which spread to many other Spanish cities into the twentieth century. Since its legal regulation in the middle of last century, it has been established as a methodology for other sectoral planning, related to urban order and development, but with a limited territorial vision. Regional planning, which has a broader scope than sectoral planning, was not fully developed until the last two decades of the twentieth century, and was linked to the rise of an environmental paradigm and to a wide political decentralization process. The Constitution of 1978 signified a great political change which brought democracy to the country and created a quasi-federal state, divided into 17 regions or autonomous communities. It conferred the departments of regional planning to the respective regional governments, and sectoral departments were shared between the state and the regions but this gave rise to difficulties in planning coordination and integration.

In the next section the relationship between sectoral and regional planning in Spain and its traditional difficulties will be examined. These difficulties are based on the long tradition and social relevance of sectoral planning, and therefore its evolution is only briefly covered, in contrast with more recent and complex regional planning. Given its central role in this work, the evolution of state water policies and hydrological planning will be depicted in its own section. The simplest and most segmented approach of sectoral planning, in contrast to wider territorial planning, must be considered, and the difficult development of holistic regional planning by a structured and compartmentalized administration with sectoral visions (Benavent, 2009). In addition, traditionally, their results would be most easily perceived by society than would territorial order and would be politically more effective.

The second section details the evolution of the hydrological sectoral planning tradition which, as in other countries, has normally been regarded as a separate management problem to spatial planning, and decisions regarding water have been made without reference to spatial planning (Woltjer & Al, 2007).

Starting from the origins of regional planning and its hydraulic considerations, the third section will demonstrate the numerous diverse experiences of regional and urban planning in Spain led by the seventeen regional governments and their territorial policies, laws and planning instruments. Specific instruments of regional planning related to hydrological aspects will be highlighted, such as the territorial sectoral plans of the Basque Country and territorial action plans within the region of Valencia.

Finally, conclusions and indications are proposed to improve the integration of hydrological and territorial planning in general.

2. Relationship between sectoral and regional planning in Spain

Traditionally there has been a long experience of sectoral planning in Spain - mainly for transport and hydraulic infrastructures - from the second half of the nineteenth century, developed by civil engineering. The country needed to be developed and all infrastructures to be made and so sectoral planning was needed to organize and order all infrastructural development. Thus, one of its main characteristics is that it has very specific and functional goals. According to these, the conventional planning process goes from the knowledge of the current situation through documentation, information, analysis and diagnosis, identifying needs based on the goals and, after an analysis of alternatives, leads to the development of proposals and action programs for the future of the plan, in order to achieve those objectives. The need to prevent and take into account the possible environmental effects of these constructions, and later its integration with urban and regional planning, would not be formally considered until the last third of the twentieth century.

Different stages in the progress of the Spanish sectoral planning can be distinguished. The initial stage, which took place between the second half of the nineteenth century and the first half of the twentieth century, was based on the need to provide infrastructure. A second phase, between approximately the 1970s and 1990s, considered adapting infrastructure to demand, and the last phase from the 1990s until today where the existing adaptation and functionality is rationalized with the new, and quality and respect for the environment and sustainability are considered.

The stage of infrastructure provision was characterized by national or state plans. For example, the development of road infrastructure required four successive General Road Plans (1860-1864-1877-1914), the National Circuit of Special Pavements (asphalted roads) (1926-1939), a Plan for Modernisation (1951), and another four successive General Road Plans (1962-1977-1984-1991), the last two fall into the second stage of adapting infrastructure to demand. The early development of the hydraulic infrastructure also progressed through five successive National Plans of Hydraulic Works, as shown in the next section. The second phase was characterized by sectoral integration plans such as the state Infrastructure Director Plan (1993-2007), which considered the two infrastructures of hydraulics and transport together. In the last and current phase, new sectoral plans have been developed, not only integrating all transport infrastructural systems but also with a strategic perspective and considering environmental sustainability, such as the last Strategic Infrastructure and Transport Plan (PEIT 2005-2020). Furthermore the new regional governments during the 1980s, which were responsible for the departments of the infrastructures in their territories, mirrored the state administrative structure and began developing the provision of sectoral infrastructural plans, such as the Roads Plan of Catalonia (1985, revised in 1995) or the General Road Plan of Andalusia (1987) also to build a new territorial identity. In the same way, they have copied the sectoral integration plans of the second phase, such as the Infrastructure Director Plan of Andalusia (1997-2007), or the Transport Infrastructures Plan of Catalonia (2006-2026) (Grindlay, 2007).

During these decades municipal urban plans were also developed, although spatial planning was implicitly effected by the sectoral actions of transport infrastructures and by the large hydraulic works and their complementary colonization actions of new irrigated lands and villages.

However with the approval of the first regional planning instruments during the 1990s, such as the Regional Planning Guidelines of Asturias (1991) or the first regional plan in Catalonia (General Territorial Plan of Catalonia, 1995), the need to integrate the new sectoral plans with these documents will be clear, as shown later. As it has been recognized, regional planning has the purpose of applying a transversal perspective to the territory, and an integrated approach to territorial issues, which will be difficult to fit into a structured and compartmentalized administration, with a sectoral perspective (Benavent, 2009).

Therefore after this long tradition of sectoral planning, the first attempts to consider environmental effects during the 1970s, and later to integrate the sectoral transport plans with the existing or new urban plans in the 1980s, proved to be very problematic. As described by McHarg (1969), sectoral plans were considered regardless of their environmental, urban or territorial effects. Since the second half of the 1980s, the environmental impact assessment of the infrastructure projects was becoming generalized (as in European Directive 85/337/CEE), and a decade and a half later it was extended to the environmental impact assessment of plans and programs (as in European Directive 2001/42/CE). Subsequently there has been a significant development of new regional plans over the second half of the 1990s and in the last decade, and a change from the traditional segmented approach of sectoral planning towards closer integration as a comprehensive element of the most complex territorial plans. A new wide-ranging and holistic perspective is currently in consideration for sectoral planning. Nevertheless despite the short experience of the last decades in regional planning, compared the long tradition of sectoral planning, later sectoral plans also consider the determinations of the holistic regional plans. Therefore, the new state infrastructure plan should consider not only all the existing territorial and sectoral plans of all the regions or autonomous communities, but also ensure new regional infrastructural plans are fully integrated in their respective territorial plans. One example would be the latest Infrastructural Plan for Sustainable Transport in Andalusia (2007-2013) (Consejería de Obras Públicas y Transportes [COPT], 2008), which develops infrastructure and is in full accord with the Spatial Development Plan of Andalusia (2006). Although these should have a holistic and complex territorial approach they are managed by a structured and compartmentalized administration, and the regional government has not created a superior department to coordinate the rest. What is more, the advantages of their proposed spatial order are largely imperceptible to society, when compared to the results of infrastructural projects. However, the integrated and territorial perspectives in planning will substantially improve the general quality of the territories and their spatial order, which is now considered as territorial capital (European Parliament [EP], 2007).

3. Evolution of hydrological planning tradition

Water planning and management is possibly one of the most critical issues in Spain due to the combination of historically cyclical droughts and a rapid increase in the number and types of water uses, which makes water policy a focal point of public intervention and a subject of political, socioeconomic, and territorial controversy (Font & Subirats, 2010). So far, water policy has risen as a strong sector policy, with an autonomous logic, fully consistent with a model of continued pressure on the hydro domain, and widely regarded as a source of production (Moral, 2009).

The background of hydrological planning in Spain is related to the solution of one of the main territorial problems in the Mediterranean area namely flooding which was addressed by the Flood Defence Projects of the rivers Jucar (1866) and Segura (1886). The key proposal of these Projects was a wide dam development along the basins. Their prime objective was to achieve defence from flooding, but with a secondary objective to increase water reserves for irrigation and supply purposes. In order to promote irrigated agriculture - at that time the main pillar of economic development - five successive National Plans of Hydraulic Works (1902, 1909, 1916, 1919 and 1933) were drawn up. Each one of these "works plans" was approved to review, revise and complete the previous dam building plans. Their purpose was to resolve the irregular spatial and temporal distribution of the essential water resources, and to remedy the shortage, considering the existing antecedents of the few Roman and medieval dams, and on the large hydraulic projects (dams and channels) poorly developed during the eighteenth century. The intense process of regulation of so-called "lazy rivers" had aimed at the promotion of irrigation for national wealth, as the principle of the "Regenerationism" movement.

Water boards, called "Confederaciones Hidrográficas" (Hydrographical Confederations), were founded at every main river during the 1920s (Ebro and Segura in 1926, Duero and Guadalquivir in 1927, Eastern Pyrenees in 1929, etc.). Their administrative area or geographical scope corresponded to the respective hydrographical basins, establishing an essential management spatial unit, and their main purpose was the exploitation of all water resources for economic development. They were public and autonomous organizations assigned to the Ministry of Public Works as organizations with full self-determination. These entities have been functioning in Spain without interruption since their origins, playing an important role in the planning, construction and maintenance of hydraulic infrastructures, in the management and supply of hydrological resources, the protection of the public water domain, the granting of private rights of usage for water, and latterly, in the formulation and monitoring of hydrological basin planning and the execution of new hydraulic infrastructure, the control of water quality, etc. (Gómez & Grindlay, 2008).

After the breakdown period of the Civil War (1936-1939), hydraulic policy followed the principles of the last National Plan of Hydraulic Works of 1933. This was a genuine hydraulic works plan, considering a comprehensive overview of water issues among the various basins, and was conceived with planning criteria. Nevertheless it was integrated as a part of a major and general infrastructure plan, the General Plan of Public Works of 1941 for the national reconstruction. According to this, the second half of the century was to bring about the construction of large reservoirs designed for the regulation of rivers, power generation, flood control (primarily in the Mediterranean area), the improvement and increase of irrigated land, and the improvement of urban and industrial supply to satisfy increasing demand. This model of intense exploitation of water resources, incrementing regulation of surface resources and groundwater exploitation was usual in most developed countries (Biswas & Tortajada, 2003). For this development and supply model, one of the most significant actions, in line with the consideration of the national hydrographical imbalance (namely that there was an over-supply of water in some areas and an under-supply in others), was the Tagus-Segura Transfer. It had been proposed in the Plan of 1933, but was built during the 1970s to connect the headwaters of the Tagus (in east-central) Spain to the Segura River. This was claimed to solve the zone's hydrological imbalances between water-rich and water-poor regions (Gómez & Grindlay, 2008; Gupta & Van der Zaag, 2008).

After reaching a certain stage of development in the 1970s, some water boards conducted studies to prepare hydrological planning, by evaluating the possibilities of increasing exploitation of water resources by new dams. National planning directives (1979) were given to develop these preliminary studies for water resources planning. However, the major impulse to hydrological planning was driven definitively by the Water Law of 1985, which replaced that of 1879, and established the necessity of a draft hydrological plan in each basin and then the nationwide National Water Plan (NWP). It would coordinate them to resolve the imbalances and water deficits between basins, and would be the latest manifestation of the hydraulic paradigm in Spain. With the rise of the environmental paradigm, a major concern of this law was also water quality, which the intensification of water use had degraded. A key aspect was the joint consideration in the public hydraulic domain of all continental waters, surface and subterranean, in establishing a unified set of measures leading to a better rationalization in the use and development of increasingly scarce water supplies.

Flooding problems also received a strong hydraulic sectoral treatment during the 1980s in the Mediterranean basins of Jucar and Segura rivers with their Flood Defence Plans (1985 and 1987). This was underlined with the development of structural measures such as large hydraulic works of new dams and channelling, which increased the flood control capacity in the basins. In addition the limitations of purely hydraulic measures led to the need to implement non-structural actions, such as the Automatic Hydrological Information Systems (AHIS) (first developed at the Jucar basin in 1989 and in the Segura basin in 1992). This is used to mitigate the effects produced by floods, and to optimize the management of water resources in the basin. Additionally the need of spatial planning measures was considered (Carmona & Ruiz, 2000; Gómez & Grindlay, 2008).

During the second half of the 1990s hydrological plans were being approved in different basins. Although these, in terms of timescale, were in a second phase of sectoral planning by adapting infrastructure to demand they were, in reality, closer to the supply oriented model. Their results encompassed the estimated hydraulic resources, all existing demands, the hydrological balance between them, and the infrastructure or solutions required to solve water deficits. The relationship with other plans, such as regional planning, was mentioned but not implemented in any way. The NWP was finally approved in 2001, and it included measures for flood control, water quality, and riverbank and wetland protection, but its leit motiv remained the development and transfer of water supplies. It estimated a very high water demand to be satisfied with water development schemes, from a supply-oriented conventional model, and a large piping scheme transfer from the Ebro River to the "in deficit" Mediterranean area was proposed (Biswas & Tortajada, 2003; Gupta & van der Zaag, 2008).

However, during these years there was a growing social concern about the environment and the consequences of the development model led to this NWP being highly contested through a broad social and environmental movement (Biswas & Tortajada, 2003; Font & Subirats, 2010; Garrido & Llamas, 2009; Jiménez & Martínez-Gil, 2005; WWF, 2002) because, among other reasons, it was contrary to the new European water policy founded in the recently approved Water Framework Directive (WFD) (European Parliament and Council [EC], 2000). The change in this water policy and its paradigms of management, was realized with its modification and substitution by the A.G.U.A. Program in 2004 ("Actuaciones para la Gestión y Utilización del Agua" – Actions for the Management and Utilization of Water). This contained a series of "Urgent Actions in the Mediterranean Basin" to increase water supplies through desalination of sea water - seeking independence from climatic influence, and to improve the management and quality of water, through water saving and reutilization, increasing efficiency in water use (Garrido & Llamas, 2009). The government chose desalination instead of large inter-basin transfers, exchanging a very tangible and immediate form of environmental impact for the less tangible environmental impact of additional energy production (Downward & Taylor, 2007).

All these new demand-oriented measures in Spanish water policy would be more in accordance with the WFD principles and, as shown later, its implementation would provide some integration between hydrological and spatial planning. In recent years new Directives related to water have been developed and approved to complement and consolidate the WFD, specifically in those aspects more relevant in Mediterranean Spain such as groundwater protection (EC, 2006) and flood management (EC, 2007).

Furthermore the administrative and departmental changes of the autonomous communities in the Spanish State in the 1980s and 1990s would increase the socio-political complexity of water planning and management in Spain. Those catchments which fall within a single region (intra-regional) have been devolved to the regional government, and six new

Regional Water Agencies have been created. In the case of rivers which cross regional boundaries (inter-regional), these are managed by the centrally controlled River Basin Authorities, the traditional "Confederaciones Hidrográficas" now linked to the Ministry of Environment. Intra-regional basins managed by the regions have reproduced the hydraulic bureaucratic structures replacing earlier state structures, as is the case of the Agencia Catalana del Agua (1999) which took over from the former Confederación Hidrográfica del Pirineo. Otherwise river basins whose territory spreads across several autonomous communities retain their state dependent administration, but the hydraulic bureaucratic structures have been expanded because of the creation of new structures at a regional level, e.g. the Agencia Andaluza del Agua (created in 2005, but it dissolved and became integrated in the new General Secretariat of Water of the Department of Environment, 2011) which coexists with the Confederación Hidrográfica del Guadalquivir. Additionally in this inter-community river arrangement the main region will claim a greater role in its administration, as in this case of Andalusia region with its new water law (9/2010) and the Guadalquivir river basin management. Therefore the existence of a greater number of autonomous communities will naturally create more difficulties in the relationship between both sets of planning – hydrological and territorial– (Grindlay et al., 2011). In accordance with the traditional water scarcity, water issues will be strongly linked to aspects of nationalism, regionalism and territorial identity. Regions are now utilising water as a means of achieving political legitimacy and to secure control over access to water both as an economic resource and as a source of territorial identity (a territorial claim), and lately the traditional state hydraulic paradigm is re-enacted at the regional government level (Lopez-Gunn, 2009).

4. The experience of urban and regional planning in Spain

From a functional point of view, traditionally in Spain, and similarly in other countries, a separation has existed between water and urban planning and territorial management and, in general, the regional functions of territorial management have been scarcely developed until recent years. Before addressing the experience of urban and regional planning in Spain a brief overview of its main background is necessary.

4.1 Origins of regional planning and its hydraulic considerations

The Scottish planner Patrick Geddes (1915) was one of the first planners to consider a broad territorial perspective beyond the city itself, including the physical environment, the need to think about its multiple zone levels, and the requirement to adopt a multidisciplinary approach. His ideas were later followed by his disciple Lewis Mumford, who founded, with other colleagues (C. Stein, P. Chase and B. MacKaye), the "Regional Planning American Association" (1923). They also considered the importance of the regional spatial unit for territorial planning, which, in the case of water processes, should naturally be the basin unit. Some of their principles, such as the basin spatial unit, were applied in one of the most relevant experiences of a complete development program of a river basin, the "Tennessee Valley Authority" (1931), in the American regional planning drive for economic development during the 1930s, described by Friedmann and Weaver (1979) in a survey of ideas concerning regions and regionalism in the United States (Weitz & Seltzer, 1998). It is also necessary to mention the original, well-known, German zoning experiences in the 1920s, and the cases of regional planning in New York (1929), and of the United Kingdom from the 1930s onwards, which adopted the zoning technique to define green areas or areas to be preserved at the regional level, many of which would be related to water processes (Hall, 2002).

During the 1960s another Scottish planner, Ian L. McHarg, clearly defined in his book "Design with Nature" (1969), the necessity to consider natural and ecological processes in planning, of course, including those related to water. He highlighted such basic considerations as to avoid development of flood plains and other areas under risk, and the aquifer infiltration areas. Moreover he developed a methodology of territorial analysis based on superimposing information map layers, originating from the current Geographical Information Systems (GIS), with their multiple spatial planning applications (e.g. Zamorano et al., 2008).

As professor Gómez Ordóñez (2006) recognizes, spatial planning has been characterized throughout its history by the concern to harmonize the natural and the artificial, with consideration for the correct relationship between what is built and what is preserved, or reserved for the future, and it is precisely in this regard that water has played, and is playing, a leading role.

4.2 Progress of hydrological considerations in Spanish urban planning

The experiences of regional planning in Spain have been relatively recent compared to those of urban planning. As mentioned earlier, Ildefons Cerdá and his first urban expansion plan ("Plan de Ensanche") for Barcelona (1859), which was accompanied of his General Theory of Urbanization (1867), represents the birth of the urban planning discipline (García-Bellido, 2000). He had a high concern for topography and drainage as the foundation for his original urban development. With his famous phrase "ruralise what is urban, urbanize what is rural" he anticipated by almost half a century the ingenious diagram of the "three magnets" of E. Howard (1898), who uses it to explain the necessity of integrating the advantages of town and country for a new urban environ in his famous proposal of the "Garden Cities" (reproduced in Wheeler & Beatley, 2009).

The urban expansion or extension plans of the second half of the nineteenth and the first half of the twentieth centuries should include a section concerning studies of the physical environment (geology, topography, hydrography, etc.) as a basis of proposed developments. Urban planning has been a municipal responsibility in its spatial administrative realm, regulated from the perspective of urban infrastructure in 1924. Additionally the land law of 1956 is the first legal text for integral spatial planning of the territory from an urban point of view, and it is considered as the beginning of Spanish contemporary urbanism. It defined the municipal urban planning instruments as the General Plan and its content had to include an analysis of the natural characteristic of the territory and it should steer later urban development (Olcina, 2007). The progress of urban municipal planning during the second half of the century, according to this law, has been described as a "story of an impossible process", referring to the results developed in many cases, so different than those originally planned (Teran, 1978). Generally successive modifications of the plans intensified developments, even building-in originally protected green areas such as riverbanks. Also buildings were often illegally constructed on unauthorized areas such as flood plains. Therefore many spatial developments have frequently been far from the spatial planning considerations regarding the hydrological environs.

One of the preliminary experiences on regional planning was a scheme for the reconstruction of Madrid (1939). Later this law also established a supra-municipal planning instrument, namely the provincial plan. Few were developed (Barcelona, 1963; Guipuzcoa, 1966; Majorca 1973 and the metropolitan plan of Barcelona 1976), all of these with an urban order perspective (Benavent, 2006).

The quasi-federal state established by the Constitution of 1978 conferred full responsibilities for urban and territorial planning to the respective regional governing bodies, and precluded the state from later establishing a national spatial plan (as dictated in 1997), making impossible a coordination between regional spatial plans. As in other European countries, the centre of gravity of spatial planning remained with the municipal land-use urban plan (Wiering & Immink, 2006). In 1998 the state land law was adapted to these changes to establish a common land legal framework but only with regards to the state responsibilities on property rights. In relation to hydrological risk, among others, land under natural risk according to sectoral planning should be categorized as (article 9.1) land with the status of "not to be developed". However it does not establish the degree of risk (Olcina, 2007), which would be required for the corresponding sectoral planning that would determine it. The last revision of the state land law was in 2008, which attempted to avoid conflict with regional departments on the matter, and among the basic criteria of land use was to address the principle of natural risk prevention and of urban and territorial sustainable development and environmental protection. It defines rural land or land not be developed with an equal status to the ground preserved by regional and town planning for its transformation through urbanization, which would include those areas with natural and technological risks, including floods. This also refers to a sectoral planning of flood risk.

As with other planning instruments, those of urban and regional planning are subject to environmental impact assessment in accordance with the provisions of the legislation to assess the effects of certain plans and programs on the environment (Law 9/2006, which supersedes Directive 2001/42/CE). In relation to this, the applicable state land law (Law 2/2008) establishes that the required environmental sustainability report of the planning instruments of urbanization actions should include a map of natural risks in the area under scrutiny. It also includes the necessity of reports from the hydrological administration on the existence of the water resources needed to satisfy the new demands, and on the protection of the public water domain. Thanks to this a major integration between the new urban planning and hydrological planning in Spain has been achieved.

A state document related to civil protection was approved, called "Basic directive of flooding" (1995), with relevant implications for regional spatial planning documents on the necessity of determination of flood areas and those at risk, considering the population, facilities and infrastructures potentially affected. It requires the analysis of flood zones and flood risk to be completed at local and regional levels but, in general, little has been developed by regional governments and hardly anything by municipalities (Olcina, 2007). It will be the new Directive on flood management (EC, 2007) and its transposition which will definitely drive these studies and plans as shown later.

4.3 The development of regional planning framed by EU policies

The development of regional planning in Spain took place on the basis of the principles of "The European Charter on Spatial Planning" (adopted in 1983), and in the progress of the decentralized state politically configured in autonomous communities. For the development of these responsibilities the respective regional governments began to develop their urban and regional planning laws, since their conformation in the 1980s, and later their planning instruments, also as a means to strengthen national identity of the regions. In consequence Catalonia (1983) was the first to set them up according to its very nationalist tradition and its own language, and similarly, later, the Basque Country (1990).

The development of the first regional laws on spatial planning of the seventeen autonomous communities in a chronological order has been: Catalonia (1983), Madrid (1984), Navarre (1986), Asturias (1987), Balearic Islands (1987), region of Valencia (1989), Cantabria (1990), Basque Country (1990), Murcia (1992), Aragon (1992), Andalusia (1994), Galicia (1995), La Rioja (1998), Castilla-La Mancha (1998), Castilla Leon (1999), Canary Islands (1999), and finally Extremadura (2001). Some of these were integrated urban and regional planning laws using the same legal stipulations as Madrid or La Rioja, and other communities set up different laws for regional and urban planning, as Andalusia (1994 and 2002 respectively). Many of them have been latterly revised and changed resulting in a complex and confused legal and planning system throughout the country demonstrated in detail by Benavent (2006).

Generally many of these new laws widely included the concepts and wording of "The European Charter on Spatial Planning", and later they were also greatly influenced by the European Spatial Development Perspective (ESDP) (European Communities, 1999). This was a broad consensus document on spatial planning matters, based on the voluntary cooperation of the EU member states, which had a long and debated process of discussion and therefore has been the most global planning policy document to date (Faludi & Waterhout, 2002). Its objectives are both simple and comprehensive, as are the processes for how these objectives should be attained, particularly referring to the conservation of natural resources and cultural heritage, sustainable development and the protection of natural and cultural heritage (Faludi, 2002).

At a European level spatial planning is an individual responsibility of the member states, and spatial planning at this level is sufficiently distinct from the spatial planning at the national and regional levels (Dühr et al., 2010). Even at this broad level it is worth highlighting that the approaches of the ESDP, whilst correct, are quite generic. It rightly sets out water resource management as a special challenge for spatial development, and the need "to co-operate across administrative boundaries in the field of water resource management". It also emphasises that "policies for surface water and ground water must be linked with spatial development policy" and, obviously, that "spatial planning can make an important contribution to the protection of people and the reduction of the risk of flood" (EU, 1999). These and other considerations around water issues will later be incorporated and broadly developed in the WFD, but divested of the spatial dimensions.

The laws on spatial planning of the autonomous communities have proposed planning or guideline instruments for the integration of the three main spatial dimensions of a territorial model, which also coincide with the elements of the policy aims and options for the territory of the EU established at the ESDP: cities and urban systems, infrastructures and access and knowledge networks, and environmental and cultural heritage.

Despite the diversity of types and designations defined in these instruments by regional governments, however, most of them have founded a similar scheme with three types of documents:

a. Integrated planning instruments at a regional level, which affect the entire territory of an autonomous community.
b. Integrated planning instruments at a sub-regional level, which affect territories with provincial, district or just supra-municipal areas, and constitutes a development of the regional level instruments.
c. Instruments attempting to integrate sectoral intervention with substantial territorial effects, such as territorial sectoral plans or regulations on sectoral plans and projects and standard assessment tools to control their territorial effects.

At the regional level there are also three types of documents: Territorial Plans (Andalusia, Cantabria, Catalonia, Madrid), Guidelines for Spatial Planning (Aragon, Asturias, Basque Country, Canarias, Castilla-La Mancha, Castilla y Leon, Extremadura, Galicia, Balearic Islands, Murcia) or Territorial Strategies (Navarre, La Rioja, region of Valencia) these last most influenced by ESDP, or European Territorial Strategy in the Spanish translation.

In the lower spatial level or sub-regional level there is a much greater diversity of documentation, as shown in the next table, but the Territorial Partial Plans, the Sub-Regional Spatial Plans or the Insular Spatial Plans for the islands can be emphasized. All of them will develop the corresponding regional level instrument in a particular territory.

However, for the purpose of this chapter, of great interest are those instruments attempting to integrate sectoral intervention with substantial territorial effects, such as territorial sectoral plans or regulations on sectoral plans. The Basque Country has been the region that has most promoted these instruments named as Territorial Sectoral Plans. These are developed by the Basque government departments with responsibilities that will produce territorial effects, and also develop the Regional Planning Guidelines (1997), which the regional territorial model established, in the sectoral areas. As, in general, it is based on the common scheme of three elements: physical environment, urban systems, and relational or infrastructural systems, the respective Territorial Sectoral Plans will develop one particular sectoral element. The approved Territorial Sectoral Plans are on the Regulation of the Margins of Rivers and Streams (1998), on Rail Network (2001/2005), on Wind Energy (2002), Public Land Creation for Economic and Commercial Equipment (2004), on Coastal Protection and Management (2007), on Roads (2008), on Wetlands (2004/2008), and on Gipuzkoa's Urban Waste Infrastructure (2009). Additionally the Territorial Sectoral Plans in preparation are pertaining to Cultural Heritage, Agro-forestry, Land for Public Housing Promotion, Ports, and Intermodal and Logistics Transport Network (Departamento de Medio Ambiente, Planificación Territorial, Agricultura y Pesca [DMAPTAP], 2011).

The next table reflects the regional planning instruments developed by autonomous communities' governments.

Autonomous Community (first Regional Planning Law)	A. Regional Level Spatial Planning Documents	B. Sub-regional Level Spatial Planning Documents	C. Other Planning Documents
Andalusia (1994)	Plan de Ordenación del Territorio de Andalucía (2006)	Plan de Ordenación del Territorio (POT) de la Aglomeración Urbana de Granada (1999-2005).	Plan Andaluz de Vivienda y Suelo 2003-2007 (2007-2011)
		Planes de Ordenación del Territorio (P.O.T.s) del Poniente Almeriense (2002); Sierra de Segura (2003); Doñana (2003); Bahía de Cádiz (2004)	Plan de Infraestructuras para la Sostenibilidad del Transporte en Andalucía (2007-2013) (2008)
		P.O.T.s Litoral Occidental de Huelva (2006); Costa del Sol Oriental-Axarquía (2006); Costa del Sol Occidental de Málaga (2006)	Plan General del Turismo Sostenible de Andalucía (2008-2011) (2007)
		P.O.T.s Levante de Almería (2009); Aglomeración Urbana de Sevilla (2009); Aglomeración urbana de Málaga (2009); Costa Noroeste de Cádiz (2011)	Plan Especial Supramunicipal del río Palmones (2010); P.E. de Ordenación de los regadíos al norte de Doñana (2011)
Aragon (1992)	Directrices Generales de Ordenación Territorial (O.T.) para Aragón (1998)	Directrices Parciales de O.T. del Pirineo Aragonés (2005-2010)	Directrices sectoriales sobre actividades e instalaciones ganaderas (2009)
		Directrices Parciales de O.T. de la Comarca del Matarraña (2008)	
Asturias (1987)	Directrices Regionales de Ordenación del Territorio (O.T.) (1991)	Directrices Subregionales de O.T. franja costera de Asturias (1993)	
		Plan Territorial Especial de Ordenación del Litoral de Asturias (2005)	
Balearic Islands (1987)	Directrices de Ordenación Territorial de las Islas Baleares (1999)	Planes Territoriales Insulares de Menorca (2003); Mallorca (2004); de Ibiza y Formentera (2005)	

Autonomous Community (first Regional Planning Law)	A. Regional Level Spatial Planning Documents	B. Sub-regional Level Spatial Planning Documents	C. Other Planning Documents
Basque Country (1990)	Directrices de O. T. de la Comunidad Autónoma del País Vasco (1997)	Plan Territorial Parcial (P.T.P.) de la Rioja Alavesa (Laguardia) (2004)	Plan de Ordenación de los Márgenes de Ríos y Arroyos (1998)
		Planes Territoriales Parciales (P.T.P.s) de Álava Central (2004); del Área Funcional de Llodio (2005);	Plan Territorial Sectorial (P.T.S.) de Red Ferroviaria en la CAPV (2001/2005)
		P.T.P.s del Área Funcional de Eibar (2005); del Área Funcional de Mondragón-Bergara (2005); del Área Funcional de Bilbao Metropolitano (2006)	P.T.S. de Creación Pública de Suelo para Actividades Económicas y Equipamientos Comerciales (2004)
		P.T.P. del Área Funcional de Zarautz-Azpeitia (2006-2008)	P.T.S. de Zonas Húmedas (2004/2008)
		P.T.P. del Área Funcional de Beasain-Zumarraga (2009)	P.T.S. de Protección y Ordenación del Litoral (2007)
		P.T.P.s de Donostia-San Sebastián (2010); del área funcional de Balmaseda-Zalla (2010)	Plan Sectorial General de Carreteras (Álava – Bizkaia) (2008)
		P.T.P.s del Área Funcional de Igorre (2010); del Área Funcional de Durango (2011)	P.T.S. de Infraestructuras de Residuos Urbanos de Gipuzkoa (2009)
Canary Islands (1999)	Directrices de Ordenación Gral. de Canarias (2003)	Plan Insular de Ordenación Territorial de Lanzarote (1991)	Directrices de O. del Turismo (2003)
		Plan Insular de Ordenación de Fuerteventura (2001)	Plan Territorial Especial de la Costa Norte de Gran Canaria (2003)
		Plan Insular de Ordenación de la Isla de El Hierro (1995-2002)	Plan Territorial Parcial Plataforma Logística del Sur (2008)
		Planes Insulares de Ordenación de Tenerife (2002-2011); de Gran Canaria (2003-2011)	Plan Territorial Especial Ordenación Turística de Tenerife (2005)

Autonomous Community (first Regional Planning Law)	A. Regional Level Spatial Planning Documents	B. Sub-regional Level Spatial Planning Documents	C. Other Planning Documents
		Planes Insulares de Ordenación de La Gomera (2011); de La Palma (2011)	Plan Territorial Especial Paisaje de Tenerife (2010)
Cantabria (1990)	Plan Regional de Ordenación Territorial (N.A.)		Plan de Ordenación del Litoral (2004)
			Plan Especial de la Red de Sendas y Caminos del Litoral (2010)
Castilla-La Mancha (1998)	Plan de Ordenación Territorial de Castilla-La Mancha (N.A.)	Plan de Ordenación del Territorio (P.O.T.) del "Corredor del Henares y zona colindante con la comunidad de Madrid" (Guadalajara) (N.A.)	
		P.O.T. de la "Zona de La Sagra" (Toledo) (N.A.)	
		P.O.T.s de la "Mesa de Ocaña y el corredor de la autovía A-3 "; del "Corredor Ciudad Real - Puertollano"; de la "Zona de influencia de Albacete" (N.A.)	
Castilla Leon (1999)	Directrices Esenciales de Ordenación de Castilla y León (2008)	Plan Regional de ámbito territorial del Canal de Castilla (2001)	Plan Regional de Ámbito Sectorial de la Bioenergía de Castilla y León (2011)
		Plan Regional de Ámbito Territorial "Valle del Duero" (2010)	Directrices de O.T. de ámbito subregional de Valladolid y entorno (2001)
		Plan Regional de Ámbito Territorial del Puerto de San Isidro (León) (2004)	Directrices de O.T. de ámbito subregional de Segovia y su entorno (2005)
		Plan Regional de Ámbito Territorial Zamor@-Duero (2010)	Directrices de Ord. de ámbito subr. de la montaña cantábrica central (2011)

Autonomous Community (first Regional Planning Law)	A. Regional Level Spatial Planning Documents	B. Sub-regional Level Spatial Planning Documents	C. Other Planning Documents
Catalonia (1983)	Pla Territorial General de Catalunya (1995)	Pla Territorial Parcial de les Terres de l'Ebre (2001-2010)	Pla director del delta de l'Ebre (1996)
		Pla Territorial Parcial de l' Alt Pirineu i Aran (2006)	Pla director de les activitats industrials i turístiques del Camp de Tarragona (2003)
		Pla Territorial Parcial de Ponent (Terres de Lleida) (2007)	Pla director urbanístic del sistema costaner (2005)
		Pla Territorial Parcial de les Comarques Centrals (2008)	Pla director urbanístic del Pla de Bages (2006)
		Pla Director Territorial de l'Alt Penedès (2008)	Pla director urbanístic (+13)
		Pla Territorial Metropolità de Barcelona (2010)	Plans director urbanístics de les àrees residencials estratègiques (+12)
		Pla Territorial Parcial de les Comarques Gironines (2010); del Camp de Tarragona (2010)	Pla comarcal de muntanya 2009-2012 (2009)
Extremadura (2001)	Directrices de Ordenación del Territorio (N.A.)	Plan Territorial de Campo Arañuelo (2008)	
		Plan Territorial de la Vera (2008)	
		Plan Territorial Embalse Alqueva (2009)	
Galicia (1995)	Directrices de Ordenación do Territorio (N.A.)	Plan de Ordenación do Litoral (2011)	
La Rioja (1998)	Estrategia Territorial de La Rioja (N.A.)		Plan de suelo de actividades económicas (2006)
Madrid (1984)	Plan Regional de Estrategia Territorial (N.A.) (Bases, 1995)	Estrategia Territorial Zona Oeste Metropolitana (1989)	Plan de Vivienda de la Comunidad de Madrid (2005-2008/2009-2012)
		Estrategia Territorial Corredor del Henares (1989)	Estrategia de Residuos de la Comunidad de Madrid (2006-2016)

Autonomous Community (first Regional Planning Law)	A. Regional Level Spatial Planning Documents	B. Sub-regional Level Spatial Planning Documents	C. Other Planning Documents
Murcia (1992)	Directrices de Ordenación del Territorio (N.A.)	Directrices y Plan de Ordenación Territorial del Litoral de la Región de Murcia (2004/2007)	Actuación de Interés Regional de Marina de Cope (2004)
		Directrices y Plan de Ordenación Territorial del Suelo Industrial de la Región de Murcia (2006)	Actuación de Interés Regional de Aeropuerto Internacional de la Región de Murcia (2004)
			Plan Regional de Vivienda (2007 – 2010/2009-2012)
Navarre (1986)	Estrategia Territorial de Navarra Directrices para la O.T. (2005)	Plan de Ordenación Territorial 1 Pirineo (2011)	Planes Directores de Acción Territorial (N.A.)
		Plan de Ordenación Territorial 2 Navarra Atlántica (2011)	Planes y Proyectos Sectoriales de Incidencia Supramunicipal
		Plan de Ordenación Territorial 3 Área Central (2011)	
		Plan de Ordenación Territorial 4 Zonas Medias (2011)	
		Plan de Ordenación Territorial 5 Eje del Ebro (2011)	
Valencia's Community (1989)	Estrategia Territorial de la Comunitat Valenciana (2011)	Plan de Acción Territorial del Entorno de Castellón (PATECAS) (N.A.)	Plan de Acción Territorial de carácter sectorial sobre prevención del Riesgo de Inundación en la Comunidad Valenciana (PATRICOVA) (2003)
		Plan de Acción Territorial del entorno Metropolitano de Alicante y Elche (PATEMAE); de la Vega Baja (N.A.); de las Huertas (Valencia) (N.A.)	Plan de Acción Territorial de carácter sectorial de corredores de infraestructuras (2005)

Autonomous Community (first Regional Planning Law)	A. Regional Level Spatial Planning Documents	B. Sub-regional Level Spatial Planning Documents	C. Other Planning Documents
			Plan de Acción Territorial Forestal de la Comunitat Valenciana (2011)

(N.A.) Not Approved

Table 1. Autonomous Communities' Spatial Planning Documents. Source: Author's own work based on Olcina (2007, pp. 161-164), and Moral (2009, table 1) and autonomous communities regional governments' web pages.

Naturally these spatial planning instruments (plans, guidelines, or strategies) will define a territorial model, in which one essential component is the environment, where the "not to be developed" lands and/or protected natural areas, and also the hydrological system are considered. Correspondingly the development model proposed should be consistent with the physical and environmental characteristics of the territory, and particularly with regard to the hydrological elements. Territorial planning must also take into account the consideration of water and the hydrological cycle in general and in particular in the distribution of the uses of the land, essentially in the future urban expansion and in the construction of infrastructures, and their conformation with the hydrological system, above all the fluvial dynamic. The traditional problem of floods and inundations in some Spanish areas has given rise to a particular concentration of attention on this question (Grindlay et al., 2011), as will be presented in the following section.

In general, as shown in Olcina (2007), most of the autonomous communities' laws on spatial planning explicitly consider the need to exclude for development those areas exposed to natural risks, such as flooding, and they establish use limitations around rivers or aquifers to protect the hydraulic public domain. However it is less common to establish the necessity for prior studies about availability of water resources to carry out the planned actions, as in the Mediterranean regions of Catalonia and Valencia, which have suffered drought effects and flood problems most intensively. All the issues referring to the hydrological system and flood risks are obviously most detailed in the sub-regional instruments, as in Catalonia and Andalusia, but especially in the territorial sectoral plans and the noteworthy experiences of the Basque Country and region of Valencia.

The first regional plan to be approved was the Territorial General Plan of Catalonia in 1994. It established the Catalonian territorial model through territorial information, analysis and diagnosis, and territorial strategies and proposals. It also had a clear economic infrastructural orientation and its physical environment analysis is only considered as a basis for future developments (Benavent, 2006; Olcina, 2007). However it is widely developed in the Territorial Partial Plans of all its counties: Terres de l'Ebre (2001-2010) I' Alt Pirineu i Aran (2006), Terres de Lleida (2007), Comarques Centrals (2008), l'Alt Penedès (2008), metropolitan area of Barcelona (2010), Comarques Gironines (2010), and Camp de Tarragona (2010).

Also very interesting is the long experience in regional planning of Andalusia that approved its Regional Spatial Plan later in 2006, but after almost ten years of development. In this

Regional Spatial Plan programs named as "Water and Territory" have also been considered, with an adequate theoretical framework, but none have been developed (COPT, 2006). The restoration of the Guadalquivir River has also been proposed as an example, but without consideration for the rest of the fluvial system (Fig.1). This was addressed later in a regional Director Plan of Riverbanks (Consejería de Medio Ambiente [CMA], 2006). The Sub-regional spatial plans were being made earlier in the most developed and complex metropolitan areas, such as Granada (1999) (Fig.2), Poniente Almeriense (2002) or Bahía de Cádiz (2004), and in those areas with a high degree of environmental protection, such Sierra de Segura (2003) or around Doñana (2003). All the most dynamic coastal areas and the metropolitan areas of Andalusia have a sub-regional spatial plan: Litoral Occidental de Huelva (2006), Costa del Sol Oriental-Axarquía (2006), Costa del Sol Occidental de Málaga (2006), Levante de Almeria (2009), Aglomeración Urbana de Sevilla (2009), Aglomeración Urbana de Málaga (2009), and Costa Noroeste de Cádiz (2011) (Fig.3). In general the hydrological environ is widely considered in all these plans, with the protection and restoration of river areas and flood risk protection, and the improvement of water supply and wastewater treatment, with an integrated management system of the water cycle (Mataran et al., 2010). However, regarding the other sectoral actions to establish the proposals for territorial organization, these plans have generally been limited to forming the projected interventions for each of the sectoral administrations, without a revision of their institution for integration with the future territorial model, and without considering their, sometimes contradictory, effects on this.

4.4 Specific instruments of regional planning related to hydrological aspects

As mentioned above, in some autonomous communities specific instruments of regional planning related to hydrological aspects have been developed, whose common base is to regulate land use and urban planning regulations in flood areas, based on different zoning and hazard maps of the fluvial territories (Berga, 2011). These documents integrate an hydrological sectoral aspect into the regional spatial planning system, and therefore they are called "territorial sectoral plans" or "territorial action plans" for regulation of river margins or flood prevention.

Traditionally the region of Valencia and the Basque Country have been two of the autonomous communities most affected by flooding in Spain, and therefore they were the first to develop planning documents with particular reference to flood risk management to reduce and/or minimize flood hazards (Olcina, 2007). Given its relevance, the rivers management plan was the first approved among all Territorial Sectoral Plans of the Basque Country. The Territorial Action Plan of the region of Valencia was developed later.

The Territorial Sectoral Plan of the Regulation of the Margins of Rivers and Streams, under revision (2008), was approved in 1998 (Cantabrian basin) and in 1999 (Mediterranean basin), developing and detailing the Regional Guidelines' chapter about physical environment in the Basque Provinces. According to this, "it has provided in these sensitive areas an integrated treatment with regards to their environmental values, hydraulic problems, and urban potential". As its name indicates, this Plan establishes the spatial order and regulates uses around rivers and streams based on their "three components: the environmental component, the hydraulic component and the urban component". From them is made a typological classification of margins, each one with specific management criteria (DMAPTAP, 2008). This constitutes an excellent example of integration in planning, between hydraulic sectoral aspects and territorial (environmental and urban) aspects (Fig.4).

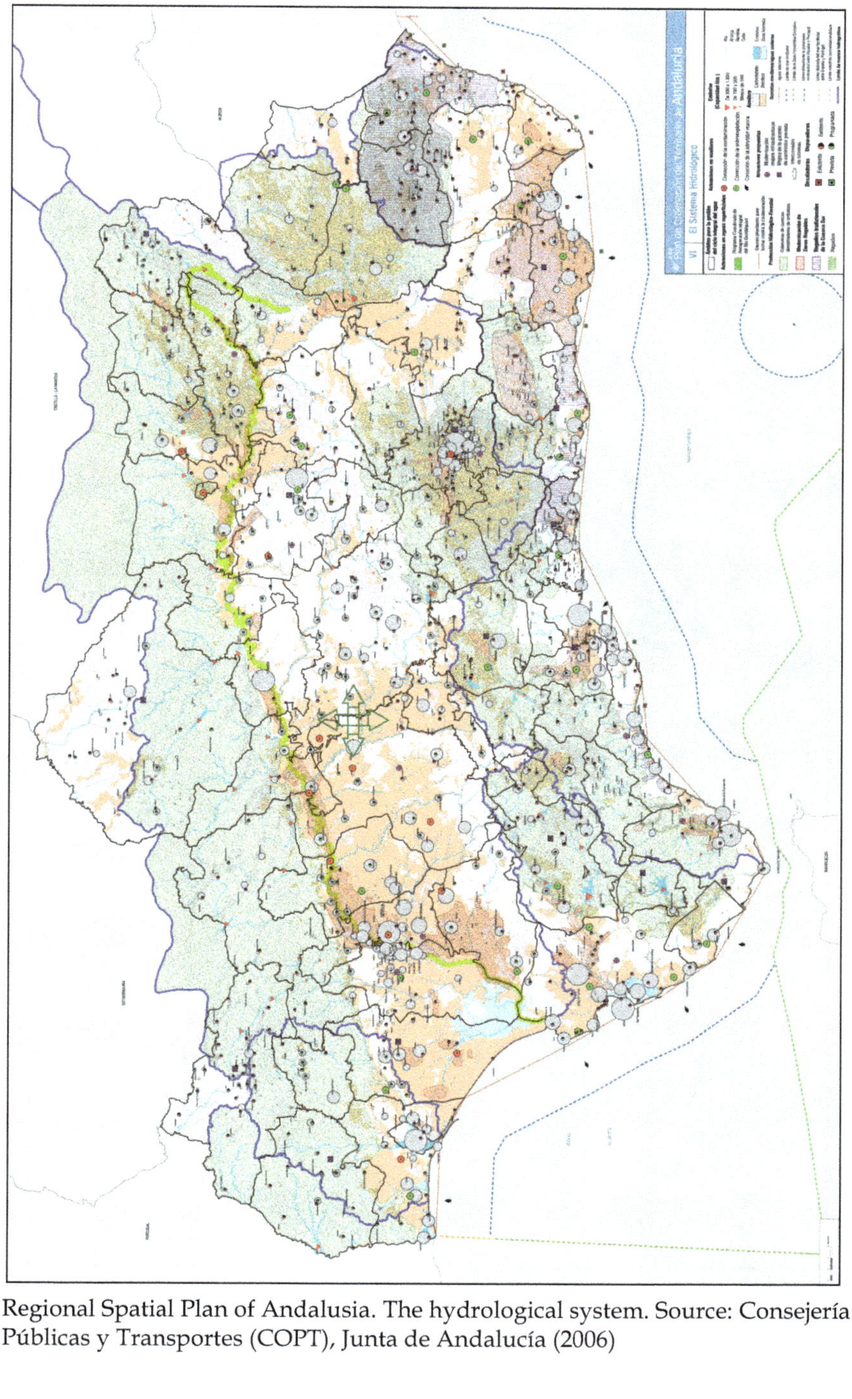

Fig. 1. Regional Spatial Plan of Andalusia. The hydrological system. Source: Consejería de Obras Públicas y Transportes (COPT), Junta de Andalucía (2006)

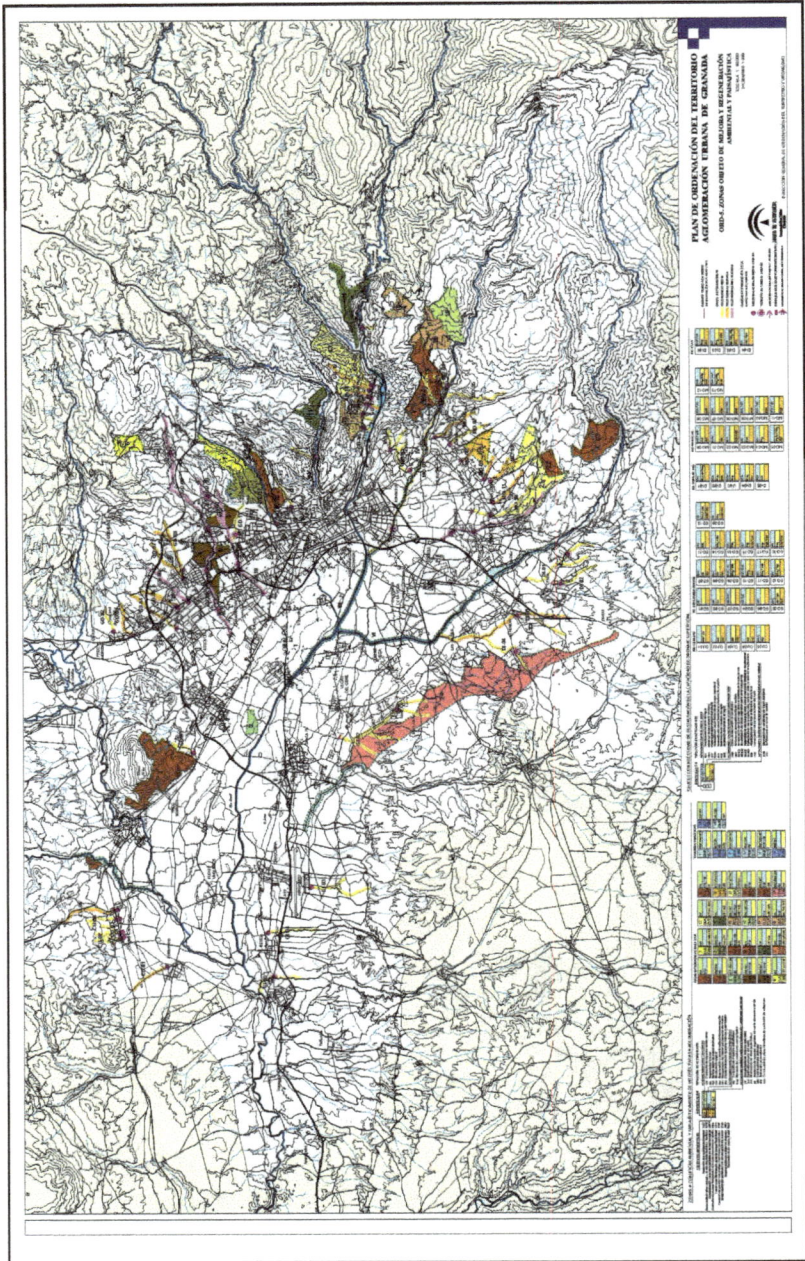

Fig. 2. Sub-regional Spatial Plan of the Agglomeration of Granada. Map of environmental and landscape regeneration areas. Source: Consejería de Obras Públicas y Transportes (COPT), Junta de Andalucía (1999)

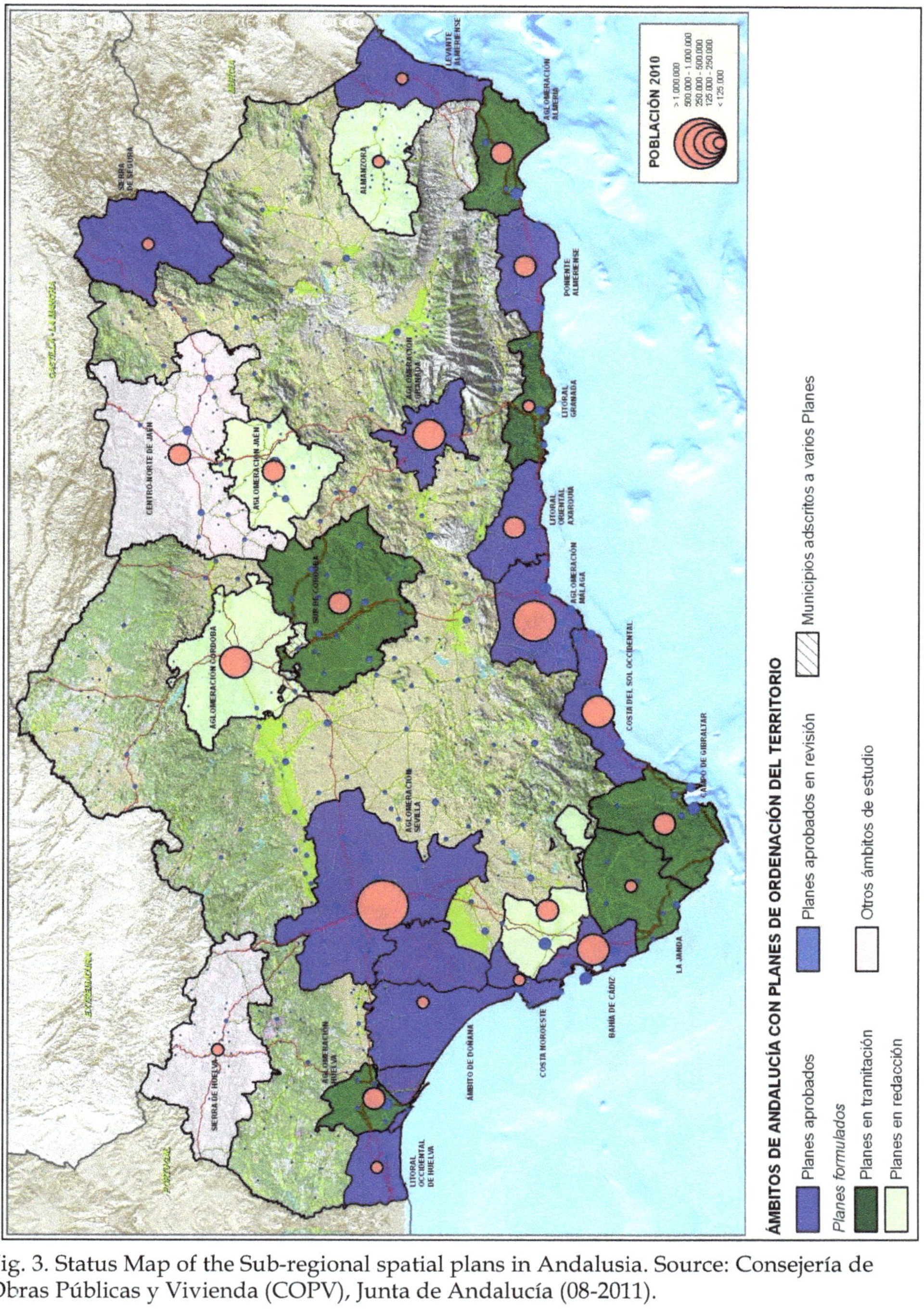

Fig. 3. Status Map of the Sub-regional spatial plans in Andalusia. Source: Consejería de Obras Públicas y Vivienda (COPV), Junta de Andalucía (08-2011).

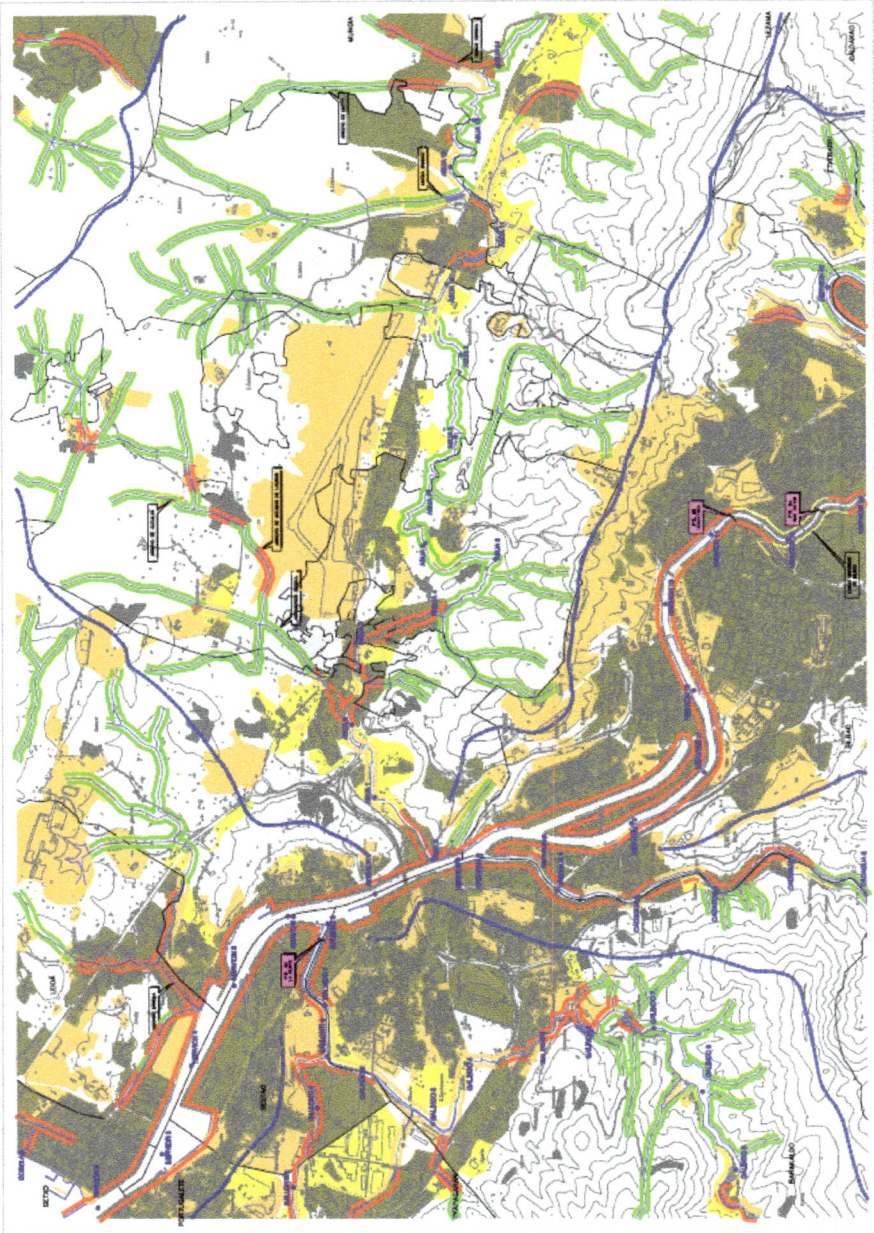

Fig. 4. Territorial Sectoral Plan of the Regulation of the Margins of Rivers and Streams. Sheet 061-II (A-4-08) urban component. Source: Departamento de Medio Ambiente, Planificación Territorial, A.P. (DMAPTAP) Gobierno Vasco (2008).

Another good example of integrated planning is the Territorial Action Plan on the Flood Risk Prevention in the Region of Valencia (2003), which was proposed in the regional planning law (1989). It was developed from flood risk cartography in the second half of the 1990s, and has later been linked to maps of existing and planned urban use. It has also added an action program and regulations for all regional rivers. The proposed actions were based on different measures: structural measures of a remedial character to reduce the actual impact of flooding, hydrological-forestry actions with their dual character of reducing risk and improving the environment, and preventive urban-regional actions which are aimed at avoiding the future impact of flood, and are constituted of land use regulations and building conditions. It defined 278 flood areas, representing 5.4% of regional territory, 73% of municipalities were affected, and 79% of the land affected corresponded to urban areas (Generalitat Valenciana [GVA], 2003) (Fig. 5). It is under revision and since its approval, according to government information (2009), it has prevented the urbanization of 1,428 hectares at risk of flooding.

From a more sectoral perspective Catalonia and Andalusia have both also developed several planning documents. The water agency of Catalonia is developing fluvial space plans, in a detailed analysis of its rivers taking into consideration all stakeholders. The agency has developed a methodology based on analysis of the natural and hydrological processes, to offer a comprehensive and set vision of hydrological-hydraulic, environmental and morphodynamic elements of river space. Their aims are to reach the WFD objectives and to reduce flood risk. There are several fluvial space plans in the process of being drawn up and processed. The Andalusian government, as a first step towards the implementation of a flood prevention plan, approved the flood prevention plan in Andalusian urban riverbeds (2002) (Fig.6). It made a diagnosis of the regional territory to locate vulnerable areas and classify them according to three degrees of risk, and immediate structural works were also proposed to reduce these risks (COPT, 2002) – many of these have been completed.

5. New hydrological planning and its integration with regional and urban planning

The physical environment has, of course, always been a key factor for urban development, and later for regional planning. However hydrological and spatial planning have been autonomous disciplines that have been developed separately, and although water management and spatial planning are clearly and inherently connected they have traditionally been separated for policymaking (Woltjer & Al, 2007). In recent years, and as a result of the implementation of WFD principles, and also the most territorial responsiveness of urban and regional planning fostered under the realization of Strategic Environmental Assessments, little integration has been achieved.

However, although it is evident that water is present in all human activities and any activity and its geographic distribution (even without a direct hydraulic objective) influences its status, generation and circulation, it is accepted today, that the discussion on water policy and planning means putting into question the forms of occupation of the territory and the model of development, and these considerations have needed some decades to be fully accepted, hence, the great complexity as well as the deep territorial significance of the debate on water (Moral, 2009). The approach to water issues still remains strongly sectoral, but softened a little through integration according to the WFD, and from their spatial dimension through urban and regional planning, and similar areas are being promoted.

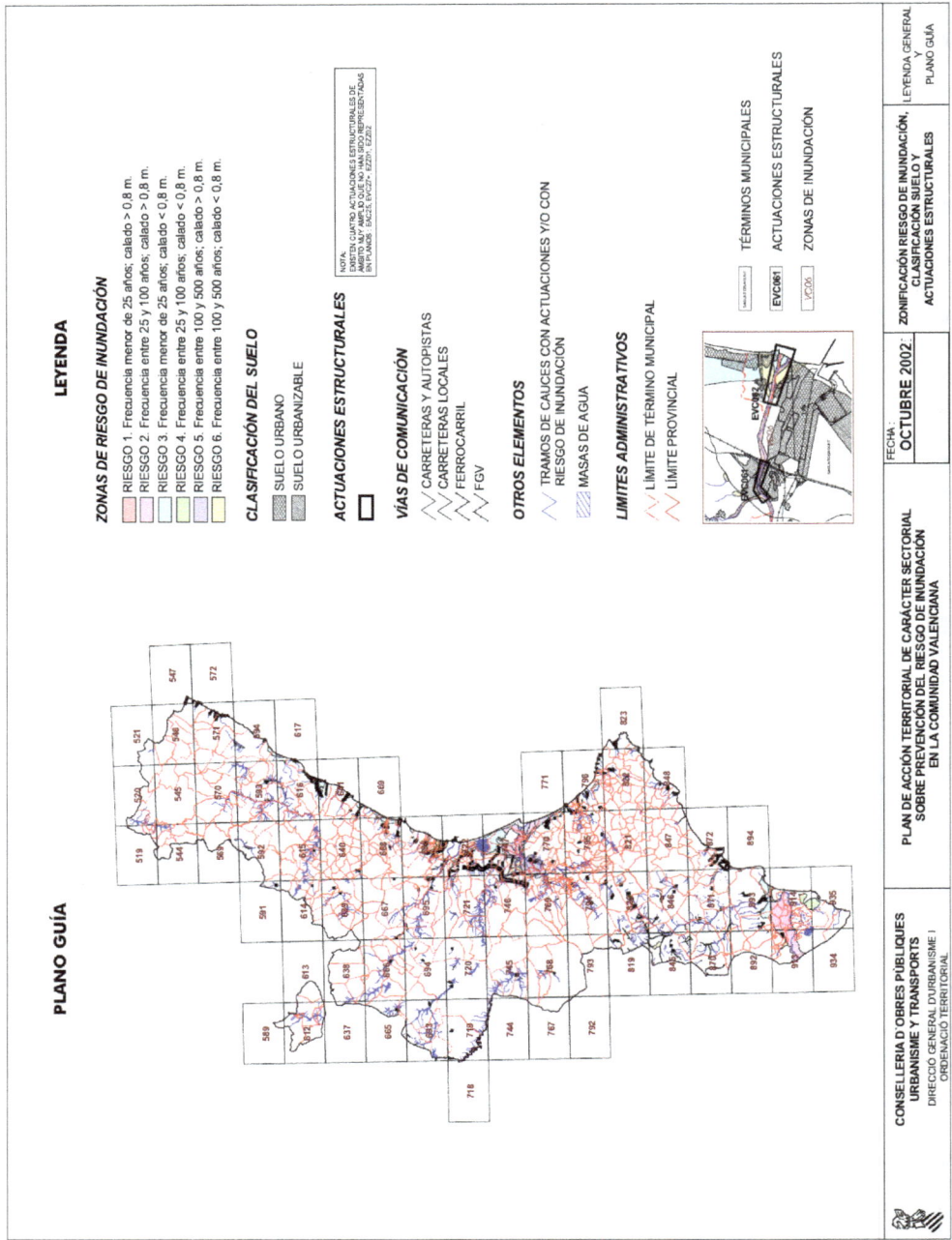

Fig. 5. Territorial Action Plan on the Flood Risk Prevention in the Region of Valencia. Guide map and general legend. Source: Generalitat Valenciana (GVA) (2003).

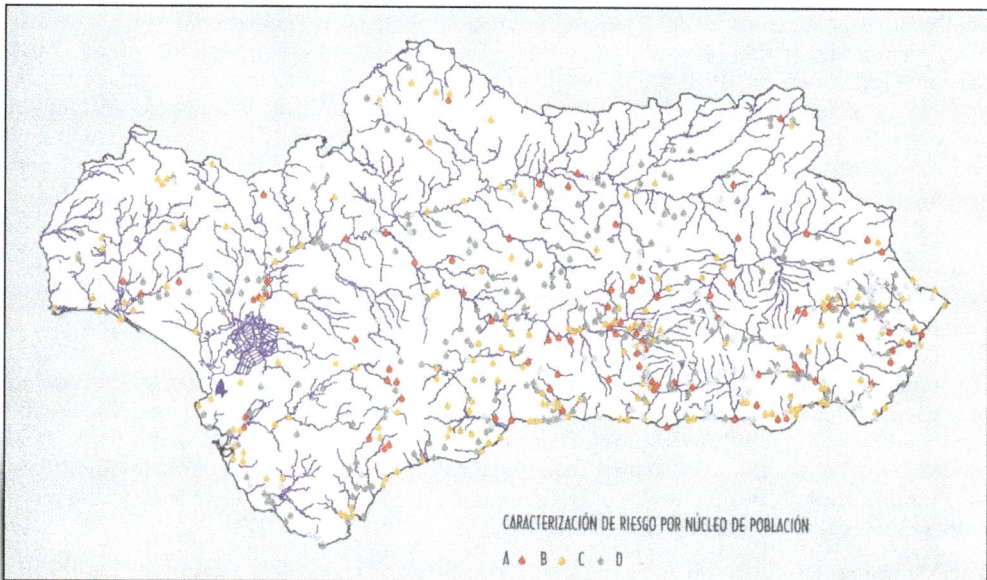

Fig. 6. Flood Prevention Plan in Andalusian urban riverbeds. Risk characterization of cities. Source: Consejería de Obras Públicas y Transportes (COPT), Junta de Andalucía (2002).

From the second half of the 1980s and in the 1990s little consideration was paid in sectoral planning to the potential urban-territorial and environmental effects, and was limited to those considered in the environmental impact assessment. Sectoral planning was prevailing over spatial planning and solely reports. However the necessity of coordination and even integration between different plans was clear.

Some accredited voices called for the strengthening of the water-territory relationship throughout an integrated management water-territory (eg. Gómez Ordóñez, 2006; López Martos 2000). Interactions and interdependencies between water management with spatial planning were obviously increasing with the new discourse of 'accommodating water', in which 'space for the river' was made, not least because of the eroding borderlines between land-use and water functions (Wiering & Immink, 2006). In fact spatial planning has been largely and clearly considered as one of the most effective measures to reduce the negative impacts of floods, restricting or if necessary preventing settlements, facilities and infrastructure in flood plains and areas potentially affected by floods (Berga, 2011; UN/ISDR, 2004; UN/WWDR, 2009). Additionally, well considered spatial planning will upgrade a region's resilience to flooding and its capacity for recovery (Woltjer & Kranen, 2011).

From a spatial planning point of view more complex methodological proposals have been put forward such a "Territorial Planning of Water" from a relational analysis of the hydrological system components and the territorial processes, to the joint planning of infrastructures and land uses, considering resource availability and reducing the environmental impact (Rodríguez Rojas, 2007), later developed to be adapted for WFD implementation (Rodríguez Rojas et al., 2008).

On the other hand, from the hydrological point of view, these reclaimed links between water management and spatial planning or land use planning will come from the Integrated Water Resource Management (IWRM) that has emerged, based on the recognition that the intrinsic complexity of interconnected biophysical, social, economic and political factors can only be addressed by combining and truly integrating social constructivist ideas of participation and empowerment with a cross-disciplinary approach ensuring that social, economic, environmental and technical dimensions are taken into account in the management and development of water resources (Global Water Partnership [GWP], 2000; 2009).

The WFD was the European response to the sustainable and integrated water management exigencies, establishing in 2000 the basic principles of sustainable water policy in the European Union and this has meant a radical change in the management of water resources (Brugge & Rotmans, 2007; Griffiths, 2002; Grimeaud, 2004; Kaika, 2003). One of the principles of the WFD is the integration of sectoral policies as a key factor for the protection of water and its related ecosystems (Moral, 2009, 2010). Its exigent objective of reaching the good status of all aquatic bodies will require significant land use changes (Carter & Howe, 2006; Volk et al., 2009). Thus the WFD is likely to influence spatial planning and prevailing land uses and, although there are no European responsibilities for spatial planning, the EU has gradually been gaining more authority over land use via directives such as this (Voogd & Voltjer, 2009).

The WFD has an approach to integrated river basin management, based on the holistic approach to water management at river basin scale: considering all waters (surface, coastal and groundwater); utilising ecological (biology, hydromorphology and physico-chemistry) and chemical standards and objectives; involving the public in management of river basins; and requiring the use of River Basin Management Planning (RBMP) through a system of analysis and planning based upon the river basin (Griffiths, 2002).

For the implementation of the WFD in Spain from 2003 the traditional river basin authorities were transformed to include, in addition to inland waterways, the coastal and transitional waters. A new administrative body has also been created, the Committee of Competent Authorities (Royal Decree 126/2007) which, in theory, is aimed at cooperation, requiring more than mere coordination between the state administration (the Ministry of Environment) and regional governments (which have responsibilities on both environmental and regional planning), where the hydraulic policy paradigm has changed in the last decades from a centred state administration to a regional government level. However, its regulation does not have an appropriate composition to address the relationships between the state and the autonomous communities, and does not provide a coordinating role, as would be consistent with the concept of integration, but only cooperation (Lopez-Gunn, 2009; Moral 2009, 2010). Thus the political-administrative complexity of the Spanish autonomic system, with some responsibilities shared between the state and the autonomous communities, has led to a lack of coordination in general but also in water policies in particular. The need to enhance and strengthen coordination between regional and state bureaucracies is at the heart of a functioning and effective federal state, and the failure to deliver effective coordination is at the heart of the tensions currently experienced in Spanish territorial politics (Romero Gonzalez 2009, cited in Lopez-Gunn, 2009).

Since the WFD approval, member states have been required to follow a rigorous timetable, establishing a river basin district structure within which demanding environmental objectives have been set, and developing a River Basin Management Plan (RBMP) by 2009 to achieve their objectives. The implementation of the WFD will be integrated with the Strategic Environmental Assessment (SEA) as required by the directives. This will aid the realisation of the WFD's objectives and ultimately could encourage the sustainable use of water resources. As Carter and Howe (2006) show, this will be by strengthening the content of RBMPs, improving the quality and availability of baseline data, the advancement of integrated consultation and public participation procedures, the development of monitoring procedures and, finally, by encouraging the sustainable management of water resources, and therefore of its territorial dimensions.

Given the relevance of achieving the demanding environmental objectives in this new hydrological planning, with particular consideration for the water environment and with attention to protected areas, clearly, a complete integration of the sectional environmental aspects within territorial planning is called for. On the other hand, the need for accurate evaluation of water demands has required close attention to the studies of the new RBMP relating to the existing territorial dynamics, and also the proposed scenarios in the current urban plans, improving the consideration of spatial dimensions on water consumption by the use of demand units and Geographic Information Systems (GIS), and thus providing more integration with spatial planning. Additionally, a higher degree of integration has been reached in the present hydrological planning process thanks to a high level of public participation in contrast with other planning processes (Grindlay et al., 2011).

As mentioned before, one of the hydrological issues where traditionally the relationship with spatial planning has been most clearly seen is in flood management. Generally its treatment has produced a transition from the merely hydraulic measures such as building dykes (separating water from land use), to considering non-structural actions such as the need to give `space for the river', from a defensive to an integrated approach (Jonge, 2009), and also usually when all possible structural actions had been made and the problem remained unsolved. Increased flooding has also been more in evidence and, above all, damages generated are due to the alteration of natural systems by humans and to the degradation and loss of hydrological integrity of catchment areas (Blackwell & Maltby, 2006). It is been developed at present, land use change and scenario modelling using GIS for flood risk mitigation and to improve the integration of hydrological and spatial planning (Feyen et. al, 2009; Barredo & Engelen, 2010).

As in other countries spatial planning instruments have traditionally been ineffective in controlling developments in floodplains. In fact, as Moss and Monstadt (2008) recognize, "planning regulations to limit urban development on floodplains have in the past generally failed to halt the loss of existing floodplains, let alone encourage the avoidance or even removal of settlements and the restoration of flood plains".

Nevertheless spatial planning policy acts in holistic terms in the sense that it analyzes phenomena and developments by using a contextual perspective, such as a landscape, a region, or an area. Therefore in relation to the treatment of safety and risk around floods, the normative discourse of spatial planners reflects the more open and flexible perspective. Risk assessment, in this view, deals with complex relations between social, physical, and spatial aspects of a particular place or context. As Wiering and Immink (2006) demonstrate, the strategy of spatial planners is focused on the establishment of collaborative planning processes in which different visions of spatial development are communicated and in which the issues are addressed by multiple actors. They are also searching for new (multi)functional and social relations between water and several types of `environments' or contexts, by seeking new links between water and other spatial functions, and considering flood hazards to be relative, because the potential impacts of these risks depend on the presence of spatial functions in a particular landscape (differentiated in urban and rural areas) as well as the perceptions of both the place and the risk by people living and working there. The traditional, more restricted sectoral perspective of water management is evolving into new comprehensive concepts such as river-basin management, as well as new environmental and rural initiatives with area-specific policies, similar to the WFD proposals.

However the flood question was not completely addressed in the WFD, and later, another Directive to complement and consolidate it was needed. The new Directive "on the assessment and management of flood risks" (EC, 2007) is focused on flood risk assessment, the preparation of flood hazard maps and flood risk maps, and establishing flood risk management plans, but gives a merely sectoral treatment of this question which then will have a direct influence on spatial planning. Naturally the importance of land use policies and their potential impact on flood risk and its management (agriculture, deforestation, urbanization, land movements, etc.) is recognized. However spatial planning is only mentioned with reference to being taken into account by the new flood risk management plans. It is also stated that these plans may "include the promotion of sustainable land use practices, improvement of water retention, as well as the controlled flooding of certain areas in the case of a flood event", but all of these actions should be put forward in a spatial plan, though this is not mentioned to avoid conflict of responsibilities. They should also focus "on prevention, protection and preparedness", and it is not only with, as mentioned, "a view to giving rivers more space" or "maintenance and/or restoration of floodplains", indicating means, where possible "to prevent and reduce damage to human health, the environment, cultural heritage and economic activity", but also taking into consideration all territorial characteristics around watercourses so it should be provided by or coordinated with, according to the case, spatial planning. However coordination is only mentioned in the Directive in relation to "the reviews of the river basin management plans" required by the WFD, which "shall be carried out with", "and may be integrated into" them.

This Directive has recently passed into Spanish legislation (R.D. 903/2010), and it recognizes the necessity of non-structural actions such as civil protection plans, implementation of early warning systems (AHIS), hydrological and forest watershed corrections, in addition to spatial planning measures, but complementing traditional structural solutions, which have proved inadequate in certain cases. It is also indicated, as mentioned above, that the state land law (2/2008) confers the legal status of rural land to any land which is susceptible to flood risk, and the prevention of such risks is required, thus the need to include maps of natural risks in the spatial planning, and reports from the hydrological administration concerning the protection of the public water domain. Some of the instruments required by the Directive had already been considered in the revised regulation of the hydraulic public domain (R.D. 9/2008), which created the National System of Cartography of Flood Zones, and also by regional water laws, such as in Andalusia (9/2010).

In this Spanish regulation, coordination is not only claimed to be implemented in the general water policy of all basins, implicitly including their RBMP, as indicated in the Directive, but is also extended to urban and regional planning as necessary for the effective prevention and protection described in the current standard. Explicitly article 15 regarding "coordination with other plans", states that "urban and regional plans in their land use regulation should not include directives that are incompatible with the contents of the flood risk management plans, and should recognize the rural character of the lands at risk of flooding or other serious accidents" (R.D. 9/2008). This will prevent discrepancies between hydrological and spatial plans in such an important aspect, and the necessary coordination between the state and the administrations of the autonomous community administrations will be definitive, at least on this point.

In addition the relationship with spatial planning has advanced as the content of future plans for flood risk management has been incorporated within urban and regional planning measures, including as a minimum: "limitations on land use for the floodable area with reference to different flood hazard scenarios, and the criteria for considering an area as not to be developed, and construction criteria required for buildings located in a flood zone". It also incorporates "the measures envisaged for adapting existing urban planning to the criteria outlined in the flood risk management plan, including the possibility to remove existing buildings or facilities that pose a serious risk, and this expropriation shall be considered as public utility" (R.D. 9/2008). Specifically this will resolve many cases of illegal and flood exposed constructions all over the country. For the first time a sectoral planning document makes explicit reference to contents regarding urban and regional planning.

With reference to urban planning, one of the fundamental coordination mechanisms is the report required by hydraulic administration, according to state and regional water laws, and later to state land law, about the water status or domain (Moral, 2009). Additionally recommendations of the regional hydraulic administration are given for new urban planning documents, and particularly on correct consideration of the hydraulic public domain and groundwater limits, and their protection areas; delimitation of the floodplains and areas at flood risk and land use regulations; adequate and assured quality and quantity of urban water supply, and availability of water resources to satisfy urban plan expectations; adequate urban wastewater treatment both current and future, taking into account expected urban growth; the consideration in urban plans of flood plains as transitional elements between the natural and built environment is also recommended, and the need to assign to them functions compatible with flood evacuation and the citizens' enjoyment of the water environment (CMA-AAA, 2009).

European spatial planning policy guidelines of the ESDP have been expanded by the Territorial Agenda (2007) (towards a European Action Programme for Spatial Development and Territorial Cohesion) to six priorities for spatial development measures, among which are "promotion of trans-European risk management including the impacts of climate change" and the "strengthening of ecological structures and cultural resources as added value for development". A study for the European Parliament about EU policies with a territorial dimension recognized that the WFD has "significant implications for spatial planning" and "restricts planning for urban development" because, "with its aim of reducing pollution, preserving protected areas and restoring and enhancing bodies of surface water it may easily conflict with urban growth strategies". But it also pointed out that "in spite of the implications for territorial development of environmental policy, overall coordination between the two does not exist, sometimes leading to policy incoherence" (EP, 2007).

It has been recommended that the territorial dimension be included in the common Strategic Environmental Assessment (SEA), which "contributes to improving and assessing different alternatives and should consequently increase the degree of positive coordination" (EP, 2007). Even better than this would be the use of Territorial impact assessment techniques for the integration of territorial aspects in sectoral policies (Golobic & Marot, 2011), such as the proposed projects on hydrological planning.

6. Conclusions and indications to improve the integration of hydrological and territorial planning

The difficulties in the relationship of a long history between sectoral and spatial planning have been seen, particularly with hydrological planning, and its evolution.

After this long history of spatial planning, however, little attention has been paid to water issues. Naturally the physical environment has been a key factor for urban development (with topography and drainage as the foundation for these developments), but hydrological and urban and regional planning have been autonomous disciplines that have been developed separately. Only in recent years, as a result of the implementation of the WFD, a certain degree of integration has been reached. Now it has been accepted that water policy and planning is directly affected by the forms of occupation of the territory and the model of development and, conversely, this is limited by the water environment and its availability.

For both hydraulic and spatial planning administrations, the need for the coordination of their respective planning documents has been clearly shown, and some mechanisms have already been established for this, such as mandatory reports, but integration means going beyond the mere requirement to issue reports on the plans of other administrations for mutual problems. This integration appears particularly visibly in the current treatment of flood risk areas, not evident in the European directive, but which have been passed into Spanish legislation. In the recent Spanish regulation on the assessment and management of flood risks, coordination is not only claimed to be implemented in the general water policy of all basins, but is also extended to urban and regional planning as necessary for effective prevention and protection. Regarding this coordination with other plans, it is explicitly stated that "urban and regional plans in their land use regulation should not include directives that are incompatible with the contents of the flood risk management plans, and should recognize the rural character of the lands at risk of flooding or other serious accidents". This will prevent discrepancies between hydrological and spatial plans in such an important aspect, and the necessary coordination between the state and the autonomous community administrations will be definitive, at least on this point. Additionally the relationship with spatial planning has advanced as the content of future plans for flood risk management has been incorporated within urban and regional planning measures, including as a minimum: "limitations on land use for the floodable area with reference to different flood hazard scenarios, and the criteria for considering an area as not to be developed, and construction criteria required for buildings located in a flood zone". It also incorporates "the measures envisaged for adapting existing urban planning to the criteria outlined in the flood risk management plan, including the possibility to remove existing buildings or facilities that pose a serious risk, and this expropriation shall be considered as public utility". In particular this will resolve many cases of illegal and flood exposed constructions all over the country.

However both disciplines remain autonomous and greater integration is required. The current implementation of the WFD ensures a certain advance in integration with the territorial dimension due to the broad consideration of environmental aspects, the uses and demands for water, public participation in the planning process and even the land uses with regards to their effects on the water environment. One of the principles of the WFD is the integration of sectoral policies as a key factor for the protection of water and its related ecosystems. In this sense, the WFD is likely to influence spatial planning and prevailing land uses in and around water environments, given its objective of reaching the good status of all aquatic bodies that will require significant land use changes. According to the relevance of achieving the demanding environmental objectives in the current hydrological planning, with particular consideration for the water environment and with attention to protected areas, clearly, a complete integration of the sectional environmental aspects within territorial planning is called for, but all land uses should be also considered. Moreover, the need for accurate evaluation of water demands has required close attention to the studies of the new hydrological plans relating to the existing territorial dynamics, and also the proposed scenarios in the current urban plans, improving the consideration of spatial dimensions on water consumption by the use of demand units and Geographic Information Systems (GIS), and thus providing more integration with spatial planning.

The recent recommendations of the hydraulic administration for new urban planning documents are a minimum basis to integrate spatial and hydrological plans, and they should be comprehensively considered, specifically: the correct consideration of the hydraulic public domain and groundwater limits, and their protection areas; the delimitation of the floodplains and areas at flood risk and land use regulations; the adequate and assured quality and quantity of urban water supply, and the availability of water resources to satisfy urban plan expectations; the adequate urban wastewater treatment both current and future, taking into account expected urban growth; and the consideration in urban plans of flood plains as transitional elements between the natural and built environment, and the need to assign to them functions compatible with flood evacuation and the citizens' enjoyment of the water environment.

Additionally, regional spatial planning now has to deal with the difficulty of establishing future scenarios with an increasing uncertainty, and it is becoming more strategic and flexible in character, adopting a broad consensus about prime objectives, as established for water issues in the WFD. However it is necessary to advance with a more strategic and broader perspective, introducing water considerations into regional planning, and vice versa, jointly considering an explicit territorial model or strategy of territorial development, for greater improvement of regional potential and territorial capital.

7. References

Barredo, J.I. & Engelen, G. (2010). Land use scenario modelling for flood risk mitigation. *Sustainability*, No.2, pp. 1327-1344, ISSN 2071-1050

Benavent Fernández de Córdoba, M. (2006). *La Ordenación del Territorio en España. Evolución del concepto y de su práctica en el S.XX*, Universidad de Sevilla, COPT, ISBN 978-84-472-0869-2, Sevilla, Spain

Benabent Fernández de Córdoba, M. (2009). Los planes de ordenación del territorio en España. De la instrumentación a la gestión. In: *Agua, territorio y paisaje: de los instrumentos programados a la planificación aplicada: V Congreso Internacional de*

Ordenación del Territorio, Sánchez Pérez-Moneo, L. & Troitiño Vinuesa, M.A. (coords.), pp. 143-158, FUNDICOT ISBN 978-84-691-9145-3, Málaga, Spain

Berga Casafont, L. (2011). Las inundaciones en España. La nueva Directiva Europea de inundaciones. Floods in Spain. The new European Directive on floods. *Revista de Obras Públicas*, Vol.158, No.3520, pp. 7-18, ISSN 0034-8619

Biswas, A.K. & Tortajada, C. (2003). An Assessment of the Spanish National Hydrological Plan. *Water Resources Development*, Vol. 19, No. 3, pp. 377–397, ISSN 0790-0627

Blackwell, M.S.A. & Maltby, E. (2006). *Ecoflood Guidelines. How to use floodplains for flood risk reduction*, European Commision, Ecoflood project, Luxembourg, 22.08.2011, Available from
http://levis.sggw.waw.pl/ecoflood/contents/Guidelines(draft_2005-10-10).pdf

Brugge, R. Van der & Rotmans, J. (2007). Towards transition management of European water resources. *Water Resource Management*, Vol.21, No.1, p. 249–267, ISSN: 0920-4741

Carmona González, P. & Ruiz Pérez, J.M. (2000). Las inundaciones de los ríos Júcar y Turia. *Serie Geográfica*, No.9, pp. 49-69, ISSN 1136-5277

Carter, J. & Howe, J. (2006). The Water Framework Directive and the Strategic Environmental Assessment Directive: Exploring the linkages. *Environmental Impact Assessment Review*, Vol. 26, Iss.3, pp. 287–300, ISSN: 0195-9255

Consejería de Medio Ambiente (CMA) (2006). *Plan Director de Riberas de Andalucía*, Junta de Andalucía, Sevilla, 22.08.2011, Available from
http://www.juntadeandalucia.es/medioambiente/web/Bloques_Tematicos/Estra tegias_Ambientales/Planes/Planes_tematicos/plan_director_riberas/riberas01.pdf

Consejería de Medio Ambiente. Agencia Andaluza del Agua (CMA-AAA) (2009). *Recomendaciones sobre el contenido mínimo de los instrumentos de planeamiento urbanístico en materia de aguas*, Junta de Andalucía, Sevilla

Consejería de Obras Públicas y Transportes (2002). *Plan de Prevención de avenidas e inundaciones en cauces urbanos andaluces*, Junta de Andalucía, Sevilla, 12.09.2011, Available from
http://www.juntadeandalucia.es/medioambiente/web/Bloques_Tematicos/agen cia_andaluza_agua/gestion/gestion_agua_andalucia/planificacion/plan_prevenci on_inundaciones/plan_avenidas.pdf

Consejería de Obras Públicas y Transportes (COPT) (2006). *Plan de Ordenación del Territorio de Andalucía*, Junta de Andalucía, Sevilla, 12.09.2011, Available from
http://www.juntadeandalucia.es/obraspublicasyvivienda/estaticas/sites/conseje ria/areas/ordenacion/documentos/POTA_WEB.pdf

Consejería de Obras Públicas y Transportes (COPT) (2008). *Plan de Infraestructuras para la Sostenibilidad del Transporte en Andalucía 2007-2013*, Junta de Andalucía, Sevilla, Available from
http://www.juntadeandalucia.es/obraspublicasyvivienda/estaticas/sites/conseje ria/areas/transportes_infraestructuras/documentos/pista.pdf

Departamento de Medio Ambiente, Planificación Territorial, Agricultura y Pesca (DMAPTAP) (2008). *Plan Territorial Sectorial de Ordenación de los Ríos y Arroyos de la Comunidad Autónoma del País Vasco*, Gobierno Vasco, Vitoria, 12.09.2011, Available from http://www.ingurumena.ejgv.euskadi.net/r49-565/es/contenidos/informacion/pts_rios_arroyos/es_pts/indice_c.html

Departamento de Medio Ambiente, Planificación Territorial, Agricultura y Pesca (DMAPTAP) (2011). *Planeamiento territorial Sectorial,* Gobierno Vasco, Vitoria, 12.09.2011, Available from http://www.ingurumena.ejgv.euskadi.net/r49-565/es/contenidos/informacion/pts/es_1161/pts_c.html

Downward, S.R. & Taylor, R. (2007). An assessment of Spain's Programa AGUA and its implications for sustainable water management in the province of Almeria, southeast Spain. *Journal of Environmental Management,* Vol.82, Iss.2, pp. 277–289, ISSN 0301-4797

Dühr, S., Colomb, C., & Nadin, V. (2010). *European Spatial Planning and Territorial Co-operation,* Routledge, ISBN 978-0-415-46773-5, Abingdon, UK

European Communities (1999). *ESDP - European Spatial Development Perspective. Towards Balanced and Sustainable Development of the Territory of the European Union,* Office for Official Publications of the European Communities, ISBN 92-828-7658-6, Luxembourg.

European Parliament and Council (EC) (2000). *Directive 2000/60/EC, of 23 October 2000, establishing a framework for Community action in the field of water policy.*

European Parliament and Council (EC) (2006). *Directive 2006/118/EC, of 12 December 2006 on the protection of groundwater against pollution and deterioration.*

European Parliament and Council (EC), (2007). *Directive 2007/60/EC, of 23 October 2007, on the assessment and management of flood risks.*

European Parliament (EP) (2007). *Follow-up of the Territorial Agenda and the Leipzig Charter: towards a European Action Programme for spatial development and territorial cohesion,* Publications office, Brussels, Available from http://www.europarl.europa.eu/meetdocs/2004_2009/documents/dv/territorial agenda-leipzig_charter_/TerritorialAgenda-Leipzig_Charter_en.pdf

Faludi, A. (ed.) (2002). *European Spatial Planning,* Lincoln Institute of Land Policy, ISBN 1-55844-153-0, Cambridge, MA; USA

Faludi, A. & Waterhout, B. (2002). *The making of the European Spatial Development Perspective: no masterplan,* Routledge, ISBN 0-415-27263-7, London, UK

Feyen, L, Barredo, J.I., & Dankers, R. (2009) Implications of global warming and urban land use change on Flooding in Europe. In: *Water & Urban Development Paradigms - Towards an integration of engineering, design and management approaches,* J Feyen, K. Shannon & M Neville (eds.), pp. 217-225, CRC Press, ISBN 978-0-415-48334-6, Balkema, NL

Font, N. & Subirats, J. (2010). Water management in Spain: the role of policy entrepreneurs in shaping change. *Ecology and Society,* Vol.15, No.2, Art.25, ISSN 1708-3087, Available from http://www.ecologyandsociety.org/vol15/iss2/art25/

Friedmann, J. & Weaver, C. (1979). *Territory and function: the evolution of regional planning,* University of California Press, ISBN 0-520-04105-4, Berkeley, USA

García-Bellido García de Diego, J. (2000). Ildefonso Cerdà y el nacimiento de la Urbanística: La primera propuesta disciplinar de su estructura profunda. *Scripta Nova. Revista Electrónica de Geografía y Ciencias Sociales,* No.61, Available from http://www.ub.edu/geocrit/sn-61.htm

Garrido, A. & Llamas, M.R. (2009). Water Management in Spain: An Example of Changing Paradigms. In: *Policy and Strategic Behaviour in Water Resource Management*, Dinar, A. & Albiac, J. (eds.), pp. 125-144, Earthscan, ISBN 978-1-84407-669-7, London, UK

Geddes, P. (1915). *Cities in evolution: an introduction to the town planning movement and to the study of civics*, Williams & Norgate, London, UK

Generalitat Valenciana (GVA) (2003). *Plan de Acción Territorial de carácter sectorial sobre prevención del Riesgo de Inundación en la Comunidad Valenciana, PATRICOVA*, Valencia, 12.09.2011, Available from
http://www.cma.gva.es/areas/urbanismo_ordenacion/infadm/publicaciones/pd f_patricova/indice.htm

Global Water Partnership (GWP) (2000). *Integrated water resources management*, TAC Background Paper 4, GWP Secretariat, Stockholm.

Global Water Partnership (GWP) (2009). *A Handbook for Integrated Water Resources Management in Basins*, Sweden. Available from
http://www.gwpforum.org

Golobic, M. & Marot, N. (2011). Territorial impact assessment: Integrating territorial aspects in sectoral policies. *Evaluation and Program Planning*, Vol.34, Iss.3, pp. 163–173, ISSN 0149-7189

Gómez Ordóñez, J.L. 2006. La Cuenca Hidrográfica y la Ordenación del Territorio. In: *Ciencia, técnica y ciudadanía, claves para una gestión sostenible del agua. Actas del IV Congreso Ibérico sobre Gestión y Planificación de Aguas*, Ibáñez Martí, C. & Prat Fornells, N. (coord.), pp. 200-208, Fundación Nueva Cultura del Agua, ISBN 978-84-7820-872-2, Zaragoza, Spain

Gómez Ordóñez, J.L., & Grindlay Moreno, A.L., (2008). *Agua, Ingeniería y Territorio: La Transformación de la Cuenca del río Segura por la Ingeniería Hidráulica*, Confederación Hidrográfica del Segura, NIPO 777-08-001-9, Murcia, Spain

Grindlay Moreno, A.L. (2007). La planificación del territorio y de las infraestructuras. In: *Organización y gestión de proyectos y obras*, Martínez Montes, G. & Pellicer Armiñana, E. (coord.), pp. 165-186, McGraw-Hill Interamericana, ISBN 978-84-481-564-11, Madrid, Spain

Grindlay, A.L., Zamorano, M., Rodríguez, M.I., Molero, E., Urrea, M.A. (2011). Implementation of the European Water Framework Directive: Integration of hydrological and regional planning at the Segura River Basin, southeast Spain. *Land Use Policy*, Vol.28, Iss.1, pp. 242–256, ISSN 0264-8377

Griffiths, M. (2002). The European Water Framework Directive: An Approach to Integrated River Basin Management. *European Water Management Online*, Available from
http://citeseerx.ist.psu.edu/viewdoc/download?doi=10.1.1.130.2753&rep=rep1&t ype=pdf

Grimeaud, D. (2004). The EC Water Framework Directive – An Instrument for Integrating Water Policy. *Review of European Community & International Environmental Law*, Vol.13, Iss.1, pp. 27-39, ISSN 1467-9388

Gupta, J. & Van der Zaag, P. (2008). Interbasin water transfers and integrated water resources management: where engineering, science and politics interlock. *Physics and Chemistry of the Earth*, Vol.33, Iss.1–2, pp. 28–40, ISSN: 1474-7065

Hall P.G. (2002). *Urban and Regional Planning*, Routledge, ISBN 0-415-21776-8, London, UK

Jiménez, N. & Martínez-Gil, J. 2005. The New Culture of Water in Spain: a philosophy towards a sustainable development. E-Water 2005-07. *European Water Association*, Available from

http://www.ewaonline.de/journal/2005_07.pdf

Jonge, V.N. de (2009). From a defensive to an integrated approach. In: *Water policy in the Netherlands: integrated management in a densely populated delta*, Reinhard, S. & Folmer, H. (ed.), pp. 17-46, Resources for the Future Press, ISBN 978-1-933115-73-3, Washington, USA

Kaika, M. (2003). The WFD: a new directive for a changing social, political and economic European framework. *European Planning Studies*, Vol.11, No.3, pp. 299–316, ISSN 0965-4313

Lopez-Gunn, E. (2009). Agua para todos: A new regionalist hydraulic paradigm in Spain. *Water Alternatives* Vol.2, Iss.3, pp. 370-394, ISSN 1965-0175

López Martos, J. (2000). Agua y territorio. *OP*, No.50, pp. 46-53, ISSN 0213-4195

McHarg, I.L. (1969). *Design with nature*. The Natural History Press, ISBN 0-471-55797-8, Garden City, NY, USA.

Marshall, T. 1993. Regional environmental planning: Progress and possibilities in Western Europe. *European Planning Studies*, Vo.1, Iss.1, pp. 69-90, ISSN 0965-4313

Mataran, A., España, M., & Fernández, A. (2010). El Planeamiento Territorial en la cuenca y su relación con el agua. In: *Infraestructuras y Políticas del Agua y su relación con el territorio del Guadalquivir*, Grindlay Moreno, A.L. & Matarán Ruiz, A. (invs. ppales.), pp. 61-93, Confederación Hidrográfica del Guadalquivir (MMAMRM), ISBN 978-84-9915-184-7, Sevilla, Spain

Moral Ituarte, L. del (2009). Nuevas tendencias en gestión del agua, ordenación del territorio e integración de políticas sectoriales. *Scripta nova Revista Electrónica de Geografía y Ciencias sociales*, Vol.13, No.285, ISSN: 1138-9788, Available from

http://www.ub.edu/geocrit/sn/sn-285.htm

Moral Ituarte, L. del (2010). Changing discourses in a modern society. In: *Water policy in Spain*, Garrido, A. & Llamas, M.R. (eds), pp. 85-93, Taylor and Francis Group, ISBN 978-0-415-55411-4. London, UK

Moss, T. & Monstadt, J. (2008). Coping with complexity: lessons for policy development, project management and research. In: *Restoring Floodplains in Europe: Policy Contexts and Project Experiences*, Moss, T. & Monstadt, J., pp. 317-337, IWA Publishing, ISBN 9781843390909, London, UK

Olcina Cantos, J. (2007). *Riesgos naturales y ordenación del territorio en España*, Fundación Instituto Euromediterráneo del Agua, ISBN 978-84-933127-4-9, Murcia, Spain

Rodríguez Rojas M.I. (2007). *Planificación Territorial del Agua en la Región del Guadalfeo. Tesis Doctoral*, Universidad de Granada, Available from

http://hdl.handle.net/10481/1747

Rodríguez Rojas, Mª I., Grindlay Moreno, A.L., Molero Melgarejo, E. (2008). Gestión Integrada del Agua y el Territorio, una propuesta metodológica para la adaptación a la DMA. In: *VI Congreso Ibérico de Gestión y Planificación del agua*, Fundación Nueva Cultura del Agua, Available from

http://www.fnca.eu/congresoiberico/documentos/c0304.pdf

Romero González, J. (2009). *Geopolítica y gobierno del territorio en España*. Tirant lo Blanch, ISBN 978-84-9876-478-9, Valencia, Spain

Terán Troyano, F. de (1978). *Planeamiento urbano en la España contemporánea.* Gustavo Gili, ISBN 978-84-252-0711-2, Barcelona, Spain

UN International Strategy for Disaster Reduction (UN/ISDR) (2004). *Living With Risk: A Global Review of Disaster Reduction Initiatives,* Inter- Agency Secretariat of the International Strategy for Disaster Reduction, New York, Available from http://www.unisdr.org/we/inform/publications/657

UN World Water Development Report (UN/WWDR) (2009). *Water in a Changing World,* World Water Assessment Programme (WWAP) 3rd UNWater, UNESCO, Available from http://www.unesco.org/water/wwap/wwdr/wwdr3/tableofcontents.shtml

Voogd, H. & Voltjer, J. (2009). Water policy and spatial planning. Linkages between water and land use. In: *Water policy in the Netherlands: integrated management in a densely populated delta,* Reinhard, S. & Folmer, H. (eds.), pp. 185-203, Resources for the Future, ISBN 978-1-933115-73-3, Washington, USA

Volk, M., Liersch, S., & Schmidt, G. (2009). Towards the implementation of the EuropeanWater Framework Directive? Lessons learned from water quality simulations in an agricultural watershed. *Land Use Policy,* Vol.26, Iss.3pp. 580–588, ISSN 0264-8377

Weitz,J. & Seltzer,E. (1998). Regional Planning and Regional Governance in the United States 1979-1996. *Journal of Planning Literature,* Vol.12, No3, pp. 361-392, ISSN: 0885-4122

Wheeler, S.M. & Beatley, T. (ed.) (2009). *The Sustainable Urban Development Reader,* Routledge, ISBN 978-0-415-45381-3, Abingdon, UK

Wiering, M. & Immink, I. (2006). When water management meets spatial planning: a policy-arrangements perspective. *Environment and Planning C: Government and Policy,* Vol. 24, Iss.3, pp. 423-438, ISSN 0263-774X

Woltjer, J. & Al, N. (2007). Integrating water management and spatial planning. *Journal of the American Planning Association,* Vol.73. Iss.2, pp. 211–222, ISSN 0194-4363

Woltjer, J. & Kranen, F. (2011). Articulating resilience in flood risk management and spatial planning. *25th ICID European Regional Conference, Deltas in Europe. Integrated water management for multiple land use in flat coastal areas.* Available from http://www.icid2011.nl/files/pdf/Paper%20III-23%20Woltjer%20et%20al.pdf

WWF (2002). *Seven reasons why WWF opposes the Spanish National Hydrological Plan, and suggested actions and alternatives,* WWF Position Paper, Available from http://assets.panda.org/downloads/sevenreasonswhywwfopposessnhp.pdf

Zamorano, M., Molero, E., Hurtado, A., Grindlay, A. & Ramos A. (2008). Evaluation of a municipal landfill site in Southern Spain with GIS-aided methodology. *Journal of Hazardous Materials,* Vol.160. Iss.2-3, pp. 473-481, ISSN: 0304-3894

Permissions

The contributors of this book come from diverse backgrounds, making this book a truly international effort. This book will bring forth new frontiers with its revolutionizing research information and detailed analysis of the nascent developments around the world.

We would like to thank Dr. Jaroslav Burian, for lending his expertise to make the book truly unique. He has played a crucial role in the development of this book. Without his invaluable contribution this book wouldn't have been possible. He has made vital efforts to compile up to date information on the varied aspects of this subject to make this book a valuable addition to the collection of many professionals and students.

This book was conceptualized with the vision of imparting up-to-date information and advanced data in this field. To ensure the same, a matchless editorial board was set up. Every individual on the board went through rigorous rounds of assessment to prove their worth. After which they invested a large part of their time researching and compiling the most relevant data for our readers. Conferences and sessions were held from time to time between the editorial board and the contributing authors to present the data in the most comprehensible form. The editorial team has worked tirelessly to provide valuable and valid information to help people across the globe.

Every chapter published in this book has been scrutinized by our experts. Their significance has been extensively debated. The topics covered herein carry significant findings which will fuel the growth of the discipline. They may even be implemented as practical applications or may be referred to as a beginning point for another development. Chapters in this book were first published by InTech; hereby published with permission under the Creative Commons Attribution License or equivalent.

The editorial board has been involved in producing this book since its inception. They have spent rigorous hours researching and exploring the diverse topics which have resulted in the successful publishing of this book. They have passed on their knowledge of decades through this book. To expedite this challenging task, the publisher supported the team at every step. A small team of assistant editors was also appointed to further simplify the editing procedure and attain best results for the readers.

Our editorial team has been hand-picked from every corner of the world. Their multi-ethnicity adds dynamic inputs to the discussions which result in innovative outcomes. These outcomes are then further discussed with the researchers and contributors who give their valuable feedback and opinion regarding the same. The feedback is then collaborated with the researches and they are edited in a comprehensive manner to aid the understanding of the subject.

Apart from the editorial board, the designing team has also invested a significant amount of their time in understanding the subject and creating the most relevant covers. They scrutinized every image to scout for the most suitable representation of the subject and create an appropriate cover for the book.

The publishing team has been involved in this book since its early stages. They were actively engaged in every process, be it collecting the data, connecting with the contributors or procuring relevant information. The team has been an ardent support to the editorial, designing and production team. Their endless efforts to recruit the best for this project, has resulted in the accomplishment of this book. They are a veteran in the field of academics and their pool of knowledge is as vast as their experience in printing. Their expertise and guidance has proved useful at every step. Their uncompromising quality standards have made this book an exceptional effort. Their encouragement from time to time has been an inspiration for everyone.

The publisher and the editorial board hope that this book will prove to be a valuable piece of knowledge for researchers, students, practitioners and scholars across the globe.

List of Contributors

Katalin Tánczos and Árpád Török
Budapest University of Technology and Economics, Hungary

Abraham Akkerman
Department of Geography & Planning, and Department of Philosophy, University of Saskatchewan, Saskatoon, SK, Canada

Torill Nyseth
Department of Sociology, Political Science and Community Planning, University of Tromsø, Norway

Olutoyin Moses Adedokun
Department Of Geography, Federal College Of Education, Zaria, Nigeria

Dickson Dare Ajayi
Department Of Geography, University Of Ibadan, Ibadan, Nigeria

Hyeon-Jeong Choi and Ju-Hyung Kim
Hanyang University, Republic of Korea

Maria Cerreta and Pasquale De Toro
University of Naples Federico II, Italy

Di Lu, Jon Burley and Pat Crawford
School of Planning, Design, and Construction, USA

Robert Schutzki
Department of Horticulture, Michigan State University, USA

Luis Loures
CIEO - Research Centre for Spatial and Organizational Dynamics, University of Algarve, Portugal

Jaroslav Burian and Vít Voženílek
Palacký University, Olomouc, Czech Republic

Beniamino Murgante and Giuseppe Las Casas
University of Basilicata

Maria Danese
Archaeological and monumental heritage institute, National Research Council, Italy

Androniki Tsouchlaraki, Georgios Achilleos and Vasiliki Mantadaki
Technical University of Crete, Greece

Zdena Dobesova and Tomas Krivka
Palacký University in Olomouc, Czech Republic

Luís Loures
CIEO - Research Centre for Spatial and Organizational Dynamics, University of Algarve,
Faro, Portugal

Jon Burley
Landscape Architecture Program; School of Planning, Design, and Construction, Michigan
State University, East Lansing, USA

Marius-Cristian Neacşu and Silviu Neguţ
Bucharest Academy of Economic Studies, Romania

Arlette Pichardo-Muñiz
The International Centre on Economic Policy (CINPE, by acronym in Spanish), Universidad
Nacional, The Costarrican Society of Urbanism and Land Use Planning, Costa Rica

Marco Otoya Chavarría
The School of Economics (ESEUNA, by acronym in Spanish), Universidad Nacional, Costa
Rica

Helena Nogueira, Cristina Padez and Maria Miguel Ferrão
University of Coimbra, Portugal

Jorge Almazán and Nakajima Yoshinori
Center for Space and Environmental Design Engineering, Keio University, Tokyo, Japan

Alejandro Luis Grindlay Moreno
Department of Urban and Regional Planning, University of Granada, Spain

www.ingramcontent.com/pod-product-compliance
Lightning Source LLC
Chambersburg PA
CBHW070715190326
41458CB00004B/986